THE THEORY

OF

SETS OF POINTS

THE THEORY

OF

SETS OF POINTS

by

W. H. YOUNG, M.A., Sc.D.
Sometime Fellow of Peterhouse, Cambridge,
Lecturer in Higher Analysis at the University of Liverpool,

and

GRACE CHISHOLM YOUNG, Phil.Doc. (Gött.),
Formerly Sir Francis Goldsmid Scholar of Girton College, Cambridge.

CHELSEA PUBLISHING COMPANY
BRONX, NEW YORK

SECOND EDITION

THE PRESENT, SECOND EDITION IS A REVISED REPRINT OF
A WORK ORIGINALLY PUBLISHED IN 1906 AT CAMBRIDGE,
WITH A NEW PREFACE AND A FURTHER APPENDIX BY
DR. R. C. H. TANNER AND DR. I. GRATTAN-GUINNESS.
THIS EDITION IS PUBLISHED IN 1972 AT THE BRONX,
NEW YORK AND IS PRINTED ON SPECIAL ALKALINE PAPER

INTERNATIONAL STANDARD BOOK NUMBER 0-8284-0259-0

LIBRARY OF CONGRESS CATALOG CARD NUMBER 75-184793

COPYRIGHT ©, 1972 BY CHELSEA PUBLISHING COMPANY

LIBRARY OF CONGRESS CLASSIFICATION NUMBER QA 248

DEWEY DECIMAL CLASSIFICATION NUMBER 511'.3

PRINTED IN THE UNITED STATES OF AMERICA

PREFACE TO THE SECOND EDITION

1. *The Preparation of the First Edition.*

ON 23 JANUARY, 1907, Georg Cantor wrote as follows from his home at 13 Händelstrasse, Halle on the Saale, to my mother Grace Chisholm Young, co-author with my father William Henry Young of *The Theory of Sets of Points:*

> I do not wish to postpone any further my thanks for your kind letter of the 16th and for the promised copy* of the book that you have written jointly with your husband, "The Theory of Sets of Points." I hope that the recognition that you will earn from this meritorious and laborious enterprise, containing so many beautiful contributions of your own to the theory, will console you for the minor oversights you write of.
>
>   *not yet arrived.

Although Cantor had not yet seen the book, he was acquainted with the original papers of the Youngs to which he alludes. These papers are indicated in the Bibliography, which also reveals the great extent to which the subject matter found its origin in Cantor's own work. The theory of sets of points, as a separate mathematical discipline, was chiefly Cantor's creation. He recognised that it was still at the building stage and testified his conviction of the Youngs' continuing share in its growth by an enthusiastic tribute he pays in a letter of 9 March, 1907, when a copy of the book at last had reached him:

> It was with great joy that I received the day before yesterday the copy you have most kindly sent me of your joint work with your husband, "The Theory of Sets of Points." My sincerest thanks to you both.
>
>   It is a pleasure for me to see with what diligence, skill and success you have worked and I wish you, in your further researches in this field as well, the finest of results, which, with such depth and acuteness of mind on both your parts, you cannot fail to attain.

Cantor entirely ignores the slips which the authors had taken so much to heart and which were inevitable in a pioneering work. Notes for a revised edition, to incorporate many improvements and new discoveries which supervened, were made by my mother in a specially bound copy of the book containing 68 additional blank pages and incorporating letters from a number of well-known mathematicians as well as an assembly of published reviews. Her notes are scattered throughout the margins and blank pages and are additional to the items in the Appendix (pp. 284-294), which would certainly have been incorporated in some form in a revised edition of the work.

This projected second edition did not eventuate, but the creative work which continued unabated produced in the seven years that followed over a hundred significant papers, mostly published under my father's sole name, but always with my mother's unremitting participation. A Cambridge Tract in Mathematics[1] by my father also falls within this period. A treatise on projective geometry was planned, but set aside.

We children were another major preoccupation for our parents, which Cantor showed himself well aware of in his letters. In September, 1903 he had visited our home in Göttingen, where *The Theory of Sets of Points* came into being, and enjoyed making the acquaintance of the three eldest of us (Frankie aged 6, myself 3, and Janet 2), as he recalled in a letter of 9 March, 1905:

> My cordial greetings to your dear children, whom I had the pleasure of seeing at Göttingen in September 1903, and I hope in the not-too-distant future to make the acquaintance of one who at that time was a new arrival[2] in the world.

The genuineness of Cantor's expressions of interest in the children is attested to by the concluding paragraph of his letter of 7 March, 1909:

> I wish to give your Frankie something that will give him pleasure, as well as food for thought, and for this purpose I have inscribed 8 cards with the numbers 1 to 255.
> With their help he will be able to guess any number be-

---

[1] W. H. Young, *The Fundamental Theorems of the Differential Calculus* (Cambridge: Camb. Univ. Press, 1910. Reprint ed., New York: Hafner, 1960).

[2] This was Helen, born 20 September, 1903.

tween these limits that another person thinks of. To do so he has only to find out on which of these cards the number appears; by adding up the initial numbers on the cards concerned he will obtain the number with unfailing certainty. The root of the matter, as you will at once perceive, is the unique representation of whole numbers in the dyadic system.

By 1908, our family was complete, with six children (the last two were boys, Laurence and Patrick). On expert advice concerning our education, the family had moved to Switzerland. But this could not shield us from the implications of the first World War. Provisionally, the revision of *The Theory of Sets of Points* was merely shelved; but in the event, the death of Frank in the Royal Flying Corps over France resulted in a complete break with pre-war undertakings.[3] The revision scheme was one of these; another was a report on university mathematical education which my father had been preparing during his period (1915-1916) as the first Hardinge Professor at the University of Calcutta.[4]

If we overlook Schönflies's unsatisfactory report on set theory for the Deutsche Mathematiker-Vereinigung which began to appear in 1900,[5] my parents' book was the first of its kind in the field. One cannot admit Lebesgue's famous *Leçons sur l'intégration*[6] of 1904 (which anticipated my father's researches on the integral), for it did not go into detail of the foundations and applications of the subject.

The Theory of Sets of Points appeared at the time when set theory began to receive text-book treatments; indeed, in the same year (1906) appeared a monograph by Hessenberg,[7] and in 1907

---

[3] See Section 15 of I. Grattan-Guinness, ''A mathematical union. William Henry and Grace Chisholm Young,'' *Annals of Science*, 29 (1972).

[4] See I. Grattan-Guinness, ''University mathematics at the turn of the century. Unpublished recollections of W. H. Young,'' *Annals of Science*, 28 (1972).

[5] A. Schönflies, ''Die Entwicklung der Lehre von den Punktmannigfaltigkeiten,'' *Jber. Dtsch. Math.-Ver.*, 8 (1900), pt. 2; and Ergb. 2 (1907). A second edition of the first part was prepared with H. Hahn as *Die Entwicklung der Mengenlehre und ihrer Anwendungen* (Leipzig and Berlin: Teubner, 1912).

[6] H. Lebesgue, *Leçons sur l'intégration et la recherche des fonctions primitives* (Paris: Gauthier-Villars, 1904). An enlarged edition appeared in 1928.

[7] G. Hessenberg, ''Grundbegriffe der Mengenlehre...,'' *Abh. Fries'schen Schule*, (2) 1 (1906), 479-706 (Reprint ed. Göttingen: Vandenhoeck und Ruprecht, 1906).

another by Nekrasov[8] and also the first edition of Hobson's treatise.[9] Hobson's work contained copious references to the literature of set theory (and my parents' papers accounted for more than that of any other author), but it was mainly a reference book, and was therefore in a different category from the living presentation of a new and emerging structure of mathematical thought. Even in the final third edition, Hobson's treatise continued to refer the reader to my parents' book for untreated subject matter.

R. C. H. TANNER.

2. *The Scope and Limitations of the Revision.*

It has not been possible to undertake a substantial revision of the text in preparing this second edition. But the Youngs' own copy of the book is still extant, and from indications in it and our own inspection of the text the following types of modification have been made.

2.1. Printing mistakes and simple errors have been corrected, including those noted in the original appendix on pages 289-294.

2.2. Consequently a few items in that appendix have been eliminated. The other items have been renumbered, and supplemented by additional items drawn from indications made (chiefly by Mrs. Young) in their copy. These comments appear below on pages 295-303 as items 15)-54). Further long notes were also prepared, but they have not been included.

2.3. Cross-references have been inserted at the appropriate places in the text to the (old and new) items in the expanded appendix.

But the book remains as a work of its time, with its deeper insights and unclarities untouched. It is in this spirit that it must be read and used today; alive and faulted, as opposed to the 'perfection' and *boredom* of the modern text-book.

I. GRATTAN-GUINNESS.

---

[8] V. L. Nekrasov', *The Construction and Measure of Linear Sets of Points* [in Russian] (Tomsk: Technical Institute, 1907).

[9] E. W. Hobson, *The Theory of Functions of a Real Variable and the Theory of Fourier's Series* (1st ed., Cambridge: Camb. Univ. Press, 1907).

PREFACE.

THE present volume is an attempt at a simple presentation of one of the most recent branches of mathematical science. It has involved an amount of labour which would seem to the average reader quite out of proportion to the size of the book; yet I can scarcely hope that the mode of presentation will appeal equally to all mathematicians. There are no definitely accepted landmarks in the didactic treatment of Georg Cantor's magnificent theory, which is the subject of the present volume. A few of the most modern books on the Theory of Functions devote some pages to the establishment of certain results belonging to our subject, and required for the special purposes in hand. There is moreover in existence the first half of Schoenflies's useful *Bericht über die Mengenlehre.* The philosophical point of view is discussed to some extent in Russell's *Principles of Mathematics.* But we may fairly claim that the present work is the first attempt at a systematic exposition of the subject as a whole.

The difficulties in arrangement which this fact suggests have been enhanced by the nature of the subject itself and by the tentative character of some of its results. The writing of the book has necessarily involved attempts to extend the frontier of existing knowledge, and to fill in gaps which broke the connexion between isolated parts of the subject. The references in the text which do not give the name of the author always refer to my own papers; in this connexion, however, I should like to point out that the citations in the text are not to be regarded as by any means complete, and are supplemented by the list of literature at the end of the volume.

On the other hand, imperfect though the book is felt to be, it is hoped that it may prove of use to a somewhat large class of readers. As far as the professional mathematician is concerned, it may be confidently asserted that a grasp of the Theory of Sets of Points is indispensable. Wherever he has to deal—and where does he not?—with an infinite number of operations, he is

treading on ground full of pitfalls, one or more of which may well
prove fatal to him, if he is unprovided by the clue to furnish
which is the object of the present volume.

In subjects as wide apart as Projective Geometry, Theory of
Functions of a Complex Variable, the Expansions of Astronomy,
Calculus of Variations, Differential Equations, mistakes have in
fact been made by mathematicians of standing, which even a
slender grasp of the Theory of Sets of Points would have enabled
them to avoid. It can scarcely be doubted that the near future
will see a marked influence exerted by our theory on the language
and conceptions of Applied Mathematics and Physics. To the
philosophical reader on the other hand and to the general public
with mathematical interests the subject presents the advantage,
as compared with other of the more recent developments of
mathematical science, that it is less technical and requires a
smaller mathematical equipment than most of them.

I should like to take this opportunity of acknowledging my
indebtedness to Professor Vivanti of Messina, who has most
carefully read all the proof-sheets, and considered various points
submitted to him; his help and criticism have throughout been
invaluable. Dr Felix Bernstein of Halle has also been so good as
to read the proof-sheets of the first eight chapters with especial
reference to the arithmetical portions of the subject. Mr Philip
Jourdain, who read Chapters VI and VII, and also looked through
the earlier proof-sheets, and Professor Oswald Veblen of Princeton,
who undertook Chapters IX and X, have also been of the greatest
help with criticisms and suggestions. Any reference to the constant
assistance which I have received during my work from my wife is
superfluous, since, with the consent of the Syndics of the Press,
her name has been associated with mine on the title-page.

In spite, however, of the greatest care to avoid error, clerical
or otherwise, mistakes are sure to have escaped notice. The
reader is recommended not to overlook the Appendix, in which
some mistakes, discovered too late for correction, and some points
in the text which seemed to require elucidation are discussed.

W. H. YOUNG.

HESWALL.
May, 1906.

TABLE OF CONTENTS.

CHAPTER IV.

POTENCY, AND THE GENERALISED IDEA OF A CARDINAL NUMBER.

CHAPTER V.

CONTENT.

CHAPTER VI.

ORDER.

CHAPTER VII.

CANTOR'S NUMBERS.

CHAPTER VIII.

PRELIMINARY NOTIONS OF PLANE SETS.

CHAPTER IX.

REGIONS AND SETS OF REGIONS.

CHAPTER X.

CURVES.

CHAPTER XI.

POTENCY OF PLANE SETS.

CHAPTER XII.

PLANE CONTENT AND AREA.

CHAPTER XIII.

LENGTH AND LINEAR CONTENT.

CHAPTER I.

RATIONAL AND IRRATIONAL NUMBERS.

1. Introductory. The student is supposed to be familiar with the ordinary theory of the natural numbers, 1, 2,..., and its extension to the fractional and negative numbers, $\frac{1}{2}$, $\frac{1}{3}$, $\frac{2}{3}$, ..., -1, $-\frac{1}{2}$, ... Both classes are grouped together under the name of rational numbers. He is also supposed to have some acquaintance with the theory of irrational numbers. This latter theory occupies, however, a fundamental position in our present subject, and we propose to give a short account of it, sufficient for the purposes in hand.

2. Sets and Sequences. Any number of rational numbers, individually given, are said to form a finite *set*. For instance,

$$1, \tfrac{1}{2}, \tfrac{1}{4}, \tfrac{3}{4}, 2, 3.$$

In this case, each being given singly, they must have a definite succession, and are said to have an *order*.

If numbers are given, not individually, but by means of some law, they are said to form a *set*, which is a *finite set*, if they can from the law be individually determined and assigned, without any numbers of the set being omitted; in the contrary case, they are said to form an *infinite set*, and to be infinite in number (more precisely, in cardinal number). If the law be such that from it the numbers are determined in a definite succession, one by one, the numbers are said to have an *order*.

Thus all the integers between 0 and 15 form a finite set, without order. The same integers in order of magnitude form a finite set in order, which is generally referred to as the *natural order*. All the rational numbers between 0 and 15 form an infinite set without order, and in order of magnitude they have an order, the natural order. All numbers satisfying the relation

$$u_n = u_{n-1} + u_{n-2},$$

where
$$u_1 = 1, \quad u_2 = 1,$$

form an infinite set in order.

A set in order is also called a *series*.

An infinite number or series of rational* numbers

$$a_1, \; a_2, \ldots,$$

is said to form a sequence, if, given any small positive quantity ϵ, a number a_n of the series can always be assigned, such that, if a_p and a_q be any numbers of the series subsequent to a_n,

$$|\,a_p - a_q\,| < \epsilon.$$

The individual numbers a_1, a_2 are called the *constituents* of the sequence.

Thus, the numbers

$$1, \quad \frac{3}{2}, \; \frac{7}{4}, \; \frac{15}{8}, \; \frac{31}{16}, \; \cdots \; \left(2 - \frac{1}{2^m}\right), \; \left(2 - \frac{1}{2^{m+1}}\right), \ldots \ldots (1)$$

for all integral values of m, form a sequence.

A series of rational numbers which constantly increase, or which constantly decrease, always defines a sequence, provided a finite number exists which in the former case is always greater, and in the latter case is always less than any number of the series.

Ex. $1, \quad 1+1, \quad 1+1+\dfrac{1}{2\,!}, \quad 1+1+\dfrac{1}{2\,!}+\dfrac{1}{3\,!}+ \ldots.$

It may now happen that the sequence is such that a rational number b exists, to which the a's continually approximate to a degree closer than any assigned quantity. More precisely it may be such that, given any small positive quantity ϵ, a number a_n of the series can always be assigned, such that a_r being any number of the series subsequent to a_n,

$$|\,b - a_r\,| < \epsilon.$$

The sequence is then said to define the number b, which is clearly unique. Thus the number 2 is defined by the sequence (1).

It is evident that the same number b may be defined by sequences whose constituent numbers differ from one another. Thus each of the sequences

$$\frac{2}{3}, \; \frac{10}{9}, \; \frac{38}{27}, \; \cdots \cdots \; \overset{n}{\underset{1}{\Sigma}} \left(\frac{2}{3}\right)^r, \; \cdots \cdots$$

$$\frac{8}{3}, \; \frac{16}{9}, \; \frac{56}{27}, \; \cdots \cdots \; \overset{n}{\underset{1}{\Sigma}} \, (-)^{n+1} \frac{8}{3^n}, \; \cdots \cdots$$

defines the number 2.

Sequences which define the same number are said to be equal.

---

* See concluding remarks on p. 5, where this restriction is removed.

3. Irrational numbers. The two most ready ways of setting up a sequence are:

(1) by means of the terminating decimal fractions which are the successive approximations to a non-terminating decimal fraction (or fraction expressed in some other scale), and

(2) by means of the successive convergents to a non-terminating "simple" continued fraction*.

The first of these forms a sequence since the constituent numbers continually increase and remain less than an assignable number: that the second forms a sequence follows from the known properties of continued fractions.

Since a rational number can be expressed in one, and only one, way (1) as a non-terminating decimal fraction, namely as a recurring decimal, and (2) as a simple continued fraction, namely a terminating one, it follows that a sequence set up in mode (1) does not define a rational number unless the decimal fraction recurs, and that (2) never defines a rational number.

In accordance with the usual law which holds in all extensions of mathematical reasoning, it is convenient still to use the word *number* when speaking of a sequence which does not define a rational number. The new numbers thus obtained are called irrational numbers; for instance the series of decimal fractions

$$\cdot 1,\ \cdot 12,\ \cdot 123,\ \cdot 1235,\ \cdot 12357,\ \cdot 1235711,\ \cdot 123571113,\ \ldots\ldots$$

where each is got from the last by appending the next prime number after the last appended, and the series of convergents

$$1,\ 2,\ \tfrac{3}{2},\ \tfrac{5}{3},\ \tfrac{8}{5},\ \tfrac{13}{8},\ \tfrac{21}{13},\ \tfrac{34}{21},\ \tfrac{55}{34}, \ldots$$

to the continued fraction

$$1 + \frac{1}{1+}\frac{1}{1+} \ \ldots\ldots,$$

are said to define irrational numbers.

4. Magnitude and Equality. In § 2 we stated that two sequences which defined the same rational number might be regarded as equal. We now give a definition of the equality of two sequences which applies when the two sequences do not define rational numbers: this definition will include the preceding as a particular case.

---

* That is to say a continued fraction of the form

$$a_1 + \frac{1}{a_2 +}\frac{1}{a_3 +} \ \ldots\ldots\ldots$$

where the a's are positive integers; Chrystal's *Algebra*, Vol. II. p. 397.

The definition is as follows :—

Two sequences $\qquad a_1, a_2, \ldots\ldots$

$\qquad\qquad\qquad\qquad b_1, b_2, \ldots\ldots$

are said to be equal, when, given any small positive quantity ϵ, *however small, an integer m can be found such that, for all values of* $n \geqslant m$,

$$|a_n - b_n| < \epsilon.$$

We may put this definition in an apparently more general, but actually equivalent, form. The two sequences are said to be equal, when, choosing out from each a partial sequence,

$$a_1', a_2', \ldots\ldots, \qquad b_1', b_2', \ldots\ldots,$$

these partial sequences satisfy the above condition of equality.

That this definition is equivalent to the former is evident from the fact that the a's and b's form sequences.

Taking any two sequences at random, and forming any partial sequences from them as above, it is easily proved that the quantities

$$a_1' - b_1', \quad a_2' - b_2', \ldots\ldots$$

form a sequence, and that all sequences so formed by means of partial sequences from two given sequences are equal. The number defined by any one of these sequences is called the *excess* of the number a, defined by the a-sequence, over the number b, defined by the b-sequence, and when taken positively is called the *difference* of a and b.

If the sequences be equal, the difference will be zero; otherwise the excess will be positive or negative, and the number a is, in the former case, said to be *greater* than the number b, $a > b$, and, in the latter case, a is said to be *less* than b, $a < b$.

An irrational number is said to be *positive or negative*, according as it is greater or less than 0.

It is evident that, (1) if $a > b$, $b < a$, and that, (2) if $a > b$ and $b > c$, $a > c$; finally that every number a is either greater than, equal to or less than any given number b. These facts are summed up by saying that *numbers may be compared as to magnitude.*

It is now easily seen that the rational numbers

$$a_1' + b_1', \quad a_2' + b_2', \ldots$$

form a sequence, whose magnitude is independent of the choice of the partial sequences. The number defined by any one of these sequences is called the *sum of the two numbers a and b.*

Similarly, $\qquad a_1' b_1', \ a_2' b_2', \ ...,$

and, *if the b-sequence do not define the number* 0,

$$\frac{a_1'}{b_1'}, \ \frac{a_2'}{b_2'}, \ ...$$

form sequences, whose magnitude is independent of the choice of the partial sequences. The former of these is said to define the *product,* and the latter the *quotient* of a and b.

Hence it follows that we can attach a definite meaning to the symbol $R\,(a,\,b,...k)$, where R is a rational function of the finite set of irrational numbers $a,\,b,...k$, provided that, in the process of calculating R by approximation, no quotient occurs whose value is, actually or in the limit, zero.

In particular, since the difference of two irrational numbers is now clearly defined, we can, in the definition of a sequence in § 2, insert the words "or irrational" after "rational," and so obtain a general definition of sequence, independent, not only of the fact whether or no the number defined be rational or not, but also of the rationality or irrationality of the constituents themselves.

5. The number ∞. We have now attached a definite number to every sequence of numbers, rational or irrational, and we saw that if

$$a_1, \ a_2, ...$$

be a given sequence, then, provided that a_ν does not become less in absolute magnitude than any assigned number,

$$\frac{1}{a_1}, \ \frac{1}{a_2}, ...$$

is also a sequence. Also

$$(a_1)\frac{1}{a_1}, \ (a_2)\frac{1}{a_2}, ...$$

forms a sequence defining the number 1, so that, by our definitions, the product of the two former numbers is unity; we therefore denote them by a and $\dfrac{1}{a}$, and say that they are the *inverses* of one another.

We shall now add another number to our list of numbers, and so remove the last restriction as to the nature of the given sequence. If the a's become less in absolute magnitude than any assigned number, the constituents of the inverse series become greater in absolute magnitude than any assigned number, and do not form a sequence as we have defined it. In other words, the number 0 is the only number which at present has no inverse. We now introduce the number ∞ as the inverse of 0. That is to say, *given a series of numbers which become and remain greater in absolute magnitude than any assigned number, we still regard this series as a sequence, and introduce the symbol* ∞ *(infinity), to denote the " number " defined by such a sequence.*

We see that this agrees so far with the preliminary definition of the term "infinite number" given in § 2.

With this convention we may add that, if in any rational function $R(\infty)$ the symbol ∞ occurs, it implies that, taking any series of numbers $a_1, a_2, \ldots$, whose inverses in order define the number 0, the quantities $R(a_n)$ form a sequence, and the number defined by this sequence is that denoted by the symbol $R(\infty)$.

We content ourselves for the present with this definition* of the symbol ∞. It will subsequently appear that we have to specialise our ideas of infinity, or more properly, of infinite numbers, for purposes which will occur in the later chapters.

6. Limit. The number defined by a sequence is said to be the *limit of* the constituent a_n of the sequence, *when n is infinite.*

It appears from what has been said, that if $R(a, b, \ldots k)$ be any rational function of the finite set of numbers $a, b, \ldots k$ (rational or irrational), then, *provided the limit of $R(a_n, b_{n'}, \ldots)$ be definite,* $R(a, b, \ldots k)$ may be defined as the limit of $R(a_n, b_{n'}, \ldots)$ when $n, n', \ldots$ are infinite, and will be a number, rational or irrational. With the above restriction, then, the rational and irrational numbers form what is called a " corpus†."

7. Algebraic and transcendental numbers. An important class of numbers is that of the algebraic numbers. These are distinguished among themselves as to rank. *An algebraic number of rank m is defined as a number satisfying an irreducible equation of degree m, with rational coefficients, and satisfying no such equation of degree less than m.*

A rational number satisfies such an equation of degree 1, and any algebraic number which satisfies an algebraic equation of degree 1 is rational; thus the algebraic numbers of rank 1 are identical with the rational numbers, and all algebraic numbers of rank higher than 1 are irrational. Methods of obtaining all algebraic numbers, *i.e.* of obtaining sequences defining them, are given in all works on the Theory of Equations.

---

* It should be noticed that in this system of fixed numbers there is no place for the symbols $+\infty$ and $-\infty$ any more than for $+0$ and -0; see however § 16.

† A corpus is a collection of objects which reproduce themselves when subjected to the simple rules of arithmetic.

All irrational numbers which are not algebraic are classed together as *transcendental*.

It is in general a most difficult problem to determine in any special case whether a given number be rational, algebraic or transcendental.   No general method and no set of necessary and sufficient conditions have at present been discovered.   There are however a few isolated theorems on the subject, among which the following is one of the most important; it enables us to write down sequences defining numbers of the last of the above classes.

Liouville's Theorem.   *If $\dfrac{p_1}{q_1}, \dfrac{p_2}{q_2}, \ldots$ be a sequence of rational fractions in their lowest terms, defining an algebraic number b of rank m, then, for every constituent $\dfrac{p}{q}$ from and after an assignable stage, we have*

$$\left| \frac{p}{q} - b \right| > \frac{1}{q^{m+1}}.$$

To prove this theorem, let the equation of degree m satisfied by b be

$$f(x) = a_0 x^m + a_1 x^{m-1} + \ldots + a_m = 0,$$

where the a's are integers; and let $\dfrac{p}{q}$ be any rational number within a certain small interval containing b, that is to say, such that the difference between that number and b, is less than a certain small positive quantity.

Then $\qquad\qquad f(b) = 0,$

therefore, $\qquad f\left(\dfrac{p}{q}\right) = \left(\dfrac{p}{q} - b\right) f'(y),$

where y is a certain number, possibly irrational, lying between $\dfrac{p}{q}$ and b; and f' is the first derived of f.

Now it is obviously possible to assign a finite number M, greater than any value of $|f'(y)|$, when y lies within the given interval.   Hence

$$\left| f\left(\frac{p}{q}\right) \right| < M \left| \frac{p}{q} - b \right|.$$

But, $\qquad\qquad f\left(\dfrac{p}{q}\right) = \dfrac{A}{q^m},$

where A is some integer; and, if we choose to assign a sufficiently small interval, so that $f(x)$ vanishes for no value of x within the interval except b,[15] A will not be zero, and therefore,

$$\left| \frac{p}{q} - b \right| > \frac{1}{M q^m}.$$

Now there are only a finite number of rational numbers whose denominators are less than M. Hence, we can determine a second interval, lying within the former interval, and containing none of these rational numbers; in this interval for every rational number $\dfrac{p}{q}$,

$$\left| \frac{p}{q} - b \right| > \frac{1}{q^{m+1}},$$

so that, given any sequence having b as limit, we can determine a stage such that for all subsequent constituents the above inequality holds.   Q. E. D.

The above property serves to determine whether a given sequence can represent an algebraic number, or a rational number, but it does not give a sufficient criterion to determine whether a given sequence actually does represent such a number.

Thus in the case of the sequence of fractions, expressed in the decimal or any other scale, such that each is got from the preceding by appending one more figure on the right, we know that the number b defined can only be rational if the figures ultimately recur in some cycle. This property is quite independent of the above, and cannot be proved from the above inequality. All that we can deduce from the inequality is that, if b be rational, then, when n is sufficiently large, n successive figures after the nth figure cannot all be noughts; while, in the more general case, when b is an algebraic number of rank m, mn successive figures after the nth cannot all be noughts, n being sufficiently large.

This property serves to define a class of transcendental numbers discovered by Liouville and called after his name, which were historically the first numbers to be proved transcendental.[16] These are the numbers

$$\frac{1}{10} + \frac{1}{10^{1.2}} + \frac{1}{10^{1.2.3}} + \frac{1}{10^{1.2.3.4}} + \ldots = \cdot 110001000000000000000000010\ldots,$$

and the decimal fractions got by replacing any 1 by any other figure.

Such numbers may of course be constructed in any other scale, and will still be transcendental.

The best known transcendental numbers are π and e: these do not belong to the class of Liouville numbers.

CHAPTER II.

REPRESENTATION OF NUMBERS ON THE STRAIGHT LINE.

8. One of the most fundamental properties of the set of rational numbers is their *order*. We shall find in the sequel that the idea of order is one of the most essential to the understanding of sets of points, and that we habitually use the order of some or all of the rational numbers as a standard of comparison.

The order of the rational numbers as a whole is such that we cannot say which is the next rational number in order of magnitude after any given one a, or before a given one c; indeed, if a and c be any two rational numbers, we can always insert a rational number b between them.

It is of assistance to the imagination that we can set up a (1, 1)-correspondence between the rational numbers and certain points of the straight line, in such a way that the order is maintained, that is to say if A_p, A_q, A_r are three of the points, corresponding to the rational numbers p, q, r, A_q lies between A_p and A_r if, and only if, q lies between p and r, and *vice versa*.

We shall now discuss shortly, how and under what assumptions with respect to the nature of the straight line, this correspondence can be extended to the irrational numbers.

In setting up the (1, 1)-correspondence referred to, measurement may be entirely avoided; in this way various difficulties which have nothing to do with the subject in hand do not come into the discussion.

The principle of the correspondence which we here choose and which is commonly referred to as the *projective scale*, is that if a, b, c, d be any four harmonic rational numbers, that is if

$$(a, b, c, d) \equiv \frac{a-b}{b-c}\frac{c-d}{d-a} = -1,$$

or, which is the same thing,

$$\frac{1}{b-a} + \frac{1}{d-a} = \frac{2}{c-a},$$

$$\quad\dots\dots\dots\dots(1)$$

the points corresponding to these four rational numbers shall form a harmonic range.

Now the equation (1) always defines a rational number d, if a, b, c be rational, unless

$$2b = a + c;$$

in the latter case there is no rational number satisfying the equation (1); but if d describe any sequence of positive or negative rational numbers, whose absolute magnitude increases without limit, $\left(a, \dfrac{a+c}{2}, c, d\right)$ approaches the limit -1. Hence in accordance with the meaning attached by us to the symbol ∞, we shall write

$$\left(a, \frac{a+c}{2}, c, \infty\right) = -1. \quad\dots\dots\dots\dots\dots(2)$$

It follows that, p being any positive integer,

$$(p-1, p, p+1, \infty) = (-p, 0, p, \infty) = \left(0, \frac{1}{p+1}, \frac{1}{p}, \frac{1}{p-1}\right) = -1.$$
$$\dots\dots\dots\dots\dots(3)$$

Now we start with any two points P, Q, and any point between them to which we attach the integer 1, or, as we shall express this more concisely, we choose this third point as the point 1. We shall see subsequently that we shall come to attach the symbols 0 and ∞ to the points P, Q, in consequence of the equations (2) and (3).

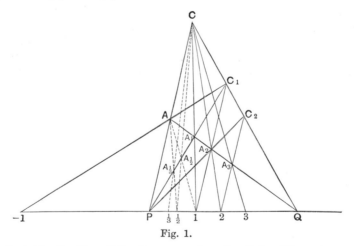

Fig. 1.

Outside the line PQ take any point C. Join CP, and on this line, between C and P, take any point A, and join AQ.

Construction of the positive integral points. The point 2 we choose to be the fourth harmonic of P with respect to 1 and Q. We therefore join $1C$, and let it meet AQ (of course between A and Q), in A_1. Join PA_1 and produce it to meet CQ in C_1, forming the quadrilateral A_11QC_1, having P and C as two of its diagonal points: the third diagonal point A_2 is obtained as the intersection of $1C_1$ with AQ. Then CA_2 will meet PQ between 1 and Q, this point we call the point 2.

The point 3 we construct as the fourth harmonic of 1 with respect to 2 and Q, precisely as we did the point 2 from P, 1 and Q. Similarly each successive integral point can be constructed.

It is evident that the integral points so constructed are in the proper order, and the greater an integer, the nearer the corresponding point lies to the point Q. It will be shewn, further on, that in this way the points corresponding to four harmonic integers are always themselves harmonic, and *vice versa*.

Construction of the negative integral points. The point $-m$ we now construct as the fourth harmonic of m with respect to the points P and Q; it can be at once obtained as the intersection of PQ with AC_m, where C_m is the second angular point on CQ of the quadrilateral used in the construction of the point $(m+1)$, just as C_1 occurred in the construction of the point 2.

It is evident that these points will again be in the proper order, the point -1 being nearest to P on the one side, or farthest from the point Q on the other side, and large negative integers nearer and nearer to the point Q. Between the positions AC and AC_1 or between two definite successive positions AC_r, AC_{r+1}, the line which is revolving round A becomes parallel to PQ, so that in the first case the points P and -1, in the second case the successive integral points $-r$ and $-(r+1)$ lie toward different ends of the straight line PQ. It will therefore be convenient to think of the whole straight line as closed, and to make the convention that any point which lies beyond the point $-r$ (or P), on the one side, or beyond the point $-(r+1)$ (or -1), on the other side, lies between the points $-r$ and $-(r+1)$ (or P and -1).

Construction of the inverse integral points $\dfrac{1}{m}$. The point $\frac{1}{2}$ we choose to be the fourth harmonic of Q with respect to P and 1; it is constructed by joining $A1$, cutting PA_1 between P and A_1

at a point which we will call $A_{\frac{1}{2}}$. $CA_{\frac{1}{2}}$ meets the line PQ between P and 1, and this is to be the point $\frac{1}{2}$.

The point $\frac{1}{3}$ is similarly constructed, by means of an auxiliary point $A_{\frac{1}{3}}$, the intersection of $A_{\frac{1}{2}}$ with PC_1, as the fourth harmonic of the point 1 with respect to P and $\frac{1}{2}$; and similarly we can construct in succession the inverse integral points, the point $\dfrac{1}{r+1}$ being the fourth harmonic of $\dfrac{1}{r-1}$ with respect to P and $\dfrac{1}{r}$.

It is again evident from the construction that these points are in the right order. The inverse integral points lie then all between P and 1, and are in the reverse order to the integral points, and they mass themselves in the neighbourhood of the point P, so that the higher the integer m the nearer to P lies the inverse integral point $\dfrac{1}{m}$.

Construction of the general positive or negative rational fractional point $\pm\dfrac{p}{q}$. The points $1+\dfrac{1}{r},\ 2+\dfrac{1}{r},\ \ldots\ldots$ can now be constructed from $P,\dfrac{1}{r}$, and Q, just as $2, 3, \ldots\ldots$ were from $P, 1$, and Q; and the point $-\dfrac{p}{q}$ from $P,\dfrac{p}{q}$ and Q, just as $-m$ was from P, m, and Q.

It is then evident that, with our convention as to points at a great distance, the points which we have constructed lie in the right order, so that if a number b lies between two numbers a and c, the point b will also lie between the points a and c, and *vice versa*.

It remains to shew for completeness, that *any four harmonic points correspond to four harmonic numbers and vice versa*.

Now it is evident from our construction:

(1)   that if we project our rational points from C on to AQ, and then project from P on to CQ, finally project from A on to PQ, we project any point x into the point $-x$;

(2)   that if we project from A_1 on to AC_1, and then from C on to PQ, we project each point x into its inverse $\dfrac{1}{x}$, and the points P and Q are interchanged;

(3)   that if we project from C_1 on to AQ, and then from C on to PQ, we change x into $x+1$.

Now by (3) we can project the four integral points $(P, m, 2m, Q)$ into $(-m, P, m, Q)$, which was by construction harmonic. Similarly $(m, 2m, 3m, Q)$ is harmonic, and so on. Thus we can project the harmonic quadruplet $(P, m, 2m, Q)$ on to the harmonic quadruplet $(P, 1, 2, Q)$; and, when we do this, the point $2m$ will be projected to the point 2, and generally the point xm to the point x, where x is any rational number (4).

By a combination of the projections (1), (2), (3) we can project each point x into the point $\dfrac{ax+b}{cx+d}$, where a, b, c, d are any given integers.

If we choose

$$c = 0,$$

$$d = (m-l)\, b,$$

$$a = \left(\frac{1}{n-l} - \frac{1}{m-l}\right) d,$$

then, if, and only if, the numbers l, m, n, p are harmonic, the points $\left(Q, \dfrac{1}{m-l}, \dfrac{1}{n-l}, \dfrac{1}{p-l}\right)$ will project into the points $(Q, 0, 1, 2)$, and will therefore be harmonic. In this case by the projection (2) the points $(P, m-l, n-l, p-l)$, and therefore by the projection (3) the points (m, l, n, p) are harmonic, so that every harmonic quadruplet of numbers corresponds to a harmonic quadruplet of points. *Vice versa*, if l, m, n, p' be the numbers corresponding to any four harmonic rational points, and p be the fourth harmonic number corresponding to m with respect to l and n, the point p will be the harmonic conjugate of m with respect to the points l and n, and will therefore coincide with p'. Thus we see that our construction is such that to four harmonic numbers correspond four harmonic points, and *vice versa*.

Extension of the correspondence to the irrational points. By construction there are an infinite number of rational points in any segment, however small, of the straight line; but we cannot assert that to every point of the straight line there corresponds a rational number, and indeed with proper assumptions as to the nature of a straight line, it can be proved that this is not the case. If, however, we choose any point x, we shall always be able to construct sequences of rational numbers, such that the corresponding points lie nearer and nearer to x, and always eventually lie between x and any other assigned point of the straight line, and these sequences will by definition all define the same number. This number we attach to the point x, and in this way every

point of the straight line has a number attached to it, the
numbers attached to the points P, Q being evidently 0 and ∞
respectively, and the numbers attached to the other points being
rational or irrational numbers such as we have defined them.

The converse of the above, namely, that *corresponding to
every irrational number such as we have defined, there is one
definite point of the straight line*, is incapable of proof, as was
first pointed out by G. Cantor, *Math. Ann.* v. 1872, p. 128.
The assumption that this is the case, which for the purposes
in hand we shall make, is therefore of the nature of an axiom,
and binds us down in the geometry of linear sets of points to
a definite conception of the straight line, considered as a continuous geometrical form, which is not *a priori* necessitated by
the axioms of projective or of Euclidian geometry. This axiom,
which has been assumed by innumerable mathematicians as
self-evident, will be found in the sequel to lead to results so
paradoxical that the student of the theory of sets of linear points
is obliged to confess that the theory of sets of numbers does not
owe more to the geometrical insight of the investigator, than
the geometrical conception of the mutual relations of points on
a straight line owes to the abstract properties of numbers.

The above axiom as to the nature of the straight line, in
whatever form it is stated, is commonly referred to as the Cantor-
Dedekind axiom, having been arrived at, in a slightly different
form, almost simultaneously with Cantor, by Dedekind.

We may now sum up the results of the present article by
saying : *the continuum of numbers and the linear continuum can
be brought into* (1, 1)-*correspondence maintaining the order.*

If we choose the point Q as the imagined intersection of the
line PQ with any parallel line, the projective scale is identical
with a scale fixed by the ordinary principle of measurement; and
for most purposes this will be in future used.

9. Interval between two numbers. The first idea which
we shall borrow from the geometry of the straight line is that of
a *segment* or *interval*, and these words will sometimes be used with
a slight difference of meaning. By the segment or interval (A, B)
of the straight line is clearly understood the finite part of the
straight line terminated at A and B, that is to say that part so
terminated which does not contain the point ∞ : and when we
speak of the segment we shall in general mean that our attention
is fixed only on points contained in that segment, while in

speaking of the interval (A, B) we shall direct our attention chiefly to the points A and B, and possibly to other points outside (A, B), but not in general to points inside the interval (A, B), or only to these in contradistinction to the others.

Having connected numbers a and b with A and B, as in § 7, every point of the segment (A, B) will have a number connected with it lying between a and b, and conversely; while every point outside the interval (A, B) will have connected with it a number not lying between a and b, and conversely. Hence we naturally consider the numerical segment (a, b) as consisting of all the numbers lying between a and b.

The segment or interval (A, B) or (a, b), must belong to one of two classes:

(1) *open* intervals or segments, *i.e.* those in which one at least of the end-points or numbers is not included;

(2) *closed* intervals or segments, *i.e.* those in which both end-points or numbers are included.

These distinctions, which for some purposes are superfluous, will be found in the course of our work to be often of fundamental importance.

A point x, which belongs to an interval unclosed at both ends, is said to be an *internal point* of the corresponding closed interval, so that a closed interval divides all the points of a straight line into *internal, external, and end-points*; or similarly for numbers.

The *length*, or *content*, or *magnitude* of the interval (A, B) or (a, b) will now be understood to mean the absolute value of the difference $(b - a)$ of the numbers corresponding to the end-points A and B.

The actual value of the difference $(b - a)$ may then be called the distance of the point B from the point A, and written AB. When the point ∞ is the intersection of the line with any parallel line this is identical with the ordinary definition of distance. The absolute value of the distance is then the same as the content; but while the content is always positive, the distance may be positive or negative.

CHAPTER III.

THE DESCRIPTIVE THEORY OF LINEAR SETS OF POINTS.

10. Sets of Points. Sequences. Limiting Points.
Given any set of numbers we have then corresponding to them
a definite set of points on the straight line. Conversely *a set of
points on the straight line, or a linear set of points, is understood to
mean any finite or infinite number of points, such that their corresponding numbers form a finite or infinite set*; for instance all the
integral points, or all the rational points, or all the points of the
segment (0, 1). This definition is to be regarded as equivalent to
the following: a linear set of points consists of points of a straight
line determined by a certain law which is such that (1) every
point of the straight line either belongs to the set or does not,
but not both, nor neither; (2) assuming that we are acquainted
with all the characteristics of the points of the straight line, given
any point we can determine whether or no it belongs to the set;
and (3) having already obtained any number or collection of the
points of the set, if there are any points of the set left, the law
permits us to determine more.

A set which is contained entirely in another set is called
a component of the latter set, and, if there are points of the latter
set not belonging to the former set, it is said to be a *proper component* of the other ; *e.g.* the Liouville points, which correspond to
the Liouville numbers (Ch. I, § 7, p. 8), form a component of the
set of all the irrational points. The remarks made at the beginning
of Chapter I with respect to the order of a set of numbers, apply,
of course, to that of a set of points on the straight line.

If the numbers corresponding to a certain set of points form
a sequence, the set of points is called *a sequence*, and, by the
Cantor-Dedekind axiom (Ch. II, p. 14), there will be a definite
point L of the straight line corresponding to the number defined
by the sequence of numbers. Since the order of points and

numbers is maintained, between this point L and any other point of the straight line we can insert an infinite number of points of the sequence, thus *the interval between L and the points of the sequence in order decreases without limit* in magnitude.* Conversely it is evident that there is no other definite point of the straight line having this property. *The point L is called the limiting point of the sequence, and we see that a sequence of points has one and only one limiting point.*

Ex. 1. The simplest example of a sequence of points and its limiting point is that got by dividing a given segment, say the segment (0, 1), into two equal parts, and then dividing the right-hand segment into two equal parts, and so on. The corresponding set of numbers is

$$\tfrac{1}{2}, \tfrac{3}{4}, \tfrac{7}{8}, \dots$$

or, using the binary notation,

$$\cdot 1, \cdot 11, \cdot 111, \dots$$

The limiting point is the point 1.

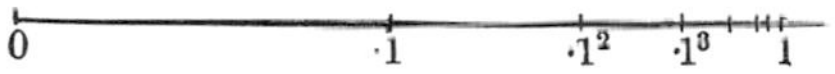

Fig. 2.

The preceding example illustrates the following theorem, which, from its fundamental importance in the theory of limiting points, deserves special attention.

THEOREM 1. *If we take any series of closed segments, each lying entirely within the preceding, and if the length of the segments decrease without limit, the end-points of the segments form a sequence, and the segments determine one and only one point L, internal to all the segments.*

COR. *Such a series of segments open at both ends determines one and only one point L, which is either internal to all the segments, or from and after a definite stage is a common end-point of them all.*

The general definition of a limiting point is as follows :—

DEF. A point L is said to be *a limiting point* of a given set if inside every interval however small containing L as internal point, there is a point of the set other than L, if L is a point of the set.

DEF. A point of a set which is not a limiting point of the set is called an *isolated point* of the set.

Thus, in Theorem 1, L is a limiting point not only of the end-points of the intervals defining it, but also of any set of points such that at least one of the points lies inside each interval.

* This phrase will be used when the constituents of the sequence are positive and decrease below any positive quantity ; they then have zero as limit.

If the point L is such that every interval however small with L as end-point, either on the right or left, always contains points of the set, the point L is said to be a limit on both sides ; otherwise, provided L is not an isolated point of the set, L must be a limit on one side only, and, every interval having L on the proper side as end-point, contains points of the set.

DEF.   A set all of whose points are isolated points of the set is called *an isolated set*.

For instance the set of Ex. 1, omitting the point 1, is an isolated set.

DEF.   A set all of whose points are limiting points of the set is said to be *dense in itself* or *concentrated*.

This is the case, for instance, with the set of all the rational points, or all the irrational points, or the Liouville points. We shall return shortly to this idea of density in itself, or concentration, which is one of the most important descriptive properties of sets of points, and give further examples.

THEOREM 2.   *A set consisting of a finite number of points only has no limiting point ; and if a set has a limiting point L there is at least one sequence belonging to the set having L as limiting point.*

The first part of this theorem is evident, since the distances of any point P from the points of the set (other than P, if P belong to the set) being finite in number, one of them is the least, and an interval of length less than this distance with P as centre contains, excluding P, no point of the set.   The second part of the theorem is also simply proved ; for drawing any interval with L as middle point, there are points of the set in this interval, and we can therefore assign one of them P_1 which is not L.   Draw an interval of length LP_1 with L as middle point, and in this take another such point P_2 and so on.   The points $P_1, P_2, \ldots$ form a sequence with L as limiting point.

We have seen that L may be *a limit on one side only, or on both sides*.   In the former case, every sequence of the given set having L as limit, will lie on the same side of L, and in the latter case, there will be sequences on each side of L, and sequences whose points lie sometimes on one side, sometimes on the other of L, having L as limit.

The converse of Theorem 2 is of great importance, and follows immediately from the definition :—

THEOREM 3.   *Any set of points, not merely finite in number, has at least one limiting point.*

For, taking any finite segment of the straight line, inside it

there are a finite or an infinite number of points of the given set.
If, in whatever way we choose our segment, there are only a finite
number of points of the set in it, then, since in the complementary
segment there are always an infinite number of points of the set,
the point ∞ itself is a limiting point of the set; indeed the set has
this and no other limiting point.   If, however, we can assign a
finite segment in which there are an infinite number of points of
the set, then we can bisect this segment, and, in one at least of the
halves, there must be an infinite number of points of the set ; if
this is the case in both, we choose the right-hand half, otherwise
we choose that half in which there are an infinite number of points
of the set.   By continued bisection, we obtain in this way a series
of intervals such as were specified in Theorem 1, and the limiting
point which they define will be a limiting point of the set.   Q. E. D.

A limiting point of a set may, but need not, belong to the set.
A set which contains all its limiting points is said to be *closed*, and
one which does not do so to be *unclosed* or *open*.   A closed interval
is a special case of a closed set, and an open interval of an open
set.[17] A closed set is in many ways easier to handle than an open
set.   We can visualise a general closed set to ourselves in a way
which is not possible in the case of an open set, by means of the
following theorem :—

THEOREM 4.   *Any closed set consists of all the external and
end-points of a set of non overlapping intervals*, and conversely.*

For either the closed set consists of every point of the straight
line, or else there is at least one point P which does not belong to
the set.   Since, however, the set is closed, P is not a limiting point
of the set; therefore we can assign an interval, containing P and
no point of the given set, and this interval we can increase at
either end until we cannot do so any more without enclosing points
of the given set; the end-points are then certainly points of the
given set, since it is closed.   If Q be any other point which does
not belong to the given set, it determines in like manner such
an interval, which either coincides with that determined by P or
does not overlap with it, the end-points of each being points of the
given set, and no internal point of either interval belonging to the
given set.   Thus the points not belonging to the given set generate

---

* These intervals will be commonly referred to as *the black intervals of the
closed set* (cp. constr. of Ex. 2, p. 20).   Cp. Du Bois Reymond, *Allg. Funktionentheorie* (1882) ;   Harnack, *Math. Ann.* XIX. p. 239 (1882) ;   Bendixson, *Acta Math.*
II. etc.   Two intervals are said to *overlap* if there is a point which is internal to
them both, and to *abut* if, without overlapping, they have a common end-point.

a set of non-overlapping intervals, whose end-points are points of the given set. Any point external to these intervals must therefore also belong to the given set. Thus we see that the first part of the theorem is true.

Conversely the set of all the external and end-points of a set of non-overlapping intervals is closed, since any internal point of one of the intervals has a definite distance from the end-points of that interval, and therefore we can assign an interval free of external and end-points and containing the point in question as internal point; thus the point in question cannot be a limiting point of external and end-points of intervals, so that the latter points by themselves form a closed set. This not only proves the second part of Theorem 4, but, since the argument is independent of whether or no the intervals overlap, gives us the following Corollary :—

COR. *The set consisting of all points of any segment which are not internal to any interval of a given set of intervals (overlapping or not), is a closed set, and consists therefore of all the external and end-points of a set of non-overlapping intervals.*

We shall see all through our work how important this point of view is, and especially what a powerful method of dealing with the properties of closed sets is afforded by Theorem 4.

DEF. A set which is both closed and dense in itself is called a *perfect set.*

The instances given of sets which are dense in themselves (p. 18) were open sets, the linear continuum is evidently an example of a set which is both dense in itself and closed, *i.e.* perfect; it is not, however, typical of the whole class of perfect sets, as the following example shews.

Ex. 2. *Cantor's typical ternary set.* Take the segment (0, 1) of the x-axis and divide it into three equal parts and blacken the middle part; this is, in the ternary notation ($\cdot1$, $\cdot2$), or, as we prefer to write it, ($\cdot0\dot2$, $\cdot2$). In each un-

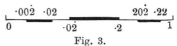

Fig. 3.

blackened segment repeat the process and so on, *ad inf.* We get in this way a definite set of non-overlapping intervals, and it is easily seen that in the neighbourhood of any end-point of one of these intervals there will be intervals of the set smaller than any assignable magnitude, so that these end-points are evidently limiting points of end-points (of course, on one side only). Similarly, any external point will be a limiting point (on both sides) of end-points; thus the set of all the external and end-points of these intervals is a set every point of which is a limiting point, *i.e.* it is dense in itself; also, by Theorem 4, it is closed, thus it is a perfect set.

The ternary numbers corresponding to this perfect set can be easily assigned. Denoting by N any combination of n figures, 0's and 1's, and by $(2N)$ the number got by multiplying N by 2, we see that any black interval obtained by our process will be of the form $(\cdot(2N)\,0\dot{2},\ \cdot(2N)\,2)$. Thus the numbers corresponding to the perfect set of points are all the terminating and non-terminating ternary fractions not involving the figure 1 (except in the equivalent form of $\dot{2}$). The limits on the left (lower) side only are those numbers which end in $\dot{2}$; the limits on the right (upper) side only are the terminating ternary fractions not involving the figure 1; all the non-terminating ternary fractions not involving the figure 1 are limits on both sides.

It will be seen that this example is typical of all perfect sets, which do not fill up any portion of the continuum. The two instances given of a perfect set are strikingly dissimilar in one particular. If we take any segment, however small, of the segment $(0, 1)$, then it either lies entirely in a black interval of Cantor's set, or else there are black intervals inside it; that is to say, inside any interval of the given segment there is always an interval free of points of Cantor's perfect set. This is in direct contradistinction to the fundamental property of the other perfect set, the linear continuum, which is such that any such interval contains only points of the latter perfect set. The example of the rational points is one in which in any such interval no interval can be found free of points of the set, though in such an interval there will be points other than those of the set, since the set is unclosed; this latter set illustrates the property under discussion therefore better than the continuum itself. This property is called *density*, and must be carefully differentiated from the property of being dense in itself *.

DEF. A set is said to be *dense in a given segment*, or *dense everywhere in a given segment*, when, assigning any interval, how- ever small, inside the given segment, there is always inside the interval a point of the set.[18]

A set is said to be *dense nowhere in a given segment*, if, assign- ing any segment, however small, in the given segment, there is an interval inside the assigned segment entirely free of points of the set.

A set which is *dense* is then certainly *dense in itself*, but not *vice versa*. Indeed we see from the above that a perfect set may be either dense. everywhere or nowhere in a given segment; it might also be dense in some segments and nowhere dense in others.

* Cantor remarks: "The expression *dense in itself* denotes a definite property of a set; on the other hand *dense everywhere* is not *a priori* a property of a set, but becomes such only when we consider the set in connection with some determinate n-dimensional portion of space, with respect to which we can say that the set is dense everywhere" (*Math. Ann.* XXIII. p. 472).

It is easily seen, however, that *a closed set which is dense in any segment and therefore perfect, consists of all the points of that segment.* The black intervals (Theorem 4) of a perfect set cannot abut anywhere, since a common end-point of two abutting intervals could not be a limit on either side. If the perfect set is dense nowhere in the given segment, the corresponding black intervals are said to be *dense everywhere* in the segment, because the set of points internal to them is dense everywhere. The black intervals of the general perfect set will not be dense everywhere; any interval, however, which has no part common with a black interval, consists entirely of points of the perfect set. Thus the black intervals together with all those intervals in which the set is dense everywhere, form a set of intervals dense everywhere. The points not internal to these intervals all belong to the perfect set. Thus *the most general perfect set consists of all the points not internal to a set of intervals dense everywhere, together with possibly all the internal points of one or more of those intervals.*

If we take a perfect set nowhere dense and omit those of its points which are end-points of the corresponding black intervals, we get a set which is dense in itself and nowhere dense. We notice too that such a set may be said to be *dense in itself on both sides,* that is to say every point of it is a limit on both sides, just as it would be if the set were dense everywhere in the segment. If, on the other hand, we took the set consisting of all the right-hand end-points of the same black intervals, we should have a set which might be said to be *dense in itself on the right only,* which is, of course, only possible when the set is nowhere dense. The numbers corresponding to a set which is dense in itself on both sides are such that there is no definite number immediately preceding or immediately following any number of the set; the same is true of the numbers of a set which is dense in itself on the right only, or of a set which is dense in itself on the left only.[19]

11. It is to be remarked that the properties of sets of points discussed in the preceding paragraph, presuppose, as is always the case in Cantor's work, the existence of an underlying continuum. It is, however, easy to frame corresponding definitions, and discuss corresponding properties when all that is presupposed is the existence of some underlying set. In speaking of a segment we shall then naturally confine our attention to points of the fundamental set lying in that segment. Theorem 2 will perhaps serve us most logically as a definition of a limiting point, and a sequence we shall define by the property of possessing one and only one limiting point. The property of a component set of being dense in itself is, as was pointed out (footnote, p. 21), independent of the fundamental set; the property of being closed is however evidently dependent on the latter, provided the fundamental set be unclosed, for the component set might in that case be an open set, and yet closed with respect to the fundamental set, if it contained all those of its limiting points which were points of the latter. Similarly the property of being dense everywhere can be so defined as to represent a general relation between a component set and the fundamental set, as follows:—

DEF. A component set is said to be *dense everywhere in the fundamental set*, when, assigning any interval containing a point of the latter set as internal point, there are always inside this interval points of the component set.

A component set is said to be *dense nowhere in the fundamental set*, if, assigning any segment containing a point of the latter set as internal point, there is always inside this segment an interval containing a point of the fundamental set and no point of the component set as internal point.

12. Derived Sets. Limiting points of various orders. It is evident from the preceding articles that, while the limiting points of a given set are always present, except in the trivial case when the number of points in the given set is finite, the limiting points may themselves be infinite in number, and may even contain every point of the given set and other points as well.

DEF. The set consisting of all the limiting points of a given set E is called *the first derived set* of E, and will be denoted by E_1. The first derived set of E_1 is called the *second derived set* of E, and denoted by E_2; and generally the first derived set of E_n, where n is any integer, is called the $(n+1)th$ *derived set of* E, and denoted by E_{n+1}.[20]

Any point of E_n which is not a point of E_{n+1} is called a *limiting point of* E *of the nth order*.

Ex. 3. In Ex. 1, p. 17, the first derived set consists of the point 1 alone, which is a limiting point of the first order, the second derived set not existing at all.

If, denoting by T_1 the set obtained, as in that example, by continued bisection of the right-hand half of a segment, we place in the segment (0, 1) a set T_1 and then in each of the intervals between consecutive points of the set so obtained we place a set T_1, the set of points so obtained will have a limiting point of the first order at each of the points of the set of Ex. 1, and this latter set will be the first derived set ; the second derived set will consist of the point 1 alone, and will be a limiting point of the second order ; the third derived set will not exist at all.

Denoting a set constructed in the above manner in any segment by T_2, we can obtain a set in which the point 1 is the single limiting point of the third order, and limiting points of higher order do not exist, by inserting between each pair of consecutive points of a set T_2 a set T_1. Similarly we can construct

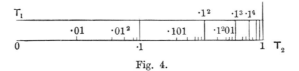

Fig. 4.

in succession sets in which the point 1 is a limiting point of the fourth, fifth and any finite order we please, the sets T_{n-1}, T_{n-2}, ... previously constructed being respectively the first, second, ... up to the nth derived sets, and all derived sets of higher order not existing at all.

Ex. 4.   Take the set T_1 in the segment $(0, 1)$; insert between the points 0 and $\cdot 1$ a set T_1, and mark these points with dots, and their limiting point $\cdot 1$ with an ordinate of height unity ; between the points $\cdot 1$ and $\cdot 11$, insert a set T_2, and mark its isolated points with dots, its limiting points of the first order by ordinates of height unity, and its limiting point of the second order $\cdot 11$ by an ordinate of height 2 ; generally between the points $\cdot 1^{r-1}$ and $\cdot 1^r$ insert a set T_r and mark its isolated points by dots, and its limiting points of various order by ordinates of corresponding height.

By this process, carried on *ad inf.* we obtain, including the point 1, a closed set of points E. If we draw parallels to the axis at heights 1, 2, 3, ... the first, at height unity, will be cut by all the ordinates in the first derived set E_1, the second in E_2, and so on. Every successive parallel is cut by the ordinates more and more to the right, the points crowding up to the point 1 which alone is common to all the derived sets.

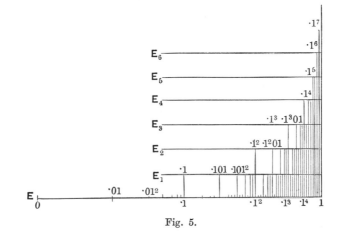

Fig. 5.

This set E possesses derived sets of every order, and each such set consists of an infinite number of points.   The binary numbers corresponding to the points of E are all those of the form $\cdot 1^m \, 01^\alpha \, 01^\beta \, 01^\gamma \ldots 01^\nu \, 01^n$, where there are at most n zeros, and the corresponding powers of 1 may be absent, or the indices a, β, $\ldots$ have any integral values whatever.

In Ex. 4 every limiting point, except the point 1, has a definite order.   If we wish to extend the idea of order so as to be applicable to the point 1, we must say that the order of the point 1 is infinite ; we shall see, however, that with our present ideas of an infinite number this terminology is undesirable and much too indefinite, since points which bear the same relation to one another as limiting points of orders 1 and 2, for instance, are in this way classified together.   In a subsequent chapter we shall see that these very considerations lead us to an extended scheme of ordinal numbers, Cantor's numbers, which enable us to differentiate completely between objects like limiting points by means of ordinal indices.

For the present it is only necessary to realise that limiting points may have these characteristics, in order to avoid mistakes which easily occur when we trust to our pristine conceptions of limiting points.

To illustrate what was said above about limiting points of infinite order, we have only to take the set constructed in any one of the preceding examples, and place between each pair of consecutive points a set of the type of Ex. 4.   In this way we get whole infinite sets of points of the type of the point 1 in Ex. 4, and the limiting points of this set, bearing the same relation to its isolated points as the limiting points of the first, second, ... orders did to the original set, ought properly to be considered as of one, two, ... higher orders.

13. Deduction. DEF. The process of taking all the points common to an infinite series of sets of points we shall call *deduction*.

Thus in Ex. 4 the point 1 is deduced from the series of derived sets.   Further we see that it is possible to construct examples on the principles already employed, in which the points deduced from the derived sets themselves form an infinite set, with derived sets of any required order, or with derived sets of every order from which we can obtain a new set by deduction, and so on.

In performing the process of deduction we are of course only concerned with those points of each set of the series which are contained in all preceding sets of the series, in general the sets from which we wish to deduce are originally given in this form (as for instance in the case of the derived sets, each of which is contained in all the preceding derived sets). When this is not the case we must reduce the given series to such a form, before investigating whether or no we can apply the following theorems.

THEOREM 5. **Cantor's Theorem of Deduction**. *If $G_1, G_2, \ldots$ be a series of closed sets of points, such that each is contained in all the preceding sets, then there is at least one point common to all the sets ; and the deduced points form a closed set.*

For, let us choose one point from each set; let these be

$$P_1, P_2, \ldots\ldots$$

Then either there are only a finite number of these points which are distinct from one another, or else they form an infinite set. In the former case there is at least one of the points such that, assigning any integer m, there is an integer $n > m$, such that P_n is identical with the point in question ; then P_n will, by hypothesis, belong to every set up to the nth inclusive. that is to say, since m could be chosen as large as we please, P_n is common to all the sets. In the second case, by Theorem 3, the points have a limiting point L, which, as a limiting point of all the points after P_m (that is, of points of G_m), is certainly a point of the closed set G_m, and therefore, as before, is common to all the sets.

To prove that the deduced points form a closed set, we have only to consider that if they are finite in number they have no limiting point, and therefore form of course a closed set; otherwise let L be any one of their limiting points, and $P_1, P_2, \ldots$ a sequence of the deduced points having L as limit, then since P_1 is a point of G_1, P_2 of G_2, and so on, L is a limiting point of a sequence of points chosen one from each successive set of the series, and is therefore, as before, a deduced point. Thus the deduced points form a closed set. Q. E. D.

When the sets of the series are unclosed, we have no security that any point at all can be obtained by deduction. Take for example the series of derived sets $E_1, E_2, \ldots$ of Ex. 4, omitting the point 1 from every one. These sets have no common point at all.

The following theorem is an immediate consequence of the definition, and is true whether or no the defining sets are closed.

THEOREM 6. *The deduced set cannot be identical with any of the defining sets* unless from and after a definite stage all sets of the series are identical.*

---

* Subject to the above condition that each set is contained in all preceding sets.

14. Theorems about a set and its derived and deduced sets.

THEOREM 7.   *A limiting point of limiting points is a limiting point.*

For let L be a limiting point of limiting points of a given set E. Let $L_1, L_2, \ldots$ be a sequence of limiting points of E, having L as limit, and all lying on the same side of L. Then since L_{2r} is a limiting point of E, therefore, by Theorem 2, between L_{2r-1} and L_{2r+1} there will be a point, say P_r, of E. These points $P_1, P_2, \ldots$ form then a sequence having L as limit, so that L is a limiting point of E.   Q. E. D.

COR.   *Every derived set is a closed set.*

THEOREM 8.   *The first derived set of a closed set is contained in that set.*

This follows at once from the definitions.

COR.   *Each derived set is contained in all the preceding derived sets.*

DEF.   From Theorem 5 together with the two corollaries just given it follows that, if the derived sets of every order exist, there is always a closed set of points common to all the derived sets, this set we call *the first deduced set of E*. As at the end of § 3 we can easily construct examples of sets whose first deduced sets have derived sets of every order, in such a case the first deduced set will itself have a first deduced set, and this we call *the second deduced set* of the original set; and generally the first deduced set of the nth deduced set, where n is any integer, is called the $(n+1)$*th deduced set of the original set.*

By the theorem and corollaries quoted, all the sets so obtained are closed and each is contained in all the sets previously obtained: it is also quite easy, in the manner indicated, to construct sets in which deduced sets of every order exist; this is done explicitly in Ex. 5. When this is the case we can deduce a closed set from the series of deduced sets, or which is the same thing, from the whole series of sets obtained by repetition of the processes of derivation and deduction in the manner indicated.

In this way we can continue in systematic manner our processes of derivation from each set obtained, and deduction from every series of sets so obtained.

It is to be noticed that, if an infinite series of sets each of which is contained in all the preceding sets have a definite *last* set, the deduced set will be the same as this last set; thus, arrived at any definite set by means of the processes of derivation and deduction, if we now deduce, we shall only get that set already obtained, and we must apply the process of derivation in order to progress a stage farther. Whenever we have an infinite series of the sets without a last one (that is a definite law by which any such set can be recognised), we progress a stage further on by deduction.

By the theorem and corollaries quoted, every set obtained in this way from a given set E is a closed set, and is contained in all the preceding sets (not including E, unless E is closed). *These sets are called the series of derived and deduced sets of E.*

Ex. 5.   Take the set E of Ex. 4 (Fig. 5), and in the interval (0, ·01) insert a set similar to E. The point ·01 becomes in this way a point of the first deduced set of the set we are constructing. In the interval (·01, ·011) insert a set similar to that already constructed between the point 0 and the left-hand end-point ·01 of this interval, that is a set similar to E.

Similarly in each interval between consecutive points of the set E, we insert a set similar to that already constructed between the point 0 and the left-hand end-point of the interval in question.

We obtain in this way a perfectly definite set, say F, which is of great assistance to us in picturing some of the possibilities in the way of derivation and deduction.

Every point of E to the left of the first limiting point of the first order, ·1, evidently belongs to the first deduced set of F; while ·1 itself belongs to the first derived of the first deduced set of F, and the same will therefore be true of every point of E between ·1 and the next limiting point of the first order ·101; this latter point will belong to the second derived of the first deduced set of F.

In this way we see that the first limiting point of the second order of E, the point ·11, will belong to the second deduced set of F, and the first limiting point of the third order of E, the point ·111, to the set deduced from the first, second, ... deduced sets of F, and so on to sets further and further on in the succession of derived and deduced sets of F.

We have then before us in the series of derived and deduced sets of an arbitrary set a never-ending vista of possibilities, with a perfectly definite property of order or succession to which we shall have reason constantly to recur, and into which we shall find we are able to introduce certain simplifications of a very important character. For the present it is only desirable to bear in mind that the series may be one of a very complicated nature, as, for instance, in Ex. 5, and we proceed to enunciate and prove certain elementary theorems with respect to the sets of the series when the given set has certain standard forms.

THEOREM 9. *The first derived set of a set which is dense in itself is perfect.* For, the set being dense in itself, it is, by definition, contained in its first derived set; any point of the latter, therefore, is a limiting point of its own points, thus the first derived, which by Theorem 7, Cor., is closed, is also dense in itself, and therefore perfect.

The next theorem does not require proof, being an immediate consequence of the definitions.

THEOREM 10. *A perfect set is identical with its own first derived; and, conversely, a set which is identical with its first derived is perfect.*

COR. *By means of the processes of derivation and deduction we obtain always the same set from a perfect set; and if by means of these processes any set is identical with any other set, this must be a perfect set.*

From the above, together with Theorem 6, the following theorem follows :—

THEOREM 11. *Unless one of the sets obtained by derivation and deduction is perfect, all the derived and deduced sets are distinct from one another.*

The next theorem again does not require proof:—

THEOREM 12. *The first derived of a set which is dense everywhere in a segment (A, B) consists of the whole continuum (A, B); and conversely, if the first derived consist of the whole continuum (A, B), the original set was dense in (A, B).*

Similarly the same is true if instead of a segment (A, B) we take any perfect set; but the theorem is not true if instead of a perfect set we substitute any set we please.

THEOREM 13. *The first derived of a set which is dense nowhere is itself dense nowhere.*

For, given any segment, we can determine in it an interval in which there are no points of the given set, and therefore, by Theorem 2, at most the end-points of this interval can be points of the first derived set, but no internal point of this interval can be such. Hence in every segment there is an interval entirely free of points of the derived set, that is, it is dense nowhere.

15. Intervals and their Limits. A set of intervals on the straight line may be (1) *overlapping*, that is such that at least two of the intervals overlap (see p. 19, footnote), or (2) *non-overlapping*, *i.e.* such that no two overlap.

In Theorem 1 we had a simple example of a set of overlapping intervals, and it was shewn that they determined a limiting point. If in that theorem we were to omit the condition that the length of the segment should decrease without limit, the left-hand end-points would have one limiting point, and the right-hand end-points another limiting point, and the intervals would determine instead of a limiting point a limiting interval, internal to all of them, whose end-points would be these two limiting points. The connection between intervals and limiting points is exceedingly important. Theorem 14 is a generalisation of Theorem 1, and the corollary stands in a certain sense in contradistinction to the latter theorem, since in it the limiting points are never internal to the intervals while in Theorem 1 the contrary was the case.

DEF. A *limiting point of a set of intervals* is one such that in any segment containing the point as internal point, there are intervals of the given set.[21]

A *limiting interval of a set of intervals* is one such that in any segment containing the interval as internal interval there are intervals of the given set containing the limiting interval, with at most one end-point common with it.

It is evident from the definition that a limiting point of a set of intervals is a limiting point both of the right-hand and the left-hand end-points of the intervals which determine it, and we shall have to distinguish, as we did before, between the case when the point is a limit on one side only, or on both sides; in the former case the segment used in the definition can be taken to have the limiting point as one of its end-points. If the point be a limit on one side only, it cannot be internal to the intervals, but may be either (1) external to all of them, or (2) an end-point of one or more of them on the side on which it is not a limit, in which case it may be said to be *semi-external to* the set of intervals, or (3) finally it may be a common end-point of a sequence of intervals as in Theorem 1, all lying on the side on which the point is a limit. If the point be a limit on both sides, it may be either internal or external to the intervals, or may be an end-point, on both sides of such sequences of intervals as above (3).

In the case of a set of non-overlapping intervals these possibilities are materially simplified. In this case a limiting point on one side must evidently be either external or semi-external, and a limiting point on both sides can only be an external point of the intervals. Further a semi-external point can be an end-point of only one of the intervals. A set of non-overlapping intervals evidently has no limiting interval.

THEOREM 14. *Any set of intervals, not merely finite in number, which has no limiting interval, has at least one limiting point.*

This follows at once from Theorem 3, the set of points there mentioned being taken to be that of the left-hand (or right-hand) end-points of the intervals. These certainly have at least one limiting point, and, having determined one such limiting point, and a sequence of the defining set having that point as limit, we only have to take the corresponding right-hand (or left-hand) end-points and determine their limiting point or points.

If this point or one of these points coincides with the point already found, this latter is a limiting point, otherwise it is an end-point of one or more limiting intervals, and we must proceed to investigate whether there is another limiting point of the left-hand (or right-hand) end-points which coincides with a limiting point of the right-hand (or left-hand) end-points of the same intervals; if this is the case we have a limiting point, and if not the set of intervals has only limiting intervals.

COR. *Any infinite set of non-overlapping intervals has at least one limiting point.*

16. Upper and lower Limit. The following remarks, though they have nothing particular to do with the main matter of this chapter, must, because of their bearing on the use of the term limit, find their place here.

A set of points must be such that it either (1) consists of points which are all internal to some finite segment (*A*, *B*) or (2) we can assign no finite segment containing every point of the given set. In the latter case, however, we can distinguish two cases, (2*a*) it may be possible to assign one definite point *P*, such that the segment (*P*, ∞) contains every point of the set, or (2*b*) this may not be possible. In the case (1), if the points *A* and *B* are not both either points of the given set or limiting points of the set, we can curtail the segment without allowing any points of the set to escape. Diminishing the segment in this way as much as possible, we must ultimately come to two definite points *A'* and *B'*, such that any further diminution would cause some point of the set to become external to the segment.

These points A' and B' may belong to the set, but this is not necessary if the set be unclosed. If they do not belong to the set however it is easily seen that they must be limiting points of the given set, and, of course, limits on one side only.

The same is true of the corresponding numbers. If of the two numbers a' and b', corresponding to A' and B', a' is that which is algebraically less than b', then a' is *called the lower limit and b' the upper limit* of the set of numbers.

The term is not altogether a happy one, since a' and b' are not necessarily limits in the sense in which we have hitherto used the term. A' and B' may be isolated points of the set, or *points of condensation,* that is, limiting points in the sense in which we have used the term ; in the latter case they will belong to the set if it be closed, but not necessarily if it be open ; in the former case they must of course belong to the set.

Similarly in case $(2a)$ we can determine a definite point P', which is either a point of the set or a limiting point, and such that every point of the set is contained as internal or end-point in the segment (P', ∞). In this case, according as the numbers corresponding to the set of points are all algebraically greater or less than p', p' is called the lower or upper limit of the set. In this case the same remarks as before apply to the meaning of the term limit, as far as it concerns p' or P'; the point ∞ must however be a point of condensation. To differentiate between the cases when the upper and the lower limits are infinite, the following convention is made as to sign*. According as p' is the lower or upper limit, the upper limit is said to be $+ \infty$, or the lower limit $- \infty$. In case $(2b)$ the upper and lower limits are $+ \infty$ and $- \infty$ respectively. Summing up we have the following statement :—

Any set of numbers has a definite upper limit L and a definite lower limit l (either of which may be finite or infinite in absolute magnitude). If the set be closed, both limits belong to the set, if open they need not do so, but if not, they certainly correspond to points of condensation of the set.

The difference $(L - l)$ is sometimes called *the oscillation* of the set of numbers.

* Cp. p. 6, footnote.

CHAPTER IV.

POTENCY, AND THE GENERALISED IDEA OF
A CARDINAL NUMBER.

17. The principle of measurement has been analysed into two primary constituents. The first is the determination of a standard object, or unit; the second is the calculation of how many times the unit has to be taken so as to be equivalent to the object to be measured in respect of size. The answer to the question—how many times?—would be given in the form of a positive integer, and would be determined by counting, that is, by setting up a (1, 1)-correspondence between the repeated unit and the several parts of the object under discussion, properly divided and arranged. Sometimes the question could not be answered accurately, and it was necessary to take a new unit. When the old unit could itself be accurately measured integrally by means of the new unit, the idea of ratio in the Euclidean sense and the introduction of fractional symbols, enabled people to do without the new unit. In the same way the generalised idea of ratio and the concept of an irrational number such as we have defined it, made the new unit superfluous, whenever the ratio could be expressed approximately by means of rational numbers forming a sequence.

This idea of measurement we are going to apply to the theory of sets of points, or numbers. We are going to answer, as far as we can, the question,—how large is a given set?—or, put more precisely,—how many points are there in a given set?

We shall regard the question as primarily answered, if we can give an integer which expresses exactly how many times we must repeat some known set, in order that we may be able to arrange the points of the given set, and those of the repeated set, or unit, in (1, 1)-correspondence. Subsequently we shall endeavour, by the introduction of appropriate numerical symbols, called *potencies*, to do away with the necessity for the use of any unit, except the simplest conceivable one, *i.e.* a single point.

DEF. Any two sets which can be brought into (1, 1)-correspondence are said to be equivalent or to have the same *potency*.

18. Countable Sets.

Starting with a single point as unit, it is evident that the only sets which we can measure are the so-called *finite sets*, that is those consisting of a finite number of points only, or, in descriptive language, those devoid of limiting points. *The simplest potencies are thus the natural numbers themselves.* Following out the idea developed in the preceding chapter, of taking the limiting points as the basis of our investigations, we naturally take as our next unit a sequence of points, that is a set having only one limiting point, and ask what sets can be arranged so that their points can be brought into (1, 1)-correspondence with those of a sequence. The following theorem shews why the term "*countable*" is applied to all such sets of points.

THEOREM 1. *A sequence can be brought into* (1, 1)-*correspondence with the natural numbers in their entirety.*

Let L denote the limiting point, this may belong to the sequence or not; if it belongs to the sequence we assign to it the number 1. Between L and any other point M we take any point A, and on the opposite side of L we take any point B. Then, since in the segment* bounded by A and B and containing the point M there is no limiting point of the sequence, the number of points of the sequence in it is finite, say n, and we can take them in any order we please and attach to them the numbers $1, 2, ..., n$, or $2, 3, ..., n, n + 1$, according as the point L is not, or is, a point of the sequence.

First let us assume that L is not the point ∞. Then bisecting (L, A) at A_1, by the same argument there are only a finite number, say n_1, of the points of the sequence in (A, A_1), and we can take them in any order, and attach to them the next n_1 integers. Then bisecting (L, B) we can do the same thing; and so we go on, bisecting alternately the segments on each side of L, and counting the points in that half segment which is not terminated in L.

In this way we attach to each point of the sequence a definite integer, and *vice versa*. Thus the sequence is brought into (1, 1)-correspondence with the natural numbers.

If L be the point ∞, we only have to alter the process so far that, instead of continually bisecting the segments on each side of

---

* The straight line being regarded, as in the preceding chapter, as divided into two segments by A and B.

L, we take the segments $(A, B), (A, A_1), (B, B_1), (A_1, A_2), (B_1, B_2), \ldots$ all of the same length.

COR.   *Any two sequences can be brought into (1, 1)-correspondence with one another.*

DEF.   Any set which can be brought into (1, 1)-correspondence with some or all of the natural numbers is said to be *countable*, and, if not a finite set, is said to be *countably infinite*.

It follows from this definition that *any component of a countable set is countable.*

THEOREM 2.   *Any set which can be divided into two parts, each of which is countable, is itself countable.*

For, let the points of the one part be denoted by

$$P_1, P_2, \ldots\ldots,$$

and those of the other part by

$$Q_1, Q_2, \ldots\ldots ;$$

then we can arrange them in the order

$$P_1, Q_1, P_2, Q_2, \ldots\ldots ,$$

and " count " them as they stand.

COR. 1.   *The set consisting of all the points of any finite number of countable sets is itself countable.*

COR. 2.   *If the set consisting of all the points of any finite number of sets is countable, each of the sets is countable.*

COR. 3.   *Any set of points whose limiting points are finite in number is countable.*

THEOREM 3.   *Any set which can be divided into a countable number of countable parts is itself countable.*

For, let us denote the points by means of a system of double indices, the first giving the number of the part to which it belongs, and the second its place in that part, when properly arranged. Arrange these pairs of numbers in the form of a wedge:

$$(1, 1) \ (1, 2) \ (1, 3) \ (1, 4) \ \ldots\ldots$$
$$(2, 1) \ (2, 2) \ (2, 3) \ \ldots\ldots$$
$$(3, 1) \ (3, 2) \ (3, 3) \ \ldots\ldots$$
$$(4, 1) \ \ldots\ldots$$
$$\ldots\ldots$$

We can then take each column in order and read it from the top to the bottom; since each column contains only a finite number of brackets, we can " count " the brackets in this order.

COR. *Any system which can be characterised by a triply infinite, quadruply infinite,... etc. system of indices is countable*.*

By the above, any set which can be brought into (1, 1)-correspondence with a sequence repeated any number of times (that is to say, any set which can be brought into (n, 1)-correspondence with a sequence), is countably infinite. On the other hand it is clear that any two countably infinite sets can be brought into (1, 1)-correspondence, thus we only get one new potency by this means, viz. the potency of the natural numbers. In accordance with the principles with which we started, we shall assign to this potency a symbol a. The symbol a, regarded as an extension of the cardinal numbers 1, 2, ... m, ... n, ... which we saw to be the simplest possible potencies, has in many ways the properties of the symbol ∞, used in an earlier part of our work: we shall see, however, that a is more precise than ∞, and that there are other potencies which have in an equal degree the properties of the symbol ∞; all such potencies are called *transfinite* and the corresponding sets are called *infinite* sets. It is clear that no proper component of a finite set can have the same potency as the whole set, but this is not true of a countably infinite set. For instance, all the even integers can be brought into (1, 1)-correspondence with all the integers, by making any even integer $2k$ correspond to k. Thus *all the even integers form a countably infinite set.* Similarly *all the odd integers form a countably infinite set.* What is true of the potencies of sets of integers follows by (1, 1)-correspondence for the potencies of components of any countably infinite set. Thus *a countably infinite set has not only components of every finite potency, but also proper components which are countably infinite.* This property—that a set can be brought into (1, 1)-correspondence with a part (proper component) of itself—has been sometimes taken as the defining characteristic of *an infinite set* in contradistinction to *a finite set*†. [22]

19. The properties of the number a already discussed are conveniently expressed in symbolic form. To do this it is necessary to give a preliminary definition of *addition* and *multiplication* of potencies. The general definitions and the discussion of the

---

* It is to be noticed that we cannot deduce as a second corollary that any set whose limiting points are countable is itself countable. This theorem will however be proved shortly.

† Dedekind, *Was sind und was sollen die Zahlen?* Cp. Russell, *Principles of Mathematics*, p. 121, seq.

validity of the processes are reserved for the chapter on Cantor's numbers.

DEF. Let G_1, G_2, ... be any countable number of sets of points without common points, and let their potencies be g_1, g_2, ... The set G consisting of every point belonging to these sets is said to have as potency the *sum* of the potencies of G_1, G_2, ...

Regarding G as objectively known, and denoting its potency by g, the process of recognising that the sets G_1, G_2, ... actually do form G in the manner indicated, is represented symbolically by the equation

$$g = g_1 + g_2 + \cdots$$

and is called *addition*.

These definitions can be generalised by using any set (instead of a countable number) of sets. The notation however must then be modified.

The set of all the common points of a set of sets is called their *sum*, and the process of forming the sum is called *addition of sets*. Unless, however, the sets are without common points, the potency of the sum is not necessarily the sum of the potencies.

DEF. If the sets G_1, G_2, ... are all equivalent, and γ be their number, whether finite, or countably infinite, or their potency in the general case, the equation

$$g = \gamma g_1$$

is substituted for the preceding equation; g is then called the product of the factors γ and g_1. The process is then called *multiplication*.

In other words *any set which can be brought into* $(g_1, 1)$-*correspondence with a set of potency* γ *has the potency* γg_1.

It is immediately evident that these definitions agree in the case of finite potencies with those given in Arithmetic, and that the addition of a finite number of potencies may be effected by the repetition of the process of adding two potencies, the commutative and associative laws holding. Multiplication, as in Arithmetic, may be extended to any *finite* number of factors by repetition of the process of multiplying two factors, and the proofs given in Arithmetic of the commutative, associative and distributive laws will still hold.

When the factors of a product are all equal, the product is called a *power* of g_1, and the usual notation is employed.

Thus, by the preceding theorems,

$$a + n = \quad a = \quad n + a,$$
$$a + a \equiv 2a = a2 = a,$$
$$na = an = a,$$
$$a \cdot a \equiv \quad a^2 = a,$$
$$a \cdot a \cdot a \equiv \quad a^3 = a,$$
$$a^n = a.$$

20. Countable Sets of Intervals. A most important example of a countable set is the following:

THEOREM 4. CANTOR'S THEOREM OF NON-OVERLAPPING INTERVALS. *Every set of intervals on a straight line is countable, provided no two overlap**.

For, let e_1, e_2, ... be any sequence of positive numbers having zero as limit, and let us consider only the case when the intervals all lie in a finite segment (A, B) of length l. There is in this way no loss of generality since we can bring the whole infinite straight line into $(1, 1)$-correspondence with (A, B)†.

The number of intervals of the given set whose magnitude lies between e_r and e_{r+1} must be finite, since the intervals do not overlap: let these be arranged in any order and denoted by G_r. Then G_r is finite and the whole set can be arranged in the order G_1, G_2, ... and "counted" as it stands; which proves the theorem.

THEOREM 5. *Given any set of intervals (overlapping in any way), we can determine a countable set from among them, such that every point internal to any interval of the given set is internal to an interval of the countable set, and vice versa*‡.

Take, first, any one of the intervals, and let us denote it by d or δ. Then, either there is no interval of the given set which overlaps with d on the left (that is, which contains the left-hand end-point of d), or else we can determine such an interval. In the latter case we denote by δ' the part of this interval which extends beyond d to the left, and by d' the interval itself.

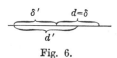

Fig. 6.

Proceeding in this way towards the left, we must ultimately either come to an interval of the given set having no interval

* Stated and proved for sets of regions in n-dimensions in precisely this way by Cantor, *Math. Ann.* xx. p. 117 (1882).

† *Infra*, § 23.   ‡ *Proc. L. M. S.* xxxv. p. 384.

overlapping with it on the left (this might, of course, have been the original interval d), or else the parts δ', δ'', δ''', ... being themselves non-overlapping intervals, must get smaller and smaller without limit, and define a limiting point P external to all of them, and therefore external to the corresponding intervals d', d'', d''', ... of the given set*. Such a point P may, however, be internal to some other interval of the given set; in this case, we choose out any one of the intervals containing P, say D. There will be only a finite number of the intervals d, d', d'', ... which do not overlap with D. Let d^i be the first which overlaps with D; then we select the intervals d, d', d'', ... $d^{(i)}$, D, and omit from consideration all the intervals $d^{(i+1)}$, $d^{(i+2)}$, ... Proceeding on these lines, we can only be stopped (1) by coming to an interval of the given set having no interval overlapping on the left, or (2) by the parts δ', δ'', δ''', ... becoming smaller and smaller and defining a limiting point Q on the left of all of them, such that no interval of the given set contains Q as internal point. In the latter case Q would be external to every one of the given intervals, unless it were semi-external, being a right-hand end-point of one or more of them. In case (1) the left-hand end-point P of the final interval might be the right-hand end-point of one or more of the given intervals, it would then be called an isolated end-point, and we should take any one of the intervals of which it was the right-hand end-point and proceed as before to the left. Otherwise in case (1) P would be external to all the given intervals except such as have it for left-hand end-point, and would again be a *semi-external* point of the given intervals.

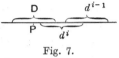

Fig. 7.

Having proceeded in like manner on the right of d, we start afresh in each of the one or two segments left over, and take again any one of the intervals and treat it as we did d. Continuing this process, we get a set of *non-overlapping* partial intervals δ', δ'', δ''', ... etc., which, by the preceding theorem, can be arranged in *countable* order δ_1, δ_2, δ_3, ... and, corresponding to these, a *countable* set of the given intervals, d_1, d_2, d_3, ... such that each δ_r coincides with part or the whole of some d_i, and has at least one end-point common with it. Any other of the given set of intervals lies entirely within one of the d_i's, or else within a set of the d_i's which overlap or abut all along.

* Ch. III. § 15.

By our construction any point which is internal to any one of the given intervals is internal to one of the c_i's, and any external point of the given intervals is external to the d_i's, any semi-external point of the given intervals is semi-external to the d_i's, and finally, any point which is an isolated end-point of intervals of the given set both on the right and on the left is an end-point of two abutting intervals of the d_i's.

Such a countable set $d_1, d_2, \ldots$ chosen in the manner indicated from among the intervals of the given set serves to classify the points of the straight line in reference to the set of intervals as completely as the whole set itself.

On account of the fundamental character of this theorem, we add an alternative proof. Yet a third proof will be found virtually given in Part II, where the corresponding theorem is stated and proved for any number of dimensions.

LEMMA*. *If G be a countable set of points, any set of intervals whose end-points are points of G, is countable.*

For, denoting the points of G by $P_1, P_2, \ldots$ and the interval whose end-points are P_1 and P_2 by (1, 2), and so on, the theorem follows by Theorem 3.

COR. *Any set of intervals whose end-points are rational is countable.*

THEOREM 5 (Alternative proof).

If P be any point internal to one of the given intervals (A, B), there is a rational point R between A and P, and one R' between P and B. There is therefore an interval (R, R') with rational end-points, containing P and lying inside (A, B). Thus the intervals with rational end-points, each of which lies inside one of the given intervals, contain as internal points every internal point of the given intervals. By the preceding corollary these intervals are countable, and may be denoted by $\delta_1, \delta_2, \ldots$.

Take an interval of the given set containing δ_1 and denote it by d_1. Let the first of the δ's not contained in d_1 be δ_i. Take an interval of the given set containing δ_i and denote it by d_2, and so on. We thus get a countable set of the given intervals, $d_1, d_2, \ldots$ containing as internal points every internal point of the δ's and therefore of the given intervals. On the other hand, the d's being intervals of the given set, every point internal to them is internal to the given set. Thus the intervals $d_1, d_2, \ldots$ satisfy the requirements. Q. E. D.

* The principle of this Lemma was first used by F. Bernstein in his proof of Theorem 19.

COR. *Given any set of intervals, a countable set of intervals with the same internal points can be found, the end-points of these intervals being rational (or belonging to any convenient set dense everywhere).*

This has been incidentally proved in the course of the preceding proof.

THEOREM 6. THE GENERALISED HEINE-BOREL THEOREM. *Given any closed set of points on a straight line and a set of intervals so that every point of the closed set of points is an internal point of at least one of the intervals, then there exists a finite number of the given intervals having the same property* *.

By the preceding theorem we may take the intervals to be countable, and denote them by $d_1, d_2, \ldots$. Then d_2 may overlap with d_1, and, if so, may be divided by d_1 into two or into three parts. Denoting d_1 by δ_1, and the non-overlapping whole, part, or parts of d_2 by δ_2, or by δ_2, δ_3, and similarly denoting the parts of each successive interval not common to it and preceding intervals, we get a set of non-overlapping partial intervals $\delta_1, \delta_2, \ldots$, and we may assume that we have omitted any which do not contain points of the given set as internal points or as isolated end-points. If the chosen intervals $\delta_1, \delta_2, \ldots$ have no external or semi-external point they are finite in number (§ 15), and it is clear that any such limiting point would be a limiting point of the given closed set and therefore a point of that set. But this is impossible, since no internal point of the given intervals can be external or semi-external to the non-overlapping intervals $\delta_1, \delta_2, \ldots$.[23] Therefore there are only a finite number of the intervals $\delta_1, \delta_2, \ldots$, and the corresponding intervals d, of which they form parts, are also finite in number, and contain every point of the given closed set as internal points. Q. E. D.[24]

A special case of the above is the following:—

THEOREM 7. THE HEINE-BOREL THEOREM. *Given any set of intervals such that every point of a given closed segment (A, B) is internal to some interval of the set, we can choose out a finite number of the intervals having the same property†.*

THEOREM 8. *Given any set of intervals (overlapping in any way), we can determine from them in one and only one way, an equivalent set of non-overlapping intervals, such that every internal, external or semi-external point, as well as every isolated end-point of the one set bears the same relation to the other set.*

* *Proc. L. M. S.* xxxv. p. 387.

† Borel, *Ann. de l'École Norm.* (3) xii. p. 51 (1895) ; *Leçons sur la Théorie des Fonctions,* p. 42 (1898). The theorem was stated by Borel only for a countably infinite set of segments. Heine, *Journ. f. Math.* lxxiv. p. 188 (1872).

The partial intervals $\delta_1, \delta_2, \ldots$ determined as in the proof of the preceding theorems, are not determined uniquely, but they have the same external and semi-external points as the given set, and any isolated end-point of the given set is an isolated end-point of the δ's, hence, if we make all the isolated end-points of the δ's, except those which are isolated end-points of the given set, into internal points, by amalgamating all the corresponding abutting intervals δ, we get a set of intervals which is the same, however the δ's were constructed, and depends only on the given set, and clearly has the properties mentioned in the enunciation.

COR. *The semi-external points and the isolated end-points of any set of intervals are countable.*

21. The following theorem follows simply from Cantor's Theorem of non-overlapping intervals (§ 20).

THEOREM 9. *Any isolated set of points is countable.*

For, if P be any point of the set, then, since P is not a limiting point of the set, we can assign an interval containing P as internal point, and no other point of the set inside it or bounding it. Doing this with all the points, and, whenever two such intervals overlap, shortening one or both of them till they only abut, we get a set of non-overlapping intervals, each containing one and only one point of the given set. Since, by Cantor's Theorem the intervals are countable, the points are so also.   Q. E. D.

Now, since any set consists of certain of its limiting points together with its isolated points, the following can be deduced as corollaries.

COR. 1. *Those points of a set E which do not belong to the first derived set E_1 are at most countable.*

COR. 2. *If the first derived set E_1 be countable, the set E is itself countable.*

This latter follows from Theorem 2, while from Theorem 3 we have the more general form of the same corollaries :

COR. 3. *Those points of a set E which do not belong to every derived set E_n are at most countable.*

COR. 4. *If any derived set E_n be countable the set E is itself countable.*

22. More than countable Sets. If any set can be brought into (1, 1)-correspondence with a part or the whole of a countable set, its potency will be either a finite number or a. If, however we arrange the set, when we set up a (1, 1)-correspondence between the points of a countable set and points of the set, there are always points of the set left over, the set may be said to be "more than countable." That such sets exist, is seen by the following theorems.

THEOREM 10. *The continuum in any segment is more than countable.*

This may be proved in a variety of ways; the following is Cantor's second proof*.

Suppose that the contrary were the case, and let the corresponding numbers, arranged in countable order, be $a_1, a_2, \ldots$. Let a be the number corresponding to the left-hand end-point of the segment, and l its length. Using the decimal notation let

$$\frac{a_1 - a}{l} = 0 . a_{11} a_{12} a_{13} \ldots\ldots,$$

$$\frac{a_2 - a}{l} = 0 . a_{21} a_{22} a_{23} \ldots\ldots,$$

$$\frac{a_3 - a}{l} = 0 . a_{31} a_{32} a_{33} \ldots\ldots,$$

and so on.

If we now define a number b, by means of the equation

$$\frac{b - a}{l} = 0 . b_1 b_2 b_3 \ldots\ldots,$$

where b_r is never the same as a_{rr}; then the number b will certainly be different from all the numbers a_r. But, since $\dfrac{b - a}{l}$ lies between 0 and 1, b is certainly a number corresponding to a point of the given interval. Thus there is at least one number left over, and the continuum is more than countable.

Cantor's first proof† was of precisely the form we shall now use to prove the following theorem.

THEOREM 11. *A countable set is never perfect.*

For let the set arranged in countable order be

$$P_1, P_2, \ldots\ldots$$

and let us suppose, if possible, that it is perfect.

Then if we take any interval d_1 with P_1 as middle point, since P_1 is a limiting point of the set, there will be an infinite number of points of the set inside d_1. Let P_i be the first point in the countable order, other than P_1, which lies inside d_1, then all the other points of the set which lie inside d_1 have indices higher than i. P_i is again a limiting point, since the set is perfect, so that we can treat it as we did P_1. Let us then describe an interval d_2, less than half as long as d_1, lying entirely inside d_1,

---

* *Jahresbericht d. d. m. Ver.* I. p. 77.

† *Jour. f. Math.* LXXVII. p. 260 ; *Math. Ann.* XV. p. 5. The proof is applicable to n-dimensions.

having P_i as middle point and not containing P_1. Let P_j be the first point in the countable order, other than P_i, which lies inside d_2, so that $j > i$, and all other points of the set inside d_2 have indices higher than j. We then proceed with P_j as we did with P_i. The process can evidently be carried on *ad infinitum*.

The infinite series of intervals $d_1, d_2, \ldots$ lying each inside the preceding and of less than half its length, defines a single limiting point (Ch. III, Theorem 1), L, which, being a limiting point of the given set, belongs to the set, since it is closed. Let it be P_k. Then since P_k lies inside the interval d_2, therefore k must be greater than 1; similarly it is greater than i, and so on. Hence the series of integers

$$1, i, j, \ldots \ldots$$

(each of which is greater than the preceding) has a definite upper limit k, so that there can be only a finite number of them, contrary to the fact that, as we saw, there were an infinite number of them. Thus the assumption was inadmissible, which proves the theorem.

While a countable set cannot by the above be both closed and dense in itself, it can possess either of these properties without the other. That it could be closed we have already seen. It is evident from Theorem 8 that a closed countable set cannot contain any component which is dense in itself, it will be shewn later* that any set which contains no component dense in itself is countable.

On the other hand countable sets can be dense in themselves, provided they are unclosed; the most familiar example is the following.[25]

THEOREM 12.   *The rational points are countable.*

For they can evidently be characterised by means of a system of double indices, namely the numerator and denominator, so that this theorem follows from Theorem 3.

THEOREM 13.   *The algebraic numbers are countable.*

For these numbers can be characterised by means of a triply infinite system of indices, viz. (1) the rank m, which is the degree of the defining equation

$$a_0 x^m + a_1 x^{m-1} + \ldots \ldots + a_m = 0,$$

(2) the integer n defined by the equation

$$|a_0| + |a_1| + \ldots \ldots + |a_m| = n,$$

(3) since the number of algebraic numbers for given m and n is evidently finite, the number p which defines the place of the particular number under consideration among these, when arranged in a predetermined order, for instance the order of

* Theorem 21, Cor. p. 55.

ascending magnitude. The theorem is now a direct result of the Corollary to Theorem 3.

The rational numbers, and therefore the algebraic numbers of which they form a component set, are dense everywhere. A countable set that is dense in itself but nowhere dense is got by taking the right-hand end-points of any set of non-overlapping and non-abutting intervals, for instance those of Ex. 2, Ch. III, p. 20. This set is dense in itself on one side only; the following is an example of a countable set dense in itself on both sides and dense nowhere*.

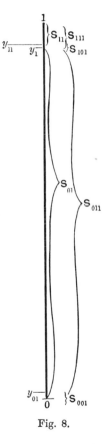

Fig. 8.

Ex. 1. Take the segment $(0, 1)$ of the y-axis, and divide it at the point y_1 into two parts, the lower s_{01}, and the upper s_{11}, so that the ratio

$$s_{01} : s_{11} = 1 + j_1 : 1 - j_1,$$

where

$$j_1 = 1 - \tfrac{1}{8}.$$

Next divide each of the two segments so obtained in precisely the same way, j_2 taking the place of j_1, where

$$j_2 = -\left(1 - \frac{1}{8 \cdot 2^2}\right),$$

and so on, j_n being defined by the equation

$$j_n = (-)^{n-1}\left(1 - \frac{1}{8n^2}\right).$$

The new points of division at the end of the second stage we denote by y_{01} and y_{11}, y_{01} lying in s_{01} and y_{11} in s_{11}; the new points at the end of the third stage by y_{001}, y_{011}, y_{101}, y_{111}, and so on. Moreover, the intervals themselves at the end of the second stage will be denoted by s_{001}, s_{011}, s_{101}, s_{111}, and y_{001} will lie in s_{001}, and so for the others. The general law of division and notation is now obvious†. The points of division are called *primary points*. Then we assert that the set of primary points is of the type required.

From the method of formation of the s's it is evident that the suffix of the maximum segment at the end of the $(2m-1)$th stage is $(01)^m$, and at the end of the $2m$th stage is $(01)^m 1$. Also, whether n be even or odd, the length of the maximum segment at the end of the nth stage is

$$\left(1 - \frac{1}{4^2}\right)\left(1 - \frac{1}{4^2 \cdot 2^2}\right)\left(1 - \frac{1}{4^2 \cdot 3^2}\right) \cdots \left(1 - \frac{1}{4^2 \cdot n^2}\right),$$

* *Proc. L. M. S.* xxxiv. p. 287.

† The indices of the new points of division introduced at the nth division are such that, prefixing to each a dot, they are all the binary fractions involving n binary places; the last figure is therefore always a 1; cf. Brodén, *Crelle*, cxviii., bottom of p. 22.

which is always greater than $\dfrac{1}{\sqrt{2}}\dfrac{4}{\pi}$, but continually approaches this value as n increases. Since each of these maximum segments lies within the preceding one, they form a *sequence*, and determine a definite interval within all of them, free of primary points, and of length $\dfrac{2\sqrt{2}}{\pi}$. The ends of this interval (which might be called $s_{\ddot{o}i}$) are, however, never reached by the primary points; they are, in fact, limiting points of the primary set, but not included in it.

Again, starting with any one of the segments left after any number, say n, of stages, we can shew in a precisely similar way, by considering the maximum segment in it obtained at each subsequent stage, that it contains within it a definite interval, free of primary points (whose length is, however, no longer $\dfrac{2\sqrt{2}}{\pi}$ of its own length).

Thus we have shewn that between every two primary points there is an interval free of primary points, possessing the property that its end-points are also not primary points. Moreover, *every primary point is approached on both sides by primary points.*

Hence it follows that any given segment of the segment $(0, 1)$ is either entirely free of primary points or contains an interval entirely free of primary points; so that *the set of primary points is dense nowhere.*

It is evident that the free intervals are the complementary intervals of a perfect set of points having the primary points as a countable set among those points of the perfect set which are limiting points on both sides.

We have purposely taken a definite numerical example, but we might equally well write

$$j_n = (-)^{n-1}\left(1 - \frac{2}{p^2 n^2}\right),$$

where p is any integer, obtaining in this way a countable set of examples of the type desired, namely, of sets of points nowhere dense and yet consisting entirely of points which are limiting points on both sides.

23. The Potency c.

We have seen that no perfect set can be brought into $(1, 1)$-correspondence with any set which we have so far measured, nor, by Theorems 2 and 3, into $(n, 1)$-correspondence, nor into $(a, 1)$-correspondence, with any such set, where n is any finite number and a the potency of a countably infinite set. We shall now see that, in order to measure a perfect set we only have to take the linear continuum itself as unit; in this way we obtain a new potency, which is more than countable, viz. the potency c of the linear continuum.

We saw (Ch. III, p. 22), that any perfect set which is dense everywhere in any segment consists of the whole continuum in that segment, and that any perfect set which is dense nowhere in any segment consists of the end-points and external points of a set of black intervals which do not abut anywhere, and which is dense everywhere. The problem divides itself therefore into two sections, which are treated separately in Theorems 14 and 15.

THEOREM 14. *Any segment, open or closed, can be brought into* (1, 1)-*correspondence with any other segment, or with the whole straight line.*

That any two closed finite segments can be brought into (1, 1)-correspondence is evident by projection. That any open segment can be brought into (1, 1)-correspondence with any closed segment can be shewn as follows:—let (A, B) be any segment, choose out any sequence having A as limit, and set up such a correspondence that every point of (A, B), except the points of this sequence, corresponds to itself, but the first point of the sequence corresponds to A, the second point to the first, and so on. In this way the segment (A, B), open at one end, is brought into (1, 1)-correspondence with the closed segment (A, B). Similarly, the segment, open at both ends, can be brought into (1, 1)-correspondence with the same segment closed.

That the whole infinite straight line can be brought into (1, 1)-correspondence with any finite segment can be shewn as follows:—take the segment $(-2, 2)$, and for all the points x outside the segment $(-1, 1)$ let us take the inversion

$$xx' = 1 ;$$

for all points x from 0 to 1, both inclusive, let

$$x' - 1 = 1 - x,$$

and for all points x from -1, inclusive, to 0, not inclusive, let

$$x' + 1 = -1 - x.$$

In this way the whole infinite straight line is brought into (1, 1)-correspondence with the segment $(-2, 2)$ open at the point -2.

It follows that any segment, closed or open, finite or infinite, has the potency of the linear continuum, and this we denote by c. By Cantor's Theorem (§ 20) we may denote the intervals of any non-overlapping set by $d_1, d_2, \ldots$. Making d_i, whether open or closed, correspond point for point to the segment $(\cdot 1^{i-1}, \cdot 1^i)$ for all values of i (cp. Ex. 1, p. 17), the points of any set of intervals with or without some or all of their end-points, are brought into (1, 1)-correspondence with those of the segment $(0, 1)$. Thus *the points of any set of non-overlapping intervals have the potency c.* Also, since any finite or countably infinite set of end-points may be omitted, *the continuum plus or minus any countable set of points has the potency c.*

THEOREM 15.   *Any perfect set, dense nowhere, has the potency c.*

Let (A, B) be the smallest segment we can find such that the whole set is internal to it. Then A and B are certainly points of the set, since the set is closed, and limiting points, since the set is dense in itself. It is plain, therefore, that A and B cannot be end-points of black intervals of the set, and are therefore external limiting points of those intervals.

Divide (A, B) into three equal parts at C and D. Then since the black intervals are dense everywhere, either (C, D) forms part of a determinate black interval or else there is a black interval inside (C, D) and possibly coinciding with it. If the latter be the case as in Cantor's typical perfect set where we have the black interval $(\cdot 0\dot{2}, \cdot 2)$, we choose some particular black interval, for instance the largest possible, and denote it by d_1. If the former be the case, we denote by d_1 the black interval of which (C, D) forms a part.

Since A and B are not end-points of any black interval, there will be two segments left over, one on each side of d_1. The end-points of these segments will again be limiting points of the black intervals in those segments, but not end-points of any of those intervals; hence we may repeat the process in each of these two segments, and choose out two new black intervals, which we shall call d_{01} and d_{11}. In Cantor's typical set, these will be $(\cdot 00\dot{2}, \cdot 02)$ and $(\cdot 20\dot{2}, \cdot 22)$. We are now left with four such segments, in each of which we can repeat our process, and choose in each a black interval. These we denote by $d_{001}, d_{011}, d_{101}, d_{111}$ in order from left to right. Proceeding thus, we use the terminating binary fractions (omitting the point), as a general system of indices for our countable set of black intervals. In the typical set the index of any black interval is the binary fraction corresponding to its middle point. This sets up a $(1, 1)$-correspondence between the black intervals of the general perfect set nowhere dense and those of Cantor's typical perfect set, of such a kind that the order of the intervals with respect to the continuum is maintained, and is the same as that of the binary fractions; that is to say if any binary fraction y lies between two binary fractions x and z, d_y will lie between d_x and d_z, and *vice versa*.

One consequence of the mode adopted for determining the indices is that, given any positive quantity e, we can determine an integer m such that, for all values of $n > m$, $d_{N1} < e$ (n being the number of places in the binary fraction N). For, by construction, the two segments left on each side of d_1 are each less than or equal to $\frac{2}{3}(A, B)$, and at each stage a similar statement can be made as to the length of the segments left over. Thus we have only to determine m so that $(\frac{2}{3})^m (A, B) < e$, and this m will certainly satisfy our requirements.

Since the order is maintained, it now follows that any sequence of intervals of the given set, defining a single limiting point, will correspond to a sequence of intervals of the typical set, defining a single limiting point, and *vice versa*. We can most easily express this correspondence between the limiting points by denoting the left- and right-hand end-points of any black interval d_{N1} by $P_{(2N)0\dot{2}}$ and $P_{(2N)2}$, the indices being the numbers corresponding to those end-points themselves in the typical case, and by denoting any external point by P_x, where x is the ternary number belonging to the limiting point of the corresponding intervals of the typical set. In this way we have set up a (1, 1)-correspondence between the points of the general and typical perfect sets dense nowhere.

At the same time we have set up a correspondence between the points of our set and the binary fractions, such that any point $P_{(2N)}$ corresponds to the binary fraction N (either terminating or non-terminating). This correspondence is (1, 1) with a countably infinite series of exceptions, namely both end-points of any black interval will correspond to the same binary fraction. This correspondence can be easily turned into one that is (1, 1) without exception, since the continuum in any segment can be brought into (1, 1)-correspondence with itself plus or minus any countable set of points.

We thus see that the points of the perfect set have the potency c; indeed, since the end-points of the black intervals are like those intervals themselves countable, the points of the perfect set which are limits on both sides have by themselves the potency c. Another way of stating this last result is the following: *the external points of a set of non-overlapping and non-abutting intervals have the potency c.*

Now any perfect set being given, we can determine all the segments in which it is dense everywhere. Since the end-points of these segments are points of the set, it is evident that no two of these segments will abut, since two such would *ipso facto* constitute a single segment in which the set would be dense. If there is any segment left over, in it the set must be dense nowhere. If there are any points of the whole continuum in which we are working left over, they must be limiting points of those segments in which the set was dense everywhere or nowhere. If we now let every such limiting point correspond to itself, and every point in any of the segments in which the set is dense everywhere also correspond to itself, finally if in any segment in which the set is dense nowhere but not entirely absent, we set up a (1, 1)-correspondence between the points of the set and all the points of that segment, we evidently set up a (1, 1)-correspondence between the points of the set and those of all the segments in which the set exists, so that *the potency of any perfect set is c.*

In the course of this paragraph the following theorems have been incidentally proved:

THEOREM 16. *If a set can be divided into a set of potency c and one of potency a or n, the potency of the whole set is c.*

THEOREM 17. *If any set is part of a set of potency c, and the other part is countable, the set is of potency c.*

COR. 1. *The irrational numbers in any segment have the potency c; and so have the transcendental numbers in any segment.*

COR. 2. *The Liouville numbers have the potency c.*

For the Liouville numbers can be at once brought into (1, 1)-correspondence with the irrational numbers together with some of the rational numbers, by taking as corresponding to a Liouville number

$$\frac{e_1}{10} + \frac{e_2}{10^{1.2}} + \frac{e_3}{10^{1.2.3}} + \ldots\ldots$$

(where $e_1, e_2, \ldots$ are any of the figures 1, 2, ... 9), the number

$$\frac{e_1}{10} + \frac{e_2}{10^2} + \frac{e_3}{10^3} + \ldots\ldots.$$

24. These theorems can be expressed symbolically by means of the following equations:

$$c + n = c,$$
$$c + a = c.$$

The theorem proved about the equivalence of the potencies of any two sets of non-overlapping intervals, may be expressed by means of the equations:

$$nc = c,$$
$$ac = c.$$

The following correspondence between the non-terminating binary and ternary fractions shews that

$$c \cdot c \equiv c^2 = c,$$

and at the same time brings out clearly the meaning of this equation.

The non-terminating binary fractions have the form

$$0 \cdot e_1 \, e_2 \, e_3 \, \ldots \ldots$$

where the e's are either 0's or 1's, and there are an infinite number of 1's. These numbers represent the continuum from 0 to 1, and have therefore the potency c. The terminating binary fractions, being rational, are countable. The same is true with respect to the potencies of the terminating and non-terminating ternary fractions. These latter may be divided up into five classes:

(A) those not involving the number 2 ;

(B) those not involving the number 1 ;

(C) those involving the number 2 only a finite number of times ;

(D) those involving the number 1 only a finite number of times ;

(E) those involving both 1 and 2 an infinite number of times.

The numbers of class A are none other than the non-terminating binary fractions, while B consists of the same fractions with 2 substituted for 1.

If y be a number of class C, it is completely characterised by giving the non-terminating binary fraction x got by omitting all the 2's in y, and the terminating binary fraction E, got from y by changing all the figures except the 2's into zeros and all the 2's into 1's. Given y, there is one and only one corresponding x; given x, however, E being at our disposal, there are a countably infinite set of y's. Thus we have a $(1, a)$-correspondence between the non-terminating binary fractions and the numbers of class C.

Similarly there is a $(1, a)$-correspondence between the non-terminating binary fractions and the numbers of class D; while the correspondence of the former with the numbers of class E is $(1, c)$, since E is in this case a non-terminating binary fraction.

Thus in each non-terminating binary fraction there corresponds one number from each of the classes A and B, a countably infinite set from each of C and D, and a set of potency c from class E, in all a set of potency

$$1 + 1 + a + a + c = c.$$

Thus we have set up a $(1, c)$-correspondence between the non-terminating binary and ternary fractions, so that, the potency of the binaries being c, that of the ternaries must be denoted by $c \cdot c$. Thus

$$c \cdot c \equiv c^2 = c.$$

In a similar manner the non-terminating fractions with base n can be brought into $(1, c)$-correspondence with those of base $(n+1)$. Thus, by induction,

$$c^n = c.$$

The only potencies at present known on the straight line are a and c together with the finite integers. In the case of a closed set it will be proved (§ 27, Theorem 26, Cor. p. 56), that no other potencies are possible, as well as in a very general type of open sets. It has not yet been proved that an open set on the straight line cannot have any other potency, although there is a very strong presumption that this is the case. Before proceeding to the proof of these theorems, it will be advisable to go more closely into the theory of limiting points in the light of our theory of potency, and to develope a step further the theory of density in itself.

THEOREM 18. SCHEEFFER'S THEOREM[*]. *If G be a perfect set, nowhere dense, and A a countable set, both lying on the same straight line, then given any two numbers c' and c'', there is at least one number c, lying between c' and c'', both inclusive, such that if the set A be translated a distance c, no point of A will coincide with a point of G.*

Let the points of A be denoted by $A_1, A_2, \ldots$, and let the points distant respectively c' and c'' from A_1 in the same direction be P and Q. Then, since G is dense nowhere, there are black intervals of G between P and Q, or else the whole segment (P, Q) is internal to a black interval of G. In the former case, choosing any one of these black intervals, and denoting the distances of its end-points from A_1 by γ_1' and γ_1'', we have

$$c' \leqslant \gamma_1' < \gamma_1'' \leqslant c'';$$

while in the latter case the signs of equality in the preceding relation may be held to hold. In either case if we shift the set G any distance d in the direction PA_1, provided

$$\gamma_1' < d < \gamma_1'',$$

A_1 will become internal to a black interval of G.

* L. Scheeffer, *Acta Math.* v.

Similarly, taking A_2 instead of A_1, and $\gamma_1{}'$ and $\gamma_1{}''$ instead of c' and c'', we get two numbers $\gamma_2{}'$ and $\gamma_2{}''$, where

$$\gamma_1{}' \leqslant \gamma_2{}' < \gamma_2{}'' \leqslant \gamma_1{}'',$$

such that shifting the set any distance d in the direction PA_1, provided

$$\gamma_2{}' < d < \gamma_2{}'',$$

both A_1 and A_2 will become internal to black intervals of G.

Proceeding thus we get a sequence of ascending numbers $\gamma_1{}'$, $\gamma_2{}'$, ... , each less than any number of the sequence of descending numbers $\gamma_1{}''$, $\gamma_2{}''$, ... Thus the upper limit of the former sequence being denoted by γ' and the lower limit of the latter sequence by γ'', we must have

$$\gamma' \leqslant \gamma'',$$

and any number between γ' and γ'' both inclusive will satisfy the requirements of the number C in the enunciation, which proves the theorem.

25. Limiting points of countable and more than countable degree*.

DEF. If L be a limiting point, and d any interval containing L as internal point, then, if d contain only a countable number of points of the given set, the same will be true when we diminish d as much as we please, as long as it contains L; in this case L will be said to be *a limiting point of countable degree*. If this is not the case L will be said to be of *more than countable degree*.

If a limiting point of more than countable degree be also a point of the set considered we shall for shortness speak of it as *a point L'.*

THEOREM 19. *A set is countable, or not, according as it has not, or has, a point L'.*

Suppose that there is a point L'. Then it follows from the definition of such a point that the set is more than countable.

Assume on the other hand that there is no point L'. Then taking any point P of the given set E, we can determine intervals having P as internal point, and containing only a countably infinite, or a finite number of other points of E. By Theorem 5 we can replace all these intervals, determined from all the points of E, by a countable set of them d_1, d_2, ...

---

* Cantor, *Acta Math* VII.

Since in d_i there is only a countable set E_i of points of E not internal to d_1, d_2,... d_{i-1}, we have in this way divided E into a countable number of countable sets E_1, E_2, ..., so that, by Theorem 3, the set of points E is countable.   Q. E. D.

This theorem may also be proved by the principle used in the alternative proof of Theorem 5.   (See footnote, p. 40.)

THEOREM 20.   COR.   *Those points of E which are not points L' are at most countable.   If a set E has at least one point L', it has a more than countable set of them ; and the set of such points L' is dense in itself.*

For consider the set consisting of E without the set of points L'; this must by Theorem 19 be countable.   Hence, if the points L' are themselves countable, the original set, being the sum of two countable sets, is itself countable, which is impossible if it had one point L'.

It remains to prove that the set of points L' is dense in itself.

Taking any one point L', there must either in (A, L'), or in (L', B), or in both, be a more than countable number of points of E.   Suppose that this is the case in (A, L').   Divide up the segment (A, L') as follows.   Bisect it at A_1, then bisect (A_1, L') at A_2, then (A_2, L') at A_3, and so on.   We thus get a countably infinite set of intervals (A, A_1), (A_1, A_2), ... such that no last interval exists, and such that L' is the sole external point in the segment (A, L').

In one at least of these intervals there must be a more than countable number of points of E, otherwise as before, there would only be a countable number of points of E in (A, L').   Let (A_i, A_{i+1}) be such a segment, then by the preceding theorem, there must be a point L' in it.

Treating the segment (A_{i+1}, L') as we did the segment (A, L') it follows from the definition of a point L' that it must contain a second point L', lying, say, in the interval (A_j, A_{j+1}).   We next take the segment (A_{j+1}, L'), and so on.   In this way we obtain a whole sequence of points L', having as limit that point L' with which we originally started.   Thus any point L' is a limit for points of the same kind, that is to say the set of points L' is dense in itself.   Q. E. D.

These results may be summed up in the following manner :

THEOREM 21.   OF THE NUCLEUS.   *Every set which is more than countable has a component, called the nucleus, consisting of all* *its points L'. The nucleus is dense in itself and more than count-able, while the points of the set which are not contained in it are at most countably infinite.*[26]

Cor. *Every set which contains no component dense in itself, is countable.*[27]

This, which is true for all sets, is the fundamental theorem in the case of an open set ; in the case of a closed set, we can go a step further.

26. Closed and perfect sets.

Theorem 22. *A perfect set is its own nucleus.*

For, if P be any point of a perfect set E, then inside any interval containing P there are points of E, which, with possibly the two end-points, evidently form a closed and therefore a perfect set, which, as such, is more than countable. Thus P is a point L', which proves the theorem.

Cor. *Every perfect set contained in a given set is contained in its nucleus.*

Theorem 23. *The nucleus of a closed set E is a perfect set, and contains every component of E which is dense in itself.*

For it is clear that any limiting point of points L' is itself a point such that in any segment, however small, containing it, there is a more than countable set of points of E ; also, since E is closed, each such limiting point is a point of E, and therefore, by definition, a point L'. Thus the nucleus is a perfect set.

Further, any component of E dense in itself is contained in the perfect set got by closing it, and since E is closed, this latter is contained in E, and therefore, by the preceding corollary, in the nucleus. Q. E. D.

27. Derived and deduced sets.[1]
We now proceed to shew how a closed set may, by means of the familiar processes of derivation and deduction, be analysed into its countable part and its nucleus.[2]

Theorem 24. *All derived and deduced sets have the same nucleus.*

For, by derivation and deduction, a perfect set is unaltered ; therefore, denoting by E^* the nucleus of the first derived set E', E^* will be contained in every derived and deduced set, and therefore in the nucleus of any one of them. That E^* must be identical with this latter set follows, since any derived or deduced set is a component of E', so that any point L' of the former is a point L' of E'.

Cor.   *If E be closed, E^* is the nucleus of E itself.*

Theorem 25.   *The sets of derived and deduced sets which are distinct from one another are at most countably infinite.*[1]

For, as has been remarked, no two can be identical unless they are perfect, in which case, by Theorem 22, each would be its own nucleus, that is E^*, and no further process of derivation or deduction would introduce a new set.   Hence, as long as the sets are distinct, we can assign one point of E', not belonging to E^*, to each successive set, and insure that no two sets shall correspond to the same point.   That is we set up a (1, 1)-correspondence between the different derived and deduced sets and points of a set, which, by Theorems 19 and 21, is known to be countable, therefore these sets themselves can be arranged in countable order.

Now as long as we do not arrive at a finite number of points, the series of derived and deduced sets cannot leave off.   If the first derived set E' be more than countable, this cannot be the case, since E^* being perfect, is by Theorem 22 always present. In this case, then, we must, after a countable series of operations, arrive at a perfect set, containing E^*, and, by the corollary to Theorem 22, contained in E^*, that is we arrive at E^* itself. Summing up, we have the following theorem and corollary.

Theorem 26.   *Every closed countable set can be reduced after a countable series of derivations and deductions to a finite number of points, and every more than countable closed set to its nucleus.*

Cor.   *Every closed set is either countable or has the potency c.*

For unclosed sets the theorem corresponding to Theorem 25, which has been also proved in the course of the preceding paragraphs, may be stated as follows.

Theorem 27.   *Every set, whose first derived is countable, leads after a countable series of derivations and deductions to a finite number of points, and every set, whose first derived is more than countable, to a perfect set E^*.*

E^* is the nucleus of any derived or deduced set of E, and contains every component dense in itself of E, in particular the nucleus of E.

This latter theorem, which for some purposes is sufficient, does not give us, as in the case of Theorem 25, an analysis of the given set into components, since neither the finite number of points in the former case, nor the set E^* in the latter case, are necessarily components of E. Nor does it enable us in the latter case to draw any precise conclusion as to the potency of an open set. In the following section we proceed to develope the method devised by Cantor for analysing open sets in a manner precisely analogous to that used above for closed sets.

28. Adherences and coherences.[1] The term *adherent* has been adopted by Cantor to denote any isolated point of a set, and *coherent* to denote any limiting point which is also a point of the set; the set of all the adherents he calls the *adherence* and that of all the coherents the *coherence*, and denotes these symbolically by the addition of a small a or c respectively to the symbol used for the set itself; thus,

$$E = Ea + Ec.$$

The adherence Ea is thus an isolated, and therefore a countable set. If the set E is not dense in itself, there is at least one adherent. Every component of E which is dense in itself is a component of Ec. The coherence Ec may however contain other points, each of which is then a limiting point of Ea. If, and only if, E is closed, Ec is the same as the first derived set E'; in any case it belongs to E'. It will for some purposes be convenient to speak of Ec as the *first derived coherence*.

If there be no second derived set, Ec will consist solely of its own adherents, and the process comes to an end. Also if Ec be dense in itself, it consists solely of its own coherents, and the process cannot by repetition lead to any new set. If, however, there be a second derived set, and Ec be neither an isolated set, nor dense in itself, it will possess an adherence and a coherence, and we may proceed in our analysis a stage further:

$$Ec = Eca + Ecc ;$$

whence　　　　　　　$$E = Ecc + Ea + Eca.$$

Here Ecc is contained in Ec', the first derived of Ec, but Eca is not. Thus Ecc is contained in E'', the second derived of E, and we shall therefore sometimes speak of it as the *second derived coherence*; it does not, however, follow that Eca is not contained in E'', or that it has not at least some points common with E'', or with some subsequent derived or deduced set; all we know is that Eca certainly is contained in E', and we shall therefore sometimes speak of it as the *first derived adherence*.

Ex. 2.   Referring to Ex. 3, Ch. III, p. 23 we construct our example by taking the set T_3 in the segment (0, 1), omitting all the limiting points of the first order, that is all the points

$$\cdot 01, \quad \cdot 011, \quad \cdot 0111, \ldots\ldots$$
$$\cdot 101, \quad \cdot 1011, \quad \cdot 10111, \ldots\ldots$$
$$\cdot 1101, \cdot 11011, \cdot 110111, \ldots\ldots$$

and so on ; generally omitting every point $\cdot 1^m 01^n$.

Here Ec consists of Eca (viz. the limiting points of the second order,

$$\cdot 1, \cdot 11, \cdot 111, \ldots\ldots)$$

and Ecc, which is the point 1, the single limiting point of the third order. Thus Eca is entirely contained in the second derived set, which, as remarked, is possible, although in the general case we can only assert that it is contained in the first derived set.

Similarly, by adding to the above set a finite number of limiting points of the first order of T_3, we get an example where Eca has some, but not all, of its points contained in E''.

If, on the other hand, we also omit all the limiting points of the second order of T_3, we get an example where Ecc is itself the third derived set, consisting of the point 1 alone.

On the same principle, using the set T_n, we can easily construct examples in which Eca is wholly or in part contained in any of the derived sets up to the nth.

If now Ecc be neither an isolated set nor dense in itself, we continue our process, and so on.   Each successive coherence and each successive adherence, will be contained in each successive derived set, so that we may speak generally of *the nth derived coherence or adherence,* and the set E will consist of the nth derived coherence together with all the derived adherences, up to the $(n-1)$th, and Ea.

Continuing the process, either (1) we get to a stage at which the series of derived sets leaves off (*i.e.* the first derived E' is countable), and therefore the series of adherences leaves off, and no coherence is left at the end, or else we get to a stage at which the latter series comes to an end but not necessarily the former, leaving us either (2) with no coherence over, or (3) with one which is dense in itself, or, finally, (4) all the derived sets exist, and corresponding to each we have a definite adherence.

In case (4) we examine whether, or no, we can " deduce " a set from the infinite series of coherences

$$Ec, \; Ecc, \; Eccc \quad \ldots\ldots\ldots\ldots\ldots\ldots(1).$$

Since these sets are not necessarily closed, they may have no common point, although the corresponding derived sets, being closed, certainly determine a deduced set; if this be the case, the original set E consists only of the countably infinite series of adherences

$$Ea, \; Eca, \; Ecca \quad \ldots\ldots\ldots\ldots\ldots\ldots(2),$$

and is therefore countable and has no component dense in itself.

Ex. 3. Consider the set constructed in Ex. 4, Ch. III, p. 24 and from this set omit the point 1. The first deduced set still consists of the point 1 alone, but although for every value of n there is an nth derived adherence and coherence, we cannot deduce a set from the coherences, since they have no single point common.

If the deduced set of the coherences does exist, it consists entirely of limiting points of every preceding coherence, and therefore is contained in the first deduced set, just as the nth derived coherence was in the nth derived set. We may therefore properly designate it as *the first deduced coherence.* Any component of E which is dense in itself will be contained in it.

If the first deduced coherence does not exist, it does not follow that the set E has no component common with the first deduced set.

Ex. 4. Take the closed set referred to at the beginning of Ex. 3, and omit all its limiting points except the point 1, which, as we saw, constitutes by itself the first deduced set. Ec is identical with Eca and with the component common to E and its first deduced set, viz. it consists of the point 1 alone.

If the first deduced coherence exists, E consists of it, together with Ea and the derived adherences of every order; the former alone of these is not necessarily an isolated set, and, in particular, contains every component of E dense in itself. We can then analyse the first deduced coherence precisely as we did the set E, forming, on the one hand, a series of adherences and coherences, and, on the other, a series of derived sets, and continuing the process, if possible, by deduction.

Proceeding in this way, either (1) the series of derived and deduced sets comes to an end, and therefore, the series of adherences comes to an end, and no coherence is left over, or else the series of adherences comes to an end, leaving us either, (2) with no coherence, or (3) with one which is dense in itself, or, finally, (4) the series of derived and deduced sets does not come to an end (E' being more than countable), but leads, after a countable series of steps, to the set E^*, while the series of adherences does not come to an end, and there is a definite coherence Ec^*, a component of E^*, left over at the final stage.

In the following example Ec^* consists of a finite number of points.

Ex. 5. Let Y_1 denote the set consisting of the middle points of Cantor's ternary set (Ch. III, Ex. 2, p. 20), together with a finite number of points of the latter set.

The set Y_1^* is then Cantor's set, and is the first derived of Y_1.

The points common to Y_1 and Cantor's set constitute the first derived coherence of Y_1, and at the same time the set Y_1c^*.

Similarly in each of the following examples Ec^* exists, and consists of the common points of E and E^*, the set Ec^* being in the successive cases more and more complicated in character.

Ex. 6. Let Y_2 denote the set obtained by inserting in each of the black intervals of Cantor's ternary set a set similar to Y_1 and adding to these points a finite number of points of Cantor's set itself.

Here the first derived coherence is identical with Y_2c^*, which is the component common to Y_2 and Y_2^*.

Forming on this principle sets Y_3, Y_4 ...... in succession, we see that for Y_n the first derived coherence is identical with Y_nc^*, which is the component common to Y_n and Y_n^*.

It is evident that by using the sets Y_1, Y_2, ... as we did the sets T_1, T_2, ... we can construct examples of a more and more complicated nature, illustrating the possibilities of the case ; for instance, the following :

Ex. 7. In the largest black interval of Cantor's set introduce a set similar to Y_1 ; in the next two largest a set similar to Y_2, and so on ; finally, take a finite number of points of Cantor's set itself. This set we call Y.

The set Y^* is reached at the first derivation, simultaneously with Yc^*, which consists of the common points of Y and Y^*.

These examples can be varied by taking instead of Y_1, a set having, in addition to a finite number of points of Cantor's set, in each black interval of Cantor's set a set T_1, or T_n, or any of the more complicated sets constructed by means of these. The set E^* is then not reached till the second derivation, or the nth, or some more remote stage of proceedings.

Again it is clear that Ec^* may not exist, and yet E may have a component common with E^*, as the following example shews.

Ex. 8. Take one of the last examples in which E^* is not reached at the second derivation, and omit all the limiting points except the finite number of points belonging to Cantor's ternary set. Here the first derived coherence and adherence are identical with the common points of E and E^*, and the process of forming the successive adherences and coherences comes to an end at the second stage, while the process of derivation leads to new sets, and ultimately to E^*. Similarly, by inserting suitably a sequence of points with its limiting point in one of the free intervals of the set last constructed, we can form an example in which some, but not all, of the points of the final adherence are points of the E^*, while Ec^* does not exist.

If Ec^* exist, we can recommence our parallel processes with it as basis, analysing it into its adherents and coherents, and subjecting it to the processes of derivation and deduction. If neither set of processes lead us to a conclusion, we start afresh with Ec^{**}, and so on.

In this way we have before us a vista of adherences and coherences, extending possibly beyond the range of the distinct derived and deduced sets, as for instance in the examples (5) and (6). The derived and deduced sets can, we know, be arranged in countable order; we will now shew that the same is true of the series of adherences in the most general case possible.

29.   The ultimate coherence†.[1] By Theorem 21 and the Corollary we know that a set is either countable, or has a nucleus. In the latter case, by what has been pointed out, the nucleus, being dense in itself, forms part of every possible coherence formed in the manner indicated, and those points of the given set which do not belong to the nucleus are countable.

Choosing then one point from each adherence we get a set of points which is at once countable and in (1, 1)-correspondence with the adherences, thus *the adherences are themselves countable.*

Summing up, we have the following theorem :

THEOREM 28.   *Every set E which has no component dense in itself, can be analysed into a countably infinite series of isolated sets (adherences); and every set E which has a component dense in itself, into such a series together with a single component U (the ultimate coherence), dense in itself, and containing every component of E which is dense in itself.*

In particular, as already remarked, if E is more than countable, U will contain a nucleus. If U does not coincide with the nucleus it can be shewn, without difficulty, to consist of two components, (1) the nucleus, and (2) another component, dense in itself, no point of which is a point L'. If E be closed, the ultimate coherence, the set E^* and the nucleus are all identical. If E is not closed, the ultimate coherence is a component of E^*, but, as pointed out in the examples, does not necessarily contain every point common to E and E^*.

† Cantor uses the term *total inherence.*

30. This analysis of Cantor's is graphically illustrated in the form of a " tree," where F denotes any coherence obtained from E, and U is the ultimate coherence.

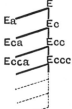

It is of course allowable to start the tree at any point of ramification we like, for instance at F. Thus we see that any theorems which are true of Ea and Ec with respect to the subsequent adherences and coherences, or the derived and deduced sets of E, are true of Fa and Fc with respect to the adherences and coherences subsequent to them in the natural order, and the derived and deduced sets of F. As to the relations of Fa and Fc and the derived and deduced sets of F to the adherences and coherences preceding F in the natural order, we have pointed out in the examples that the possibilities are of the most varied nature.

Thus, if G denote any derived or deduced set of E, and F the coherence corresponding to it, we have the following relations :

(1) F is contained in G ;

(2) F' contains Fc, but not Fa, or any point of any adherence preceding F in the natural order ;

(3) G' contains F'', and therefore Fc, but may contain some or all points of Fa, or of any adherence preceding Fa in the natural order except Ea. The above statements are then still true, if, instead of E, we start with any coherence preceding F in the natural order.

Fig. 9.

THEOREM 29.   *Each adherence consists entirely of points which are limiting points of every adherence preceding it in the natural order* [28].

That this is true for the derived adherences is easily seen, for Eca being a component of Ec, consists of limiting points of E, (that is, of Ea together with Ec), but no point of Eca is a limiting point of Ec, hence every point of Eca must be a limiting point of Ea. Similarly each derived adherence consists of limiting points of the adherence immediately preceding it in the natural order, and therefore, by induction (since limiting points of limiting points are limiting points), of every adherence preceding it in the natural order.

* *Quarterly J. of M.* cxxxviii. p. 115.

That it is true of the first deduced adherence we may prove as follows.

Let F denote the first deduced coherence. If P be a point of F, but not a limiting point of Ea, it must be a limiting point of Ec; and, not being a limiting point of Ea, it cannot be a limiting point of Eca, by the above, thus P must be a limiting point of Ecc and so on. Thus P, being a limiting point of every one of the derived coherences, is a limiting point of F, and therefore a point of Fc and not of Fa. Similarly, if we had started with any derived adherence instead of E, as that of which P was not a limiting point, it would follow that P could not be a point of Fa. Thus every point of the first deduced adherence Fa is a limiting point of every derived adherence. The above method of proof is perfectly general, and may be extended, by induction, to any adherence whatever, which proves the theorem.[28]

31. Ordinary Inner Limiting sets. As already remarked the theory of open sets is still so far imperfect that we cannot say whether or no potencies other than the finite integers, a and c, exist on the straight line. The theory of density in itself, however, as developed by Cantor, and given in the preceding paragraphs, enables us to make the same assertion with respect to a very general type of open sets, that we were able to make with respect to closed sets, namely that potencies other than those mentioned do not exist*. The sets in question are defined as follows.

DEF. Given a series of sets of intervals, the set of all those points such that each is internal to at least one interval of every set of the series is called *the inner limiting set of the series of sets of intervals, or an ordinary inner limiting set*.

THEOREM 30. *Any inner limiting set can be defined by means of a series of sets of non-overlapping intervals such that each interval lies inside an interval of the preceding set of the series, possibly coinciding with this latter.*

Such a series of sets of intervals will be called *normal intervals*. We have seen (Ch. III, Theorem 4, Cor. p. 20), that the internal points of any set of intervals determine a set of non-overlapping intervals of which they are the set of internal points. Having, in this way, replaced the given sets of intervals by sets of non-over-

---

* "Zur Lehre d. nicht abgeschlossenen Punktmengen," *Leipziger Bericht*, 1903.

lapping intervals, there may be an interval of the second set which does not lie entirely in a single interval of the first set; in this case, however, we only have to take, instead of that interval, the parts of it which lie inside the different intervals of the first set; the internal points common to the first and second sets do not in this way suffer any alteration. In this way we can modify the intervals of each successive set till we get normal intervals.

LEMMA. *If an ordinary inner limiting set is such that the greatest length of its normal intervals at each successive stage decreases without limit*, and also that we can assign a series of constantly increasing integers $r_1, r_2, \ldots$ and corresponding to these an interval of the r_1th normal set which contains two intervals of the r_2th normal set entirely internal to them, each of which contains two intervals of the r_3th normal set entirely internal to them, and so on, then the potency of the inner limiting set is c.*

To prove this,[29] let us denote the normal interval of the r_1th set by d_{01}, and the two intervals of the r_2th set which it contains by d_{001} and d_{011}, and, continuing this system of notation, let us denote by d_{N01} and d_{N11}. the two normal intervals referred to in the enunciation, which lie inside that already denoted by d_{N1}, where N is any combination of zeros and ones with n figures. Since the greatest length of the intervals at each successive stage decreases without limit, every series of these intervals, one lying entirely inside the other, defines a point of the inner limiting set. To each such series of intervals, however, by our system of notation, corresponds one and only one non-terminating binary fraction, which, if d_{N1} be any interval of the series, begins with the figures denoted by N; conversely to each non-terminating binary fraction there corresponds such a series of intervals. Thus there is a (1, 1)-correspondence between all or some of the points of our inner limiting set and the non-terminating binary fractions; since the potency of these latter is c, so is the potency of the former.

THEOREM 31. *An ordinary inner limiting set which has a component dense in itself has the potency c; otherwise it is countable.*

By the theorem of the nucleus (Theorem 21), any set which has no component dense in itself is countable, therefore we only have to prove the first part of the theorem.

Suppose the given ordinary inner limiting set to have a component U which is dense in itself. Let P be any point of U,

* See footnote, p. 17.

then there is a normal interval d_{01} of the first normal set containing P as internal point, and since P is a limiting point of U, there will be another point Q of U inside d_{01}. If at every stage the normal interval containing P contains also Q, then every point between P and Q will be internal to a normal interval at every stage, and will therefore belong to the ordinary inner limiting set, so that the latter, containing all the points of an interval, has the potency c. For the same reason if the length of the normal interval containing P does not decrease without limit as the stages advance, the potency will certainly be c, and the same is of course true of Q or of any other point of the inner limiting set. We may therefore now assume that the lengths of the normal intervals decrease without limit. This being so, we can assign a stage at which the normal intervals containing P and Q are distinct, and both are entirely internal to d_{01}. Denoting these by d_{001} and d_{011}, there will be a component of U in each, the argument can be repeated, and thus the conditions of the Lemma are satisfied. Thus the potency of the ordinary inner limiting set is c if it has a component dense in itself. Q. E. D.

COR. 1. *Since the component of an ordinary inner limiting set internal to any interval is itself an ordinary inner limiting set, it follows that any point of an ordinary inner limiting set is of degree c in the set, if it is not of countable degree.*

COR. 2. *The necessary and sufficient condition that a countable set G should be an ordinary inner limiting set is that G should contain no component dense in itself.*[29]

32. It has been asserted that the class of ordinary inner limiting sets embraces a large class of open sets, it might have been added that it includes all closed sets. It remains to shew that this is the case, and to give examples ; we shall also be able to give what is almost an equivalent statement of Theorem 31, but in a form for some purposes more convenient (Theorem 34).

It is often convenient to start with a given set of points E, and describe intervals round each point of E, giving us a set of intervals, and then to let the length of each interval converge towards zero according to any law, giving us a sequence of sets of intervals whose inner limiting set will, by Theorem 1, Chap. III, contain E. The following theorem is then almost self-evident :—

THEOREM 32. *The inner limiting set consists of the chosen set E together with possibly certain points of the first derived set E_1.*

Suppose P to be a point of the inner limiting set not contained in E. Then we can assign an interval from each successive set containing P as internal point, and each of these intervals contains a point of E. Since the length of these intervals decreases without limit, it follows that P is a limiting point of points of E.   Q. E. D.

THEOREM 33.   *Any closed set is an inner limiting set.*

Let us [30] choose as set E any component set dense everywhere in the given closed set (Chap. III, § 11, p. 23), and take all the intervals in each set of equal length, this length diminishing without limit for the successive sets; also let the interval round any point of E be described with that point as centre. Then, since there are points of E as near as we please to any limiting point of E, every point of E_1 will be internal to intervals of any particular set, and will therefore belong to the inner limiting set; also by the preceding theorem none but points of E or E_1 can belong to the inner limiting set, so that the latter will be the given closed set. [30]

THEOREM 34.   *The potency of the inner limiting set is the same as that of E_1, the first derived of the chosen set E, unless E contains no component dense in itself while E_1 is more than countable.*

This theorem is to all intents and purposes merely a restatement in other words of Theorem 31 ; it is only not *a priori* obvious that a set, such as that mentioned at the end of the enunciation, could lead to a countable inner limiting set. In the following simple case, however, it is easily seen that this is the case. Take the middle points of the black intervals of Cantor's perfect set (Chap. III, Ex. 2, p. 20) ; and describe round each such point as centre an interval of length e_n times that of the corresponding black interval, where

$$e_1, \ e_2, \ \ldots\ldots$$

is any sequence of positive proper fractions decreasing continually with zero as limit.

The points of the first derived set, that is of Cantor's set, then lie from the first outside all the intervals, which, as here constructed, are normal intervals. Hence the inner limiting set contains no points of E_1, and consists only of the chosen countable set.

The following theorems shew that the above is true, not only in the special case there considered but in the most general case.

THEOREM 35 *a*. *If E contain no component dense in itself, we can so arrange the intervals that the inner limiting set consists of E alone, and if E_1 is more than countable, we can so arrange the intervals that the inner limiting set may be either countable or have the potency c.*

THEOREM 35 *b*. *In general we can so arrange the intervals that those points of the inner limiting set which are not points of E are limiting points only of the greatest component of E which is dense in itself (i.e. the ultimate coherence of E, § 29).*

The proof of this latter theorem includes that of Theorem 35 *a* as a special case, viz. when the ultimate coherence U is non-existent.

Having arranged the adherences in countable order (Theorem 28), let us arrange the points in each adherence also in countable order, and let P_{ij} denote the *j*th point of the *i*th adherence A_i.

Let us assume any finite positive quantity *l*.

Since A_i contains none of its limiting points, we can assign a definite largest interval, of length not greater than *l*, say $2d_{ij}$, having P_{ij} as centre, such that in it there is no point of A_i, except P_{ij}; one or both of its end-points may be points of A_i, or limiting points of A_i.

The law of intervals which we take is that, round each point P_{ij} as centre, we describe an interval of length d_{ij} for the first set of intervals, and $e^n d_{ij}$ for the *n*th set of intervals, where *e* is any assigned positive quantity, less than unity. The law of intervals for points of U (if it exist) may be any we please.

Now, if there be any point of the inner limiting set not belonging to E, it is, by Theorem 32, a point of the first derived set E_1.

Let Q be any point of E_1 which is not a point of E. If Q be not a limiting point of U, it must, by Theorem 29, be a limiting point of Ea, and may be a limiting point of other adherences. Let A_i be any adherence of which Q is a limiting point; then, by Theorem 29, Q is a limiting point of every adherence preceding A_i in the *natural* order.

Hence, by the construction, Q is external to all the intervals d_{ij} whose centres belong to A_i, or to any adherence preceding A_i in the natural order.

Thus, if Q be a limit for every A_i, Q cannot be a point of the inner limiting set without remaining always internal to intervals described round points of U, which it cannot do, if it be not a limiting point of U. If Q be not a limit for every A_i, we must, by Theorem 29, be able to assign an integer h, such that Q is a limiting point of A_h, but not of any adherence subsequent to A_h in the natural order.   Let A_i denote the next adherence after A_h in the natural order.   Then we can assign a definite largest interval, of length say D, with Q as centre, such that inside this interval there is no point of A_i, and therefore, by Theorem 29, no point of any adherence subsequent to A_i in the natural order.

Let us now determine an integer m, so that

$$e^m l < D,$$

then, for all values of i and j,

$$e^m d_{ij} < D ;$$

therefore Q is external to all the intervals $e^m d_{ij}$ of the mth set of intervals whose centres P_{ij} belong to adherences subsequent to A_h in the natural order.   Also, since Q was shewn to be external to all the remaining intervals of the mth set whose centres are not points of U, it follows that Q cannot be a point of the inner limiting set without being a limiting point of U; which proves the theorem.

In reference to the above theorems it is to be remarked that although we can so arrange the intervals as to include all the limiting points, and, under certain circumstances, so as to include none of the limiting points, there are a great variety of other possibilities.   When the intervals of the individual sets have external limiting points, or when a normal interval from each successive set has a common end-point, these points will certainly be limiting points of the given set without belonging to the inner limiting set, and we can use this property to exclude any particular limiting point from the inner limiting set.

Ex. 9*.   Take as set E all the rational points $\dfrac{p}{q}$ of the segment $(0, 1)$, and describe round each point $\dfrac{p}{q}$ as centre an interval of length $\dfrac{2e}{nq^3}$, n being any integer and e any positive quantity less than $\dfrac{1}{M}$, where

$$M = 2 \sum_{q=1}^{q=\infty} \frac{q-1}{q^3} = \frac{\pi^2}{3} - 2 \sum_{q=1}^{q=\infty} \frac{1}{q^3}.$$

The sum of the lengths of the non-overlapping intervals (Chap. III, Theorem 4, Cor. p. 20), which these intervals exactly fill up, being certainly less than the sum of all the intervals, is still less than

$$\sum_{q=1}^{q=\infty} (q-1)\frac{2e}{nq^3}, \quad \text{or} \quad \frac{eM}{n},$$

a fortiori is less than 1.

* Borel, Théorie des Fonctions (1898), p. 44.

From this it is easy to see that there is a more than countable set of points external to these intervals. For, if there were only a countable set,

$$p_1, p_2, \ldots\ldots$$

we could assign any positive quantity ϵ less than $\left(1 - \dfrac{eM}{n}\right)$ and describe round each point p_i an interval of length $\dfrac{\epsilon}{2^{i+1}}$; adding these intervals to the original intervals, every point of the segment $(0, 1)$ is internal to these intervals; therefore the equivalent non-overlapping intervals must amalgamate into the whole segment $(0, 1)$, whose length would however be greater than the sum of the equivalent overlapping intervals, which is impossible.

The set E of the rational points being dense everywhere, we see that this shews that a more than countable set of the limiting points of E does not appear in the inner limiting set. On the other hand, by the general theory, we know that the inner limiting set has the potency c, while E is countable; hence a set of limiting points of potency c is introduced into the inner limiting set of the intervals described round the points of E.[31]

We can, in fact, shew that all the Liouville points (Chap. I, § 7, p. 8), which, as we know, have the potency c (p. 50), belong to the inner limiting set. For, x being a Liouville number, there are an infinite number of rational numbers $\dfrac{p}{q}$, such that

$$\left| x - \frac{p}{q} \right| < \frac{1}{q^4} \, ;$$

we can therefore determine such a rational number with denominator greater than $\dfrac{n}{2e}$, in which case

$$\left| x - \frac{p}{q} \right| < \frac{2e}{nq^3} \, ,$$

so that the point x will lie inside the interval of length $\dfrac{2e}{nq^3}$ described round $\dfrac{p}{q}$ as centre. Thus, corresponding to each integer n, we can assign one of the intervals in which x lies, so that x is a point of the inner limiting set, and this latter includes all the Liouville points.

33. Generalised inner and outer limiting sets*.

DEF. If $G_1, G_2, \ldots$ be a series of sets of points such that, for all values of n, G_n contains G_{n+1} and G be the set of all the points common to all the sets G_n, G is called *the inner limiting set of the series*, or shortly *a generalised inner limiting set*.

* The terms *greatest common factor* and *least common multiple* of sets, used by Cantor and others for the set of all the common points and all the points respectively of any set of sets, have been purposely avoided as misleading. The term *sum*, which is occasionally used for all the points of any set of sets, is restrained by authors who use the above terms for the case when the sets have no common points.

The process of forming a generalised inner limiting set has already been used and called deduction, and the set itself the deduced set.

Here if G_n consists of a set of *open* intervals (each interval being considered as not including either of its end-points), G is an *ordinary* inner limiting set.

If G_n consist of a set of intervals, each of which may be closed, or may include one or neither of its end-points, G differs only by a countable set from the inner limiting set Γ of the same intervals considered as including none of their end-points. For it is clear that G will contain Γ, and, if P be any other point of G, P must, after some definite stage, be a common end-point of a normal interval at each succeeding stage; since the number of normal intervals at each stage is countable, and also the number of stages is countable, the number of such end-points is countable, and therefore the points of G which are not points of Γ, if any, are countable.

DEF. If G_1, G_2, ... be a series of sets of points such that, for all values of n, G_n is contained in G_{n+1}, and G be the set such that every set G_n is contained in G, while every point of G belongs to some definite G_n, G is said to be *the outer limiting set of the series*, or shortly *a generalised outer limiting set*.

DEF. If G_1, G_2,... be all closed sets, G is said to be *an ordinary outer limiting set*.

Any countable set is an ordinary outer limiting set, since G_1, G_2, may be taken to be finite sets.

An ordinary outer limiting set has clearly the potency c if it is not countable, since either one of the sets G_n is of potency c, or else the difference $(G_n - G_{n-1})$ being countable, and $G = G_1 + (G_2 - G_1) + ... + (G_n - G_{n-1}) + ...$, G is countable.[32]

34. Sets of the first and second category.

In connection with the terms used in the preceding article it may be pointed out that *Baire's sets of the first category* are none other than generalised outer limiting sets, the defining sets G_1, G_2, ... being subject to the sole condition that each of them is dense nowhere. Any set which is not of the first category is said by Baire to be of the *second category*.

It follows from the definitions that *the sum of any finite or countably infinite number of sets of the first category is a set of the first category*.

The following fundamental theorem and corollaries with respect to these sets are due to Baire*.

THEOREM 36.   *If G be a set of the first category, then in every interval, however small, there is a point not belonging to G.*

For let G be a set of the first category, and G_1, G_2, ... the defining sets.   Then in any segment d of the continuum there is a segment δ_1 containing no point of G_1, since G_1 is dense nowhere. Inside δ_1, since G_2 is dense nowhere, we can assign an interval δ_2, containing no point of G_2, and so on.   The intervals δ_1, δ_2, ..., lying entirely one inside the other, contain at least one point Q internal to all of them, which point Q, not being a point of any G_n, is not a point of G.   Thus G has the property that in every segment d of the continuum there is a point *not* belonging to G.

COR. 1.   *The continuum is not a set of the first category.*

COR. 2.   *The complementary set of a set of the first category with respect to the continuum (or any other set of the second category) is a set of the second category.*

This follows from Theorem 36, since the sum of two sets of the first category is a set of the first category.

COR. 3.   *The complementary set of a set of the first category with respect to the continuum is of potency c.*

For since G_n is nowhere dense, so is the set $G_n{}'$, got by closing G_n, and the complementary set of $G_n{}'$ is a set of open intervals D_n. The inner limiting set of the sets of intervals D_n is contained in the set in question, and by Theorem 36 has a component dense everywhere and therefore dense in itself; hence by Theorem 31 the set in question has the potency c.   Q. E. D.

Taking as fundamental set F any perfect set instead of the continuum, we may define sets of the first and second category with respect to F, and shall have properties of these sets and of F precisely analogous to those enunciated and proved for sets of the first and second category and the continuum.   Thus for instance, corresponding to Theorem 36, we have the theorem, that in every interval, however small, containing a point of F there is a point of F not belonging to the set of the first category with respect to F.

All the properties in question can be most easily proved by using the correspondence between a perfect set and the continuum employed in the proof of Theorem 15, p. 49.[33]

---

* Baire, *Sur les fonctions de variables réelles*, 1899, pp. 65—67. The term *set of the second category* is used as equivalent to complementary set of a set of the first category by Schoenflies, *Bericht*, p. 108, and Bernstein, *Dissertation*, § 11.

35. The following theorems shew how very general are the classes of ordinary inner and outer limiting sets.

THEOREM 37 *a*. *The set G of all the points of any finite number of ordinary inner limiting sets is an ordinary inner limiting set.*

THEOREM 37 *b*. *The set of all the points of any finite number of ordinary outer limiting sets is an ordinary outer limiting set.*

In case (a) we only have at each stage to take together all the intervals corresponding to all the given sets. This gives us a new series of sets of intervals whose inner limiting set contains G. If P be any point not belonging to G, we can, since the number of the given sets is finite, assign a stage after which P is not internal to any of the defining intervals. Therefore P is not a point of the new inner limiting set, which is therefore G.

Similarly (b) can be proved, since the set consisting of all the points of any finite number of closed sets is clearly* a closed set.

THEOREM 38 *a*. *An inner limiting set of a sequence of inner limiting sets is an ordinary inner limiting set.*

THEOREM 38 *b*. *An outer limiting set of ordinary outer limiting sets is an ordinary outer limiting set.*

Let the sets defining G_n be G_{n1}, G_{n2}, ..., for all values of n. For shortness, let me use the symbol < to mean "is contained in," and > to mean " contains."

Then, when G is a generalised inner limiting set $G_1 > G_2$. Hence, if $G_{1,r} \leqslant G_{2,r}$, we can remedy this by taking, instead of $G_{2,r}$, the common part of $G_{1,r}$ and $G_{2,r}$, which is clearly* also a closed set and contains G_2. Doing this for all values of r, $G_{2,r} > G_{2,r+1}$, and $G_{1,r} > G_{2,r}$.

Doing this in succession for the sets defining G_3, G_4, ..., we get the following table:—

$$G_{1,1} > G_{1,2} > G_{1,3} > G_{1,4} > \ldots\ldots > G_1$$
$$\lor \qquad \lor \qquad \lor \qquad \lor \qquad\qquad \lor$$
$$G_{2,1} > G_{2,2} > G_{2,3} > G_{2,4} > \ldots\ldots > G_2$$
$$\lor \qquad \lor \qquad \lor \qquad \lor \qquad\qquad \lor$$
$$G_{3,1} > G_{3,2} > G_{3,3} > G_{3,4} > \ldots\ldots > G_3$$
$$\ldots\ldots\ldots\ldots\ldots\ldots\ldots\ldots\ldots\ldots\ldots\ldots \lor$$
$$\ldots\ldots\ldots\ldots\ldots\ldots\ldots\ldots\ldots\ldots\ldots\ldots \vdots$$
$$G.$$

This being so, consider the sequence of closed sets $G_{1,1}$, $G_{2,2}$, $G_{3,3}$, ..., and let their inner limiting set be denoted by G'.

* See Chap. V, Theorem 7, p. 84.

If P be a point of G', it belongs to $G_{m,m}$, for all values of m, and therefore to $G_{n,m}$, for all values of $n > m$, and therefore to G_n, for all values of n ; that is, P is a point of G.

If, on the other hand, P is a point of G, we can assign an integer m such that P is a point of G_m, and therefore of $G_{m,m}$, for all values of m; that is, P is a point of G'. *Thus G is identical with G', and is, as was asserted, an ordinary inner limiting set.*

Next, if G be a (generalised) outer limiting set of outer limiting sets, $G_1 < G_2$. If $G_{1,r} > G_{2,r}$, we can remedy this by taking, instead of $G_{2,r}$, the set consisting of all points belonging to at least one of $G_{1,r}$ and $G_{2,r}$, which is clearly* also a closed set and contained in G_2. Doing this for all values of r, $G_{2,r} < G_{2,r+1}$ and $G_{1,r} < G_{r,1}$.

Doing this in succession for all the sets defining G_3, G_4, ..., we get the following table :—

$$G_{1,1} < G_{1,2} < G_{1,3} < G_{1,4} < \ldots\ldots < G_1$$
$$\wedge \qquad \wedge \qquad \wedge \qquad \wedge \qquad\qquad \wedge$$
$$G_{2,1} < G_{2,2} < G_{2,3} < G_{2,4} < \ldots\ldots < G_2$$
$$\wedge \qquad \wedge \qquad \wedge \qquad \wedge \qquad\qquad \wedge$$
$$G_{3,1} < G_{3,2} < G_{3,3} < G_{3,4} < \ldots\ldots < G_3$$
$$\ldots\ldots\ldots\ldots\ldots\ldots\ldots\ldots\ldots\ldots \wedge$$
$$\ldots\ldots\ldots\ldots\ldots\ldots\ldots\ldots\ldots\ldots \vdots$$
$$G.$$

This being so, consider the sequence of closed sets G_{11}, G_{22}, G_{33}, ..., and let their outer limiting set be denoted by G'. If P be any point of G', we can assign an integer m such that P is a point of $G_{n,n}$, and therefore of G_n, for all values of $n > m$; that is, P is a point of G.

If, on the other hand, P be any point of G, we can assign an integer m such that P is a point of G_m. Then, since G_m is an outer limiting set, we can assign an integer r such that P is a point of $G_{m,r}$. If now $m \leqslant r$, P is a point of $G_{m,m}$; but, if $m > r$, P is a point of $G_{r,r}$: in either case, P is a point of G'.

Thus G is identical with G', and is, as was asserted, an ordinary outer limiting set.

THEOREM 39. *The difference of two closed sets is both an ordinary outer and an ordinary inner limiting set.*

First, to prove it is an ordinary outer limiting set. Enclose the smaller closed set in a finite number of open intervals each of length less than e. The points of the larger closed set left over form a closed set. This closed set, as e describes a sequence

* See Chap. V, Theorem 7, p. 84.

with zero as limit, generates the difference of the two given closed sets.   Q. E. D.

Next to prove that it is an ordinary inner limiting set. Enclose the larger closed set in intervals each of length less than e. By Theorem 6, Chap. IV, a finite number of these suffice, which therefore cover up a finite number of non-overlapping segments. Let d be any one of these segments : then the points of the smaller closed set which lie in d form a closed set, *inside* the black intervals of which lie all the points of the set in question which lie inside d. Taking all such black intervals in all the segments d, we have a set of intervals containing the whole set in question. As we diminish e, we get a series of sets of intervals each lying inside the preceding, and each containing the set in question. The inner limiting set of this series will therefore certainly contain the set in question ; but, since each such set of intervals lies inside the corresponding finite number of segments, this inner limiting set is a component of the larger closed set, and contains no point of the smaller closed set; so that the set in question contains this inner limiting set. Thus the set in question is none other than this ordinary inner limiting set.   Q. E. D.

THEOREM 40. *If we subtract a closed set from either an ordinary outer or an ordinary inner limiting set, we still get an ordinary outer or an ordinary inner limiting set.*

In the former case the theorem is a direct consequence of Theorems 38 and 39. In the latter case the difference of the two sets is the ordinary inner limiting set of the parts of the defining intervals of the ordinary inner limiting set that are internal to the black intervals of the closed set.

THEOREM 41. *If we subtract an ordinary outer limiting set from an ordinary inner limiting set containing it, the difference is an ordinary inner limiting set ; and, if we subtract an ordinary inner limiting set from an ordinary outer limiting set containing it, the difference is an ordinary outer limiting set.*

The first part of the theorem is proved in precisely the same way as the second part of the preceding theorem, only that, instead of a single closed set, we have a sequence of closed sets each containing the preceding, and therefore a sequence of sets of black intervals each containing the succeeding.

To prove the second part we proceed as follows :—

Let D_1, D_2, ... denote the successive sets of intervals defining the inner limiting set D, and let P_1, P_2, ... denote the closed sets of which D_1, D_2, ... are the black intervals; also let G_1, G_2, ... denote the closed sets defining the outer limiting set G. The points common to G_n and P_n clearly* form a closed set, say K_n, contained in G and having no point common with D; further, given any point of G, not belonging to D, we can assign an integer m such that, for all integers n greater than m, that point is a point of P_n (since it is not a point of D), and an integer m' such that, for all integers n greater than m', it is a point of G_n (since it is a point of G); therefore, if m'' denote the larger of m and m', the point is a point of K_n, for all integers n greater than m''. Thus the outer limiting set of the series of closed sets K_n, each one of which evidently contains the succeeding, is the difference $G - D$.   Q. E. D.

COR.   *If we subtract any countable set from an ordinary inner limiting set containing it, we get an ordinary inner limiting set.*

* See Chap. V, Theorem 7, p. 84.

CHAPTER V.

CONTENT.

36. In this chapter we come to the second important metrical idea in the theory of sets, that of Content. The potency of a set is a metrical relation between one set and another in which the individuals of the sets are regarded as indistinguishable. The content of a set is one in which the individuals are regarded as having a characteristic by means of which they become of varying importance which must be taken into account; it is determined by the relative position of these individuals but is independent of their actual situations in the fundamental region.

The idea is a natural one when we start with intervals instead of points or numbers. The distinguishing characteristic is here apparent, the *length* of the individual intervals. The potency of a set of intervals is instinctively felt to be an affair of minor importance*; what interests us more is the relation of the intervals to the linear continuum, not regarded as a collection of points (a one-dimensional variety), but as a whole, capable of division into parts comparable by means of finite numbers with itself, a variety of zero dimension.

Recalling the description and properties of a perfect set dense nowhere, given in Chap. III, we recognise that *the parts into which the continuum may be divided are not exclusively segments.* In other words, if we begin by blackening out of the continuum a set of intervals and continue till no more intervals are left, we shall not necessarily in this way have exhausted the whole continuum, the part left over is a set of points nowhere dense, and, if we regard the intervals as open intervals, it is a closed set of points. From this point of view the idea of the content of a *closed* set of points

* And is rigidly shewn by Cantor's theorem to have the minimum infinite range of variation, viz. if not a finite number, the potency is that of the natural numbers.

is an immediate extension of that of a set of intervals, and we shall obviously so frame our definitions that *the content of a set of intervals together with that of the complementary closed set of points is equal to the content of the segment of the continuum in which we are operating.*

37. We define *the content I_δ of a finite number of intervals to be the sum of their lengths.* With this definition we see at once that

(1) *The content I_δ is positive and less than or equal to the length l of the segment (A, B) (supposed finite), of the straight line in which the set lies.*

(2) If the content I_δ be less than l, then there exists a complementary set of intervals, whose content I_P is equal to the excess of l over the content of the given set, *i.e.*

$$I_\delta + I_P = l.$$

(3) If the content I_δ is equal to l, there are no complementary intervals, and *vice versa* ; in this case

(i) There are no end-points (except, of course, A and B), which do not belong to two intervals.

(ii) There are no points of (A, B) external to all the intervals.

38. Next consider an infinite number or set of non-overlapping intervals in the segment (A, B). What are the analogous theorems to those just enunciated ? By Cantor's Theorem the intervals are countable, and, from the proof given in Chap. IV, § 3, it is evident that they can be arranged in order of magnitude, δ_1, $\delta_2, \ldots$ and that, if (A, B) be a finite segment, given any positive quantity ϵ, we can assign an integer m, so that, for all values of $n \geqslant m$, $\delta_n < \epsilon$.

39. Now the sum of any number of the intervals cannot be greater than l. There must, therefore, be an upper limit I_δ, less than or equal to l, such that the sum of any finite number of the intervals is always less than I_δ; it can be made as near as we please to I_δ by taking sufficient of the intervals. That is to say, given any ϵ, we can find an integer m, such that, for all values of $n \geqslant m$, $I_\delta - \epsilon < \sum_1^n \delta_r < I_\delta$.

In the usual manner we express this fact in other words by saying that the series

$$\sum_1^\infty \delta_s \equiv \delta_1 + \delta_2 + \ldots\ldots \text{ ad inf.}$$

is convergent, and has I_δ for its sum. *I_δ we call the content of the set of intervals.* This evidently agrees with the definition of the content in the case when the set contains only a finite number of intervals.

It now follows that, *given any small positive quantity σ, we can assign a small positive quantity ϵ, such that the sum of all the intervals of the set which are less than ϵ is less than σ.* Denoting this sum by $R(\epsilon)$, we have $R(\epsilon) < \sigma$.

40. Having so defined the content, we see that (1) of § 37 holds as it stands for any set of non-overlapping intervals.

When $I_\delta = l$, it follows from the meaning of this equation that no complementary interval can exist (3).

That the converse (*i.e.* (2) of § 37) is not necessarily true, is well shewn by Example 1 (see below). It is however true, and is proved in § 41 that *if the content I_δ is less than l, there are always points in more than countable number, external to the intervals,* and this fact is the proper generalisation of the statement (2) of § 37. In Example 1 it is evident that this is the case, since the closed set consisting of the external and end-points of the intervals is perfect (Theorem 15, Chap. IV, p. 48).

Ex. 1. Divide the segment (0, 1) into three equal parts, and blacken the central part (·1, ·2) using the ternary notation. The ternary numbers corresponding to points in the black interval will then be those having the figure 1 in the first place.

Next divide each of the $(3-1)$ unblackened segments into 3^2 parts, and blacken the central part in each, viz. (·011, ·012) and (·211, ·212). The ternary numbers corresponding to points in either of these blackened intervals will then be those having the figure 1 in the second and third places.

Similarly if we next divide each of the $(3-1)(3^2-1)$ unblackened segments into 3^3 parts, and blacken the central part in each, the numbers corresponding to the points in these black intervals will be those which have the figure 1 in the 3 places after the first $(1+2)$ places.

Proceeding in this manner, at the (nth) stage we have $(3-1)(3^2-1)\dots$ $(3^{n-1}-1)$ unblackened segments, each of which we divide into 3^n parts, and we blacken the central part. The numbers corresponding to the points in these black intervals will then be those having the figure 1 in the n places after the first $\dfrac{n(n-1)}{2}$ places.

This process, carried out *ad infinitum*, defines a set of black intervals of content I_δ which is clearly given as follows :—

$$I_\delta = 1 - \frac{3-1}{3} + \frac{3-1}{3}\left(1 - \frac{3^2-1}{3^2}\right) + \frac{3-1}{3}\cdot\frac{3^2-1}{3^2}\left(1 - \frac{3^3-1}{3^3}\right) + \dots$$

$$= 1 - \underset{n=\infty}{\text{Lt}}\left(1 - \frac{1}{3}\right)\left(1 - \frac{1}{3^2}\right)\dots\left(1 - \frac{1}{3^n}\right).$$

Denoting the above infinite product by I_P,

$$\left(1-\frac{1}{3}\right)\left(1-\frac{1}{3^2}-\frac{1}{3^3}-\ ...\right)<I_P<\left(1-\frac{1}{3}\right),$$

$$(1-\tfrac{1}{3})\,(1-\tfrac{1}{2}\cdot\tfrac{1}{3})<I_P<(1-\tfrac{1}{3}),$$

$$\tfrac{5}{9}<I_P<\tfrac{2}{3}.$$

Thus $\tfrac{1}{3}<I_\delta<\tfrac{4}{9}.$

More accurately,

$$\left(1-\frac{1}{3}\right)\left(1-\frac{1}{3^2}\right)...\left(1-\frac{1}{3^{n-1}}\right)\left(1-\frac{1}{2}\cdot\frac{1}{3^{n-1}}\right)<I_P<\left(1-\frac{1}{3}\right)\left(1-\frac{1}{3^2}\right)...\left(1-\frac{1}{3^n}\right).$$

Whence putting $n=3$,

$$\frac{2}{3}\cdot\frac{8}{9}\cdot\frac{17}{19}<I_P<\frac{2}{3}\cdot\frac{8}{9}\cdot\frac{26}{27}<\frac{2}{3}\cdot\frac{8}{9}\cdot\frac{17\frac{1}{3}}{9},$$

$$\tfrac{136}{243}<I_P<\tfrac{139}{243}.$$

In this way I_P and I_δ can be calculated to any desired degree of accuracy. Using the ternary notation,

$$I_\delta=\cdot1\quad 02\quad 212\quad 2000\quad 01001\quad 01\ ....$$

There is however no interval of the segment $(0, 1)$ which does not lie wholly, or in part, inside one of the intervals so constructed. Thus by Theorem 15, Chap. IV, p. 48, the external and end-points form a perfect set dense nowhere. Adopting the convention already used in Chap. III, Ex. 2 (Cantor's typical ternary set), that ternary fractions having a terminal 1 shall be denoted in the alternative manner with a terminal $0\dot{2}$, we may say that the corresponding perfect set of numbers consists of all the numbers between 0 and 1, with the exception of those which, expressed in the ternary scale, have the figure 1 in each of the n places after the first $\frac{1}{2}n\,(n-1)$ places, for any and every integral value of n.

Ex. 2. Suppose that we are given any small positive quantity e, we can construct on the principle of the preceding example a set of intervals dense everywhere in the fundamental segment and of content less than e. To do this, we determine an odd integer m such that

$$\frac{1}{m-1}<e,$$

and we use the base m in our division in place of the base 3*.

Ex. 3. Instead of taking a definite base as above, we may take any convergent infinite product

$$0<I_P=(1-\lambda_1)\,(1-\lambda_2)\,(1-\lambda_3)\,........<1,$$

---

* The principle employed in the construction of Exx. 1 and 2 was first used by H. J. S. Smith, *Proc. L. M. S.* Vol. vi. 1875, who, however, always took the right hand interval instead of the middle interval, so that the external and end-points do not form a perfect set. The sets of points with which H. J. S. Smith, and his successors before Cantor, were interested, were always countable sets, and the object of their investigations was to find a countable set of points which could not be shut up in a finite number of intervals, the sum of whose lengths is as small as we please. The end-points of the above intervals, for instance, satisfy these requirements.

and, in the segment (0, 1), take first an interval d_1, of length λ_1, with neither of its end-points at the points 0 or 1 ; then in each of the segments left on each side of d_1 perform a similar division, the new intervals having the ratio $\lambda_2 : 1$ to the lengths of the segments in which they lie, and so on. We then have for the content of this set of intervals

$$I_\delta = 1 - (1-\lambda_1) + (1-\lambda_1)\{1-(1-\lambda_2)\} + (1-\lambda_1)(1-\lambda_2)\{1-(1-\lambda_3)\} + \ldots = 1 - I_p.$$

41. In these examples, the complementary set of points, being perfect, has the potency c, and Theorem 15, Chap. IV, p. 48 shews that this will always be the case when the intervals are non-abutting, whether or no the content be less than that of the fundamental segment. On the other hand the following theorem gives us a connection between the potency of the complementary set and the content of the set of intervals, without reference to the descriptive properties.

THEOREM 1. *If the content of a set of non-overlapping intervals be less than that of the fundamental segment, the potency of the complementary set of points is c.*

Since the set consisting of all the external and end-points is closed (Theorem 4, Chap. III, p. 19), while the end-points, like the intervals themselves, by Cantor's theorem, are countable, it is evidently immaterial whether some or all the intervals be regarded as open or closed, the potency will be c, if the points are more than countable. (§ 23.)

Suppose, if possible, that they are countable, then we can enclose the first point in an interval of length $\frac{1}{2}e$, the second in one of length $\frac{1}{4}e$, and generally, the nth in an interval of length $\frac{e}{2^n}$.

In this way, choosing e sufficiently small, since the content of the given intervals was less than that of the fundamental segment, we should have enclosed every point of the fundamental segment in a set of intervals the sum of whose contents would be less than that of the segment itself; by the Heine-Borel Theorem, however, there would be a finite number of these intervals enclosing every point of the fundamental segment ; this is evidently impossible, since the sum of their contents would be less than that of the fundamental segment, whereas it must be greater, since they certainly overlap. Hence the assumption that the points are countable is inadmissible, which proves the theorem.

COR. *However a segment be divided up into non-overlapping intervals, so that no internal point of the segment, or only a countable number of such, is external to the small intervals, the sum of the lengths of the small intervals is equal to the length of the segment.*

42. Summing up then we have the following definition and properties of the content as defined for sets of intervals.

DEF. *The content I_δ of a set of non-overlapping intervals is the sum of their contents.*

(1) *The content I_δ is positive and less than or equal to the length l of the fundamental segment (A, B).*

(2) *If the content I_δ be less than l, then there is a complementary set of points of potency c.*

(3) *If the content I_δ is equal to l, there are no complementary intervals; the converse is not true, since sets of intervals can be constructed having no complementary intervals, and of content I_δ less than any assigned positive quantity e.*

It is to be noticed that the content of a set of intervals is the same, with this definition, whether, or no, some or all the intervals are regarded as closed, or open at one or both ends. Also *the content of a given set of intervals is the lower limit of the content of a set of intervals containing the given set.*

43. When the sum of two sets of non-overlapping intervals is a set of non overlapping intervals, it is evident from the definition that its content is the sum of their contents. The sum of two sets of non-overlapping intervals may however be a set of overlapping intervals. Now by the Corollary to Theorem 4, Chap. III, p. 20, the portion of the fundamental segment covered by a set of overlapping intervals is identical with that covered by a certain set of non-overlapping intervals. Since, in the theory of content, we are dealing with division of the continuum into parts, it is convenient to define *the content of any set of intervals to be that of the equivalent set of non-overlapping intervals.* With this convention, however, we see at once that the content of the sum of two sets of intervals is not in general the sum of their contents. The contents of sets of intervals, however, obey quite a simple addition theorem, which is as follows:—

THEOREM 2. ADDITION THEOREM FOR THE CONTENT OF SETS OF INTERVALS.

Given two sets of intervals of content I_1 and I_2, and calling the content of their sum I, and the content of the set of intervals consisting of all their common parts I',

$$I_1 + I_2 = I + I'.$$

Let us call the equivalent sets of non-overlapping intervals G_1, G_2, G and G'. Take any small positive quantity e, and determine a finite number of the intervals of G_1 and G_2 so that the sum of the remaining intervals of G_1 and G_2 respectively is less than e. Then the sum of the remaining intervals of G_1 and G_2 being less than $2e$, the same is true of the parts which the chosen intervals of G_1 and G_2 have in common with the remaining intervals of the other set G_2 or G_1. Thus the sum of these two finite sets is a set whose content differs from I by less than $2e$, and the set consisting of the common parts of these two finite sets is a set whose content differs from I' by less than $2e$. Denoting then by θ_1, θ_2, θ, θ', four quantities lying between 0 and 1, we may denote the contents of the two finite sets by $I_1 - e\theta_1$, $I_2 - e\theta_2$, that of their sum by $I - 2e\theta$, and that of their common parts by $I' - 2e\theta'$.

Now consider the finite number of intervals which are parts of the chosen intervals of G_1 but not of the finite intervals of G_2: when these are added to the common parts we get the chosen intervals of G_1; hence their content is $I_1 - 2e\theta_1 - I' + 2e\theta'$.

But the same intervals added to the chosen intervals of G_2 give the sum of the two finite sets of intervals, whence their content is $I - 2e\theta - I_2 + e\theta_2$.

Equating these two values, since e is as small as we please, the theorem follows.

44. In accordance with the principle enunciated in § 35, we regard the set of points left over when we have blackened out a set of intervals from the fundamental segment, as a part of the continuum comparable in regard to content with the other part and such that the sum of the contents of the two parts is the content of the whole. This leads us to the following definition, referring for the present to closed sets only, that is, regarding the remaining part of the fundamental segment as consisting of a set of open intervals. The definition must however be regarded as equally applicable when some or all of the intervals are closed, in which case we get the content of an open set; as, however, we do not in this way get the most general open set, we shall deal with the content of open sets separately, this special case will then arise in its proper place.

DEF. *The content I_p of a closed set of points is the difference between the content of the fundamental segment and that of the black intervals* of the set.*

---

* See footnote, p. 19.

With this definition the content of a closed set of points is always positive or zero, and less than the content of the fundamental segment, and it has two important properties, contained in the following theorems.

THEOREM 3. *The content I_p is the upper limit of the content of closed components of the closed set.*

For the black intervals of the given set will be contained in those of the component, while, on the other hand, we can form a closed component of the given set having content as near as we please to I_p in the following way :—

Take any point of the given set, enclose it in a small interval of length less than some assigned small quantity e; the points which are not internal to this interval nor to the black intervals of the given set form a closed set (by Theorem 4, Cor., Chap. III, p. 20) contained in the given set, and the content of its black intervals differs from that of the given set by at most the content of the added interval, that is by less than e.

THEOREM 4. *The content I_p is the lower limit of the content of a set of intervals containing the given closed set as internal points.*

For such a set of intervals together with the black intervals of the given set contain every point of the fundamental segment, so that, by the Heine-Borel Theorem, a finite number from among these two sets cover every point of the fundamental segment, and have in consequence content greater than that of the fundamental segment. Therefore the sum of the contents of the set of intervals and the set of black intervals is not less than the content of the fundamental segment, that is the content of the set of intervals is not less than that of the closed set which they enclose.

On the other hand we can construct a set of intervals containing the given set of points and having content as near as we please to I_p as follows :—

Given e, find a finite number of the black intervals of the given set of points such that the sum of the remaining black intervals is less than e. These black intervals, in number finite, leave over a finite number of intervals containing every point of the given set of points, and of content less than $I_p + e$, since together with the finite number of black intervals in question they fill up the fundamental segment.

45. THEOREM 5. *If G_1 and G_2 be two closed sets of points having no point common, the set consisting of G_1 and G_2 together is a closed set of content I equal to the sum of their contents $I_1 + I_2$.*

For the points of G_2 must, in this case, be internal to a finite number of the black intervals* of G_1. Let these be $d_1, d_2, \ldots d_r$. Then the black intervals of G consist of $d_{r+1}, d_{r+2}, \ldots$ together with the black intervals of G_2 not counting the parts external to $d_1, d_2, \ldots d_r$. That is

$$l - I = l - I_2 - (l - d_1 - d_2 \ldots - d_r) + d_{r+1} + d_{r+2} + \ldots = l - I_1 - I_2,$$

whence
$$I = I_1 + I_2.$$

THEOREM 6. *If a closed set of content I contain a closed component of content J, it contains a closed component of content $I - J - e$ (where e is as small as we please), having no point common with the former component.*

By the preceding lemma no closed component could have content greater than $I - J$.

Let e be any assumed small quantity, and let us shut up all the points of the given closed component in a finite number of intervals of content lying between J and $J + e$. The points of the given set which are not *internal* to these intervals form, as is easily seen, a closed set; if the content of this latter set were less than $I - J - e$, we could enclose all its points in a finite number of intervals of content less than $I - J - e$, which together with the intervals first described would form a set of a finite number of intervals of content less than I, enclosing all the points of a closed set of content I; which is impossible. Thus the content of the closed component in question is not less than $I - J - e$; which proves the lemma.

THEOREM 7. ADDITION THEOREM FOR THE CONTENT OF CLOSED SETS.

If G_1 and G_2 be two closed sets of points of contents I_1 and I_2, (a) the set consisting of all the points common to G_1 and G_2 is a closed set, say G' of content I', and (b) the set consisting of all points belonging to one or both of G_1 and G_2 is a closed set, say G of content I. Further, (c) $I_1 + I_2 = I + I'$.

For (a), if P be a limiting point of G', it is a limiting point both of G_1 and G_2, and therefore a point of both, that is a point of G'; so that G' is closed.

(b) If P be a limiting point of G, it must be a limiting point of one at least of G_1 and G_2, and is therefore a point of that one, and therefore a point of G; so that G is closed.

* See footnote, p. 19.

(c) By Theorem 6, G consists of the closed set G_1 and a complementary component containing closed sets of content as near $I - I_1$ as we please, but not any whose content exceeds $I - I_1$. Since this complementary component is also the complementary component of G' with respect to G_2, by the same theorem, it contains closed sets of content as near as we please to $I_2 - I'$, but none whose content exceeds $I_2 - I'$. Hence $I - I_1 = I_2 - I'$, which is equivalent to the statement to be proved.

46. THEOREM 8. *The content of a countable closed set is zero.*

This follows from Theorem 4, since we can enclose the first point in an interval of length $\frac{1}{2}e$, the second in one of length $\frac{1}{4}e$, and so on. Thus the whole set can be enclosed in a set of intervals of content less than e, where e is as small as we please.

COR. *The potency of a closed set of positive content is c.*

Similarly we can prove the following theorem.

THEOREM 9. *The content of a closed set is the same as that of any one of its derived or deduced sets.*

For the closed set consists of the derived or deduced set in question, say G', together with a countable set of points. These latter can as before be enclosed in a set of intervals of content less than e, and the former in a set of intervals of content less than I' and e, where I' is the content of G'. Hence the whole set can be enclosed in a set of intervals of content less than $I' + 2e$. But it cannot be enclosed in a set of intervals of content less than I', because G' is a part of the closed set (Theorem 8, Chap. III, p. 27). Hence by Theorem 4 the theorem follows.

COR. *A closed set which is more than countable has the same content as its nucleus.*

The preceding theorem is a special case of the following :—

THEOREM 10. *The contents of two closed sets are equal if the points of either which are not common to both sets are countable.*

For, by Theorem 7, the common points form a closed set, whose content, by an argument precisely similar to that used in the proof of the preceding theorem, is equal to that of either of the given sets.

47. The two fundamental properties of the content of closed sets enunciated and proved as Theorems 3 and 4, could either of them be taken as definition of the content, the other would then follow, as well as the property that the content of the closed set is the difference between the length of the fundamental segment and the content of the black intervals. The disadvantages of such a definition will appear most clearly when we come to deal with the content of open sets. Historically however the question is of interest, since the idea of content, arising as it did from the theory of integration, was originally connected with that of intervals, and indeed with a finite number of intervals.

Riemann* may perhaps be regarded as the originator of the idea of content, though it was H. J. S. Smith† who, in making Riemann's proof of the condition of integrability rigid, first expressly drew attention to the connection of the subject with the theory of sets of points and introduced the idea but not the name of content.

Hankel‡ was the originator of the term content, and his definition is as follows :—

Divide the fundamental segment into n equal parts, and take the sum of those parts which contain points of a given set of points ; the limit of this sum when n is indefinitely increased and the size of the intervals indefinitely decreased, is called the content of the given set.

It has here to be shewn that the limit in question is definite. The following proof shews at the same time that, in the case of a closed set, Hankel's definition agrees with that adopted in § 44.

Let d denote the norm at any stage. Take a definite small positive quantity e, and determine the largest integer k such that

$$2kd < e.$$

Then, as d decreases indefinitely, k increases indefinitely ; thus we may assume that d is so small that the sum of the first k black intervals of the set got by closing the given set differs from the sum of all those black intervals by less than e', where e' diminishes indefinitely with d. Now the sum of those parts which lie entirely inside any black interval d' lies between d' and $d' - 2d$. Hence also the sum of those parts that lie inside the first k black intervals lies between

$$I_\delta - e' \text{ and } I_\delta - e' - e.$$

These parts contain no points of the given set. The remaining parts, if any, which do not do so, must each lie inside one of the remaining black intervals ; this shews that their sum

* Riemann, *Ueber die Darstellbarkeit einer Funktion durch eine trigonometrische Reihe, Ges. Werke,* 2nd Edition, p. 240.

† H. J. S. Smith, *Proc. L. M. S.* VI. p. 55.[34]

‡ Hankel, *Math. Ann.* XX. p. 87.

is less than e'. Thus the sum of all those parts which do not contain points of the given set lies between

$$I_\delta - e' \text{ and } I_\delta - e,$$

where I_δ is the content of those black intervals. It follows that the sum of those parts which do contain points of the given set lies between

$$I_p + e' \text{ and } I_p + e,$$

where I_p is the content of the set got by closing the given set. Since e is at our disposal and can be chosen as small as we please, this shews that the limit is definite and is the content I_p of the set got by closing the given set.

Two things have to be noticed with respect to this definition. The first is that if, as Hankel does, we omit the word "closed" in the definition, we get as the content of the set that of the set got by closing it. Now in the case of one particular open set we have already adopted a definition of content in direct variance with the Hankel definition, viz. in the case of an infinite set of intervals. The definition we adopted of the content of a set of intervals is the most natural one, and is indeed the only one of any conceivable use ; it would certainly not be reasonable to substitute for it the content of the set of points got by closing it, which may be the whole continuum even when the content of the intervals is as small as we please. There would appear therefore to be no sufficient reason for defining the content of an open set in the way that Hankel defines it.

The second point is one to which Hankel himself was the first to direct attention : it is that if we omit the condition as to the finiteness of the number of parts into which we divide the segment, we do not, in the case of an open set, get to a definite limit, as is evident by taking any simple example, for instance that of the rational points, which, being countable, can be enclosed in a set of intervals of content as small as we please, while the limit obtained in Hankel's manner is always the content of the fundamental segment. In the case of a closed set this difficulty does not occur ; for, if the fundamental segment be divided into parts, finite or infinite in number, in any manner, each of length less than r, then, by Theorem 6, Chap. IV, p. 41, there will only be a finite number of these intervals which contain the given closed set : thus we get the same limit, when r is indefinitely decreased, as by dividing the fundamental segment into a finite number of parts. Hankel himself pointed out the reason for the occurrence of this peculiarity; when the number of intervals is finite, all the limiting points are of necessity internal to the intervals at every stage of proceedings, but, when the number of intervals is infinite, this is not the case. The question suggests itself, Which of the limiting points can be cut out by a proper choice of the intervals, and what is the effect on the content of the intervals so chosen? That all points can be cut out, except those which are points of the given set, we know to be possible in the case of what we have called an ordinary inner limiting set, and it is evident that it is only possible in this case. These questions, which were not investigated till quite lately, will recur later.

The definition of content adopted by Cantor is likewise based upon Theorem 6 of Chap. IV, and runs as follows:—

Round every point of the closed set describe a small one-dimensional sphere of radius r, and calculate the content of the set of intervals covered by these spheres ; the limit of this content when r is indefinitely decreased, is called the content of the closed set.

Here again we know by Theorem 6, Chap. IV, that the set of intervals in question will consist of a finite number of intervals only, and this is expressly stated by Cantor. If, however, we apply this process to an unclosed set, we must, at each stage of proceedings, enclose every limiting point in the one-dimensional spheres, thus we shall again, as in Hankel's case, get a definite limit, as r is indefinitely decreased, and this limit will be the content of the set got by closing the given set.

The equivalence of Cantor's definition with that adopted in § 44 may be proved in the following way.

Given r we can determine k, so that k is the greatest integer for which

$$2kr < e,$$

where e is any assigned small positive quantity.

This determines e', where e' is the sum of all the black intervals except the first k : here e' diminishes indefinitely with r, since k increases indefinitely with $1/r$.

Since the one-dimensional spheres overlap with the black intervals, their content is certainly not less than $l - I_\delta$. The parts common to the spheres and the black intervals have however a sum certainly less than $2kr + e'$ or $e + e'$. Thus the content of the spheres lies between $l - I_\delta$ and $l - I_\delta + e + e'$, and has therefore the definite limit $l - I_\delta$, the content as defined in § 44.[35]

48. The point of view adopted by Hankel and Cantor, is, as was pointed out, none other than that of regarding a closed set of points as a special case of an inner limiting set, viz. when there are only a finite number of intervals in each defining set. The following theorems give the connection between the contents of the sets of intervals defining any inner limiting set whatever, and the content of any closed component of the latter set.

THEOREM 11. *Given a countably infinite series D_1, D_2, ... of sets of intervals, each of which contains only a finite number of intervals such that each interval of D_{n+1} is contained in an interval of D_n (with possibly one or both end-points common), there is at least one point common to an interval from each set; and the common points form a closed set.*

For, since the number of intervals in D_n is finite, the internal and end-points form a closed set, and, by hypothesis, the closed set of points D_{n+1} is a component of the closed set D_n; hence, by Cantor's Theorem of Deduction (Theorem 5, Chap. III), the conclusion follows.

THEOREM 12. *If to the hypothesis of Theorem 11 we add that the content of each D_n is greater than some positive quantity $\geqslant g$, the common points form a closed set of points D' of content $\geqslant g$, so that, by Theorem 8, Cor., they have the potency c.*

For, if possible, let the content be less than g, and let the difference be greater than e. Then we can enclose all the common points in a finite set of intervals of content less than $(g - e)$. Out of the set D_n let us cut those parts which are common to D_n and the intervals just constructed : there remain over a finite number of intervals of content greater than e. The intervals so constructed for successive values of n satisfy the requirements of Theorem 11 ; so that there is at least one point common to them, and therefore to the original sets D_n, contrary to the assumption that all the common points had been cut out. The assumption was then inadmissible that the common points could be enclosed in a finite set of intervals of content less than g.   Q. E. D.

49. If we remove the restriction that the number of intervals in D_n is finite, these conclusions are inadmissible, since the points of D_n do not then form a closed set. The following simple example proves this.

Ex. 4.   Let D_{m+1} consist of all the abutting intervals between the points whose numbers in the binary scale are of the form $\cdot 1^n$ $(n \geqslant m)$ (Fig. 10).   Here

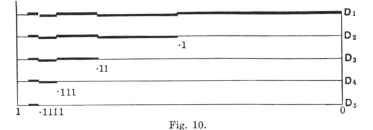

Fig. 10.

the only limiting point of intervals one from each set is the point unity, and is external to the intervals of every set.

In this example the content of the set decreases without limit; when the sets of intervals have no point common this must always be the case, as is shewn by the following theorem.

THEOREM 13.   *Given a countably infinite series D_1, D_2, ... of sets of intervals such that* (1) *each interval of D_{n+1} is contained in an interval of D_n for every value of n, and* (2) *the content I_n of each set D_n is greater than some positive quantity g, then* (a) *there is a set of points such that each is internal to an interval of D_n for every value of n, and* (b) *it contains closed components of content $> g - e$, where e is as small as we please ; so that the potency of these points is c*.*

Let the non-overlapping intervals which have the same internal points as D_n be arranged in countable order, and denoted by $D_{n,r}$, for successive integral values of r.

---

* This theorem includes Arzelà's *lemma fondamentale*, the enunciation of which is as follows :—

Sia y_0 un punto limite per un gruppo qualsiasi di numeri (y) ; e indichi $G_0 = (y_1, y_2, ...)$ una successione, comunque scelta, di numeri (y) tendenti al limite y_0. Assumendo le variabili come coordinate ortogonali di un punto nel piano, si consideri il gruppo delle rette $y = y_1$, $y = y_2$, ... ; nell' intervallo $a ... b$ sopra ciascuna si segnino dei tratticelli distinti l' uno dall' altro, in numero finito che può variare da retta a retta e anche crescere indefinitamente via via che y_s si approssima a y_0. La somma dei tratticelli $\delta_{1, s}$, $\delta_{2, s}$, ..., $\delta_{n, s}$ segnati sulla $y = y_s$ sia d_s. *Se per ogni valore $s = 1, 2, ...$ si ha sempre $d_s \geqslant d$, d numero determinato positivo, necessariamente esiste tra a e b almeno un punto x_0 tale che la retta $x = x_0$ incontra un numero infinito di tratti δ.*

In other words, assuming that the sets of intervals in the enunciation of Theorem 13, are finite (a restriction which is subsequently removed), Arzelà asserts that there is at least one point x_0 common to intervals of every set, *i.e.* either an internal or end-point of such intervals.   Arzelà, "Sulle Serie di Funzioni," Parte prima, *R. Acc. d. Sc. di Bologna*, 1899.   As to the character of Arzelà's proofs see *Proc. L.M.S.* Ser. 2, Vol. II. Part I. p. 22.

Let us determine a finite number of the intervals $D_{1,r}$ such that the content of the remainder is less than $\dfrac{e}{2^{1+1}}$; and from each end of each of these intervals (in finite number) $D_{1,r}$ let us cut off a fraction $\dfrac{1}{2^{1+3}} \times \dfrac{e}{I_1}$ of its length. The sum of these pieces is less than $\dfrac{e}{2^{1+1}}$, and therefore the finite number of curtailed intervals, which we denote by D_1', has content $> g - \frac{1}{2}e$.

The parts of intervals $D_{2,r}$ which lie inside D_1' evidently have content $> I_2 - \frac{1}{2}e$: choosing out a finite number of these so that the content of the remainder may be less than $\dfrac{e}{2^{2+1}}$, and curtailing them at each end by a fraction $\dfrac{1}{2^{2+3}} \times \dfrac{e}{I_2}$ of their length, we get a finite set of intervals D_2' of content $> g - \dfrac{e}{2} - \dfrac{e}{2^2}$ lying inside the intervals D_1'.

Proceeding thus with each successive set of intervals, we obtain from $D_{n,r}$ a finite set D_n' of content $> g - \dfrac{e}{2} - \dfrac{e}{2^2} \dots - \dfrac{e}{2^n}$, a fortiori $> g - e$, lying inside the finite set D_{n-1}, for every value of n. Applying Theorem 12 to these sets, we deduce that they have in common a closed set of points of content $> g - e$. By construction these points are internal to the original intervals; which proves the theorem.

THEOREM 14. *Given an infinite number of sets of intervals, in a finite segment (A, B) of length L, such that the content of each set of intervals is greater than some positive quantity g, then a set of points of potency c exists, which is internal to an infinite series of these sets of intervals, and contains closed components of content $> g - e$, where e is as small as we please.*

For consider the non-overlapping intervals having the same internal points as any one of the sets D_1. Their content $> g$, and therefore we can choose out a finite number of them whose content is greater than g. Suppose this done for all the sets; then in each set we have only a finite number of non-overlapping intervals. Let the integer m be determined so that

$$mg \leqslant L < (m+1)g \quad \dots\dots\dots\dots\dots\dots(1).$$

Let us consider a group of n of the sets, where n is a sufficiently large integer, later to be more particularly specified.

The points of (A, B), if any, internal to at least two of these n sets form a finite number of non-overlapping intervals, whose content we denote by $I_{1,n}$. The points which are only simply covered, therefore, form a finite set of non-overlapping intervals of content $> n (g - I_{1,n})$, whence

$$n (g - I_{1,n}) < (m + 1) g;$$

therefore $$I_{1,n} > \{1 - (m + 1)/n\} g \dots\dots\dots\dots\dots(2).$$

Let us choose an integer n' so that $(m + 1)/n' < \frac{1}{2}e$, that is,

$$n' > 2 (m + 1)/e \dots\dots\dots\dots(3);$$

then $$I_{1, n'} > (1 - \frac{1}{2}e) g \dots\dots\dots\dots(4).$$

Grouping the given sets together in distinct groups of n', and taking the corresponding sets of double non-overlapping intervals, we have conditions exactly similar to those with which we started, only that, instead of g, we have $(1 - \frac{1}{2}e) g$.

To these new sets we apply the same reasoning as before, taking, however, $\frac{1}{4}e$ instead of e, and substituting for n an integer n'' such that $n'' > 4 (m + 1)/e$ and grouping the sets of double intervals in groups of n''. The content of the double parts corresponding to any such group being denoted by $I_{2,n''}$, it follows that

$$I_{2, n''} > (1 - \frac{1}{2}e) (1 - \frac{1}{4}e) g > (1 - \frac{1}{2}e - \frac{1}{4}e) g > (1 - e) g \ \dots(5).$$

There will, therefore, certainly be such parts for every one of the groups, and they will, by the construction, be at least quadruply covered by the original sets.

In this way we can always proceed a stage further: the sets of intervals which we construct at each successive stage always have content $> (1 - e) g$. Returning to the equation (2), we see that there will be a set of intervals* in (A, B) which are covered at least doubly by the given sets and that this set of double intervals has a content I_1' greater than or equal to $I_{1,n}$ for all values of n, $I_1' \geqslant g$. Similarly, denoting by I_2' the content of the set of intervals in (A, B) which are covered at least quadruply by the given intervals, $I_2' \geqslant g$; and, generally, $I_n' \geqslant g$, where I_n' is the content of the set of intervals which are covered by at least 2^n of the given sets.

Now, since the intervals corresponding to the content I_n' certainly lie inside those of content I'_{n-1}, we can apply

---

* There might, of course, be points of (A, B) external to these intervals which belong to two or more of the given sets, but they do not affect the argument.

Theorem 13, since the content of each is certainly greater than $g - e$, which proves the theorem.

50. The following theorems shew that the intervals of the preceding sections may be replaced by closed sets of points of positive content. Our sets of intervals, when infinite in number, become open sets of points. The theorems about to be obtained will contain the earlier ones as special cases.

The analogon of Theorem 11 is none other than Cantor's Theorem of Deduction (Theorem 5, Chap. III) which may for convenience be restated here :—

CANTOR'S THEOREM OF DEDUCTION. *Given a countably infinite series of closed sets of points, $G_1, G_2, \ldots$, such that each point of G_{n+1} is also a point of G_n, there is at least one point common to all the sets, and the deduced points form a closed set.*

THEOREM 15. *If to the hypothesis of the above Theorem we add that the content of each G_n is greater than some positive quantity g, the common points form a closed set G' of content $\geqslant g$; so that they have the potency c.*

If possible, let the content I' of G' be less than g. Denote by $I_1, I_2, \ldots$ the contents of $G_1, G_2, \ldots$. By Theorem 6 we can find a closed component of G_1, all of whose points are distinct from those of G', and whose content is $I_1 - I' - e$, where e is a positive quantity, smaller than some assigned quantity. This set has in common with G_2 a closed set, whose content, by Theorem 7 is equal to $I_1 - I' - e + I_2 - K$, where K is the content of the closed set constituted by G_2 and the closed component of G_1 above found, and is certainly less than I_1. The content of this component of G_2 is therefore greater than $I_2 - I' - e$; *a fortiori*, greater than $g - I' - e$.

In other words, we have found a component of G_2 which is closed and has no points in common with G', and whose content is greater than $g - I' - e$.

We can therefore repeat the argument, and obtain in each succeeding set such a closed component, each component lying inside the one previously obtained. It follows then, by Cantor's Theorem of Deduction, that there are points other than G' common to all the given sets, contrary to the hypothesis. Therefore &c. Q. E. D.

51. **Open Sets.**

THEOREM 16. *Given a countably infinite series G_1, G_2, ... of sets of points such that the upper limit I_n of the content of closed components in G_n is greater than a positive quantity g, the same for all values of n, each set G_n being contained in the foregoing G_{n-1}, then a set of points exists of potency c, common to all the sets, and this set contains closed components of content greater than $g - e$, where e is as small as we please.*

By the definition of I_n, we can find a closed component G_1' of G_1 such that, its content being denoted by I_1', $I_1 - \frac{1}{2}e < I_1' \leqslant I_1$; and, for all values of n greater than 1, we can, in like manner, find a closed component G_n'' of G_n such that, its content being denoted by I_n'',

$$I_n - \frac{1}{2^n} e < I_n'' \leqslant I_n.$$

Those points of G_2'' which belong to G_1' form a closed set, whose content is greater than $I_2 - \frac{1}{2}e - \frac{1}{4}e$ [since the set consisting of all the points belonging to one or both of the sets G_1' and G_2'' is a component of G_1, so that its content is not greater than I_1, by Lemma 3; therefore, the content of the set common to G_1' and G_2'' is greater than

$$I_1 - \tfrac{1}{2}e + I_2 - \tfrac{1}{4}e - I_1 \quad \text{or} \quad I_2 - \tfrac{1}{2}e - \tfrac{1}{4}e].$$

Let us denote this closed component of G_2'' by G_2'. *Then G_2' is contained in G_1' and has content greater than $g - e$.*

Similarly we can determine a closed component G_3' of G_3'' and of G_2', of content greater than $g - e$; generally we determine successively closed components of each G_n'' and G'_{n-1}, of content greater than $g - e$.

Applying Theorem 15 to these sets G_n', the result follows.

THEOREM 17. *Given an infinite number of sets of points G_1, $G_2, \ldots$, components of a closed set of finite content* L, such that the upper limit of the contents of the closed components of G_n is greater than some positive quantity g, the same for all values of n, then an infinite series of these sets exists, having in common a set of points of potency c, the content of whose closed components has an upper limit $\geqslant g$.*

---

* It will be seen that it is sufficient if L is the upper limit of the content of closed sets in the whole set, which does not need to be closed; this is brought out in the re-statement of this theorem as Theorem 20.

Let us choose out a closed component of each set of content greater than g, and let these be denoted by G_1', G_2', ... Let e be any small positive quantity, and let the integer m be determined such that

$$mg \leqslant L < (m+1)g \quad\dots\dots\dots\dots\dots(1).$$

Let us consider a group of n of the closed sets G', where n is a sufficiently large integer, subsequently to be further specified.

The points common to any particular pair of the sets of the group form a closed set of points (Theorem 7); therefore, since the sum of any finite number of closed sets is a closed set, the points common to at least two of the sets of the group form a closed set: let us denote it by $G_{1,n}$, and its content by $I_{1,n}$.

The points of $G_{1,n}$ which belong to any particular set of the group G' form a closed component of $G_{1,n}$, whose content is therefore less than or equal to $I_{1,n}$; by Theorem 6, therefore, there is a closed set of content greater than $g - I_{1,n}$, consisting entirely of points belonging to no set of the group, except G'. Corresponding to each of the n sets G' we can find such a closed component, and they will have no common points; so that they form a closed set of content greater than $n(g - I_{1,n})$, by Theorem 7. Hence, by (1), $n(g - I_{1,n}) < (m+1)g$; and therefore

$$I_{1,n} > \{1 - (m+1)/n\}\, g \quad\dots\dots\dots\dots\dots(2).$$

Thus the set $G_{1,n}$ certainly exists, and has the potency c, for all values of n greater than $m+1$.

Let us determine an integer n' such that $(m+1)/n' < \tfrac{1}{2}e$, that is, $n' > 2(m+1)/e$. Then

$$I_{1,n'} > (1 - \tfrac{1}{2}e)\, g \quad\dots\dots\dots\dots\dots(3).$$

Grouping our sets G' together in distinct groups of n' sets, and taking the sets of points belonging to at least two sets of each in turn of these groups, say G_1'', G_2'', ..., we have the same conditions as before, only the content of each closed set is now greater than $(1 - \tfrac{1}{2}e)\,g$, instead of g.

Repeating on these sets the process just gone through, we obtain sets of quadruple points of the original sets whose content $I_{2,n}$ satisfies the inequality

$$I_{2,n} > \{1 - (m+1)/n\}\, (1 - \tfrac{1}{2}e)\, g \quad\dots\dots\dots(4).$$

Thus sets $G_{2,n}$, consisting of points common to at least 2^2 of the given sets, certainly exist, and have the potency c, for all values of n greater than $m+1$.

CONTENT [CH. V

As before (using $\frac{1}{2}e$ instead of e), we can then determine n'' so that

$$I_{2,\,n''} > (1 - \tfrac{1}{4}e)(1 - \tfrac{1}{2}e)\,g > (1 - \tfrac{1}{2}e - \tfrac{1}{4}e)\,g > (1 - e)\,g \ldots(5).$$

This process can be continued *ad infinitum*, and at each stage we see that *there are sets $G_{r,\,n}$ (consisting of points common to at least 2^r of the given sets), of potency c, and of content greater than $(1 - e)\,g$, where e is as small as we please.*

Now the set, in general open, consisting of *all* the points belonging to at least 2^r of the given sets is certainly contained in the set consisting of all the points common to at least 2^{r-1} of the given sets, and, by the above, these sets satisfy the other condition of Theorem 16 ($g - e$ being substituted for g). Hence, by Theorem 16, the result follows.

52. The inner content. We have seen that, in the case of an open set, the upper limit of the content of closed components plays a most important *rôle*. In the lemmas and theorems relating to open sets, enunciated and proved, this concept has to them precisely the relation that content itself has to closed sets; it is called *the inner measure of the content* or briefly *the (inner) content of the open set*.

DEFINITION. *The (inner) content* of a set is defined to be the upper limit of the content of its closed components.

The introduction of this term simplifies the statements of the preceding theorems: thus Theorems 5 and 6 can be replaced by the following simple proposition :—

THEOREM 18. *If a closed set G be the sum of two non-overlapping sets, one at least of which is closed, the content of G is the sum of the (inner) contents of the components.*

Theorems 15 and 16 are replaced by the following :—

THEOREM 19. *The (inner) content of a generalised inner limiting set is the limit of the inner content of the defining sets $G_1, G_2, \ldots, G_n, \ldots$*

COR. *The (inner) content of an inner limiting set G of sets of points $G_1, G_2, \ldots$, each contained within the preceding is the limit of the (inner) content of G_n, when n is indefinitely increased.*

Theorem 17 is replaced by the following :—

THEOREM 20. *Given an infinite number of sets of points, components of a set of finite (inner) content, the (inner) contents of these sets having a positive lower limit g, then an infinite number of these sets exists, having in common a set of (inner) content $\geqslant g$.*

53. The (inner) content, so defined, is certainly a magnitude, and, in the case of a closed set, the (inner) content is the content itself. Further it is clear that the inner measure of the content has the same connection with the potency that the content of a closed set has, viz.—*If the (inner) content be positive, the set has the potency c. The (inner) content of a set which has not the potency c is zero.*

The question arises whether the (inner) content possesses all the properties which we are accustomed to associate with the term "content" as long as this term was confined to closed sets. First, we ask, *Is the (inner) content of the sum of two non-overlapping sets always equal to the sum of their (inner) contents?*

All that has been proved in the preceding sections is that this is the case provided the sum of the two sets as well as one of the components is closed. We can, however, at once extend the result to the case when the sum is open. In other words—*Even if the sum of two non-overlapping sets be open, its (inner) content is the sum of their (inner) contents, provided one at least of the components is closed.*

For, if the content of the closed component be a, and the (inner) content of the sum $a + b$, we can, by the definition, find a closed component of content $a + b - e$, where e is as small as we please. The part common to these two closed components must have content $\geqslant a - e$, and $\leqslant a$ [since, otherwise, the remaining component of the first closed component would have (inner) content $> e$, and we could therefore find in it a closed component having no point common with that of content $a + b - e$, and these two together would form a closed component of the whole set of content $> a + b$].

In the closed component of content $a + b - e$ there must then, by Theorem 18, be another distinct component of (inner) content $\geqslant b - e$ and $\leqslant b$. This being true for all values of e, it follows that the (inner) content of the original open component is not less than b. But it cannot be greater than b, since otherwise we could find a closed component which with the first given component would form a closed set of content greater than $a + b$. Thus the second component has (inner) content b; which proves the theorem.

Summing up the result so far, we have the following:—

THEOREM 21. THE INNER ADDITION THEOREM.

The (inner) content of the sum of two non-overlapping sets, one of which is closed, is the sum of their (inner) contents.

54. Two cases remain:—The sum of two non-overlapping open sets is (1) closed; or (2) open.

If we assume that what we have, for shortness, called the (inner) addition theorem may be extended to open sets so that it holds, or, still more shortly, *that the (inner) addition theorem holds* in Case (1), it is easy to deduce that it holds in Case (2).

For, if I be the (inner) content of the sum, and a and b of the components, we can find a closed component of the sum, of content $I-e$, and this cuts out of the components two (open) sets, whose (inner) contents, by the same argument as before, lie between a and $a-e$, and b and $b-e$, respectively. The sum of these two (open) sets, being closed, has, under the supposition that the (inner) addition theorem holds in Case (1), content lying between $a+b-2e$ and $a+b$, but must be equal to $I-e$. Since this is true for all values of e, the result follows.

55. We are now left with the discussion of Case (1). By means of the theorems already proved, we can reduce the problem of determining whether in this case the (inner) addition theorem holds to the following:—*Can the sum of two open sets, each of (inner) content zero, be a closed set of positive content?*

To show this we proceed as follows :—

Suppose, if possible, we have a closed set of content $a+b+c$, and it can be divided into two open components, whose (inner) contents are b and c respectively. In these open components there exist closed components of content $b-e$ and $c-e$ respectively, where e is as small as we please; the content of their sum is then $b+c-2e$. The remaining points of the whole set form a set, in general open, whose (inner) content, by Theorem 21, is $a+2e$, and which is the sum of two non-overlapping sets, the (inner) content of each of which is, by what has been proved, not greater than e. Hence, by the usual argument, we can find a *closed* component of the whole set of content $a+e$, which is the sum of two non-overlapping sets, the sum of whose (inner) contents is not greater than e. With respect to these sets we can now repeat the argument, using $\frac{1}{2}e$ instead of e, and so on. Ultimately, by Theorems 15 and 16 we shall determine a closed set whose content is a, divided into two non-overlapping components, both of whose (inner) contents are zero.

With our present imperfect knowledge of open sets, it seems impossible to assert definitely that such a case could not arise. At present however no such case is known.

In the next section it is shewn that the (inner) addition theorem holds when one of the components is any set whatever and the other component is any one of a large class of open sets.[36]

THEOREM 22.[36] *If the (inner) addition theorem holds when one of the two components is a set of a certain type, it is also true when one of the components is the outer limit of sets of that type.*

In other words, if, for every value of n, the addition theorem holds for G_n and *any* other set O_n, it holds for G and any other set O. For let the sum of G and O be H, and let O_n be the set which added to G_n makes up H. Then, using the letters indiscriminately for the sets and their (inner) contents, $G_n + O_n = H$.

Now, as G_n increases towards G, O_n diminishes towards O, each O_n lying in the preceding O_{n-1}. Therefore, by Theorem 19, the (inner) content O is itself the limit of the (inner) content O_n. Also Lt G_n + Lt $O_n = H$; therefore Lt $G_n + O = H$.

Now the (inner) content of G is evidently not less than the limit of G_n. Therefore $G + O \geqslant H$, the letters denoting (inner) contents. But, evidently, $G + O \leqslant H$; therefore $G + O = H$, the letters denoting either sets or their contents.

COR. 1.[36] *The (inner) addition theorem holds when one of the components is an ordinary outer limiting set, by Theorem 21.*

COR. 2.[36] *An outer limiting set which is the limit of a sequence of sets each of which has the property that the (inner) addition theorem holds for it and any other set whatever has for (inner) content the limit of the (inner) contents of the sets of the sequence.*

COR. 3.[36] The theorem

Lt (inner) content = (inner) content of Lt

is true for a sequence of expanding open sets when the expansion is due *to the increase of a component for which the (inner) addition theorem holds.*

THEOREM 23.[36] *If the (inner) addition theorem is true when one of the two components is a set of a certain type, it is also true when one of the components is an inner limiting set deduced from an infinite series of sets of this type.*

In other words, if, for all values of n, $G_n + O = H_n$ (the letters being used indiscriminately for a set and its content), and each G_n is contained in the preceding set G_{n-1}, then $G + O = H$. In fact, H is itself an inner limiting set, and therefore by Theorem 19, Cor., its (inner) content is

$$\text{Lt } H_n = \text{Lt } G_n + O = G + O.$$

COR.[36] *The (inner) addition theorem holds if one of the components is an ordinary inner limiting set.*

For an infinite set of closed intervals is a special case of an ordinary outer limiting set, and therefore, by Cor. 1 to Theorem 22, the (inner) addition theorem applies when one of the components consists of the points belonging to such a set of intervals. Hence, applying our present theorem, it holds for the deduced set of a sequence of such sets of intervals.   Q. E. D.

Applying the results of this section, we see that, if we keep applying in any order Theorems 22 and 23 to any series of ordinary outer or inner limiting sets, the sets so obtained must always have the property in question.

We have thus already obtained a large class of open sets possessing the property in question; we can, however, extend this class still further.   The class of sets for which the (inner) addition theorem holds may, for shortness, be called *the (inner) additive class.*

THEOREM 24.[36] *If each of two sets which do not overlap belong to the (inner) additive class, their sum also possesses the property in question.*

Let G_1 and G_2 be two such sets, and H any other set whatever. Also let G be the sum of the two sets. Let the (inner) contents of G_1, G_2, and G be denoted by I_1, I_2, and I, and that of H by J. Then, since G_1 belongs to the class, we have at once $I_1 + I_2 = I$. For the same reason $I_1 + J$ is the (inner) content of $(G_1 + H)$. Hence also, since G_2 belongs to the class, the (inner) content of $(G_1 + H) + G_2$ is $I_1 + J + I_2$, i.e., it is $I + J$. In other words, the (inner) content of $G + H$ is $I + J$. Therefore, &c.   Q. E. D.

THEOREM 25.[36] *If each of two sets one of which is a component of the other belong to the (inner) additive class, so does their difference.*

Use the same notation as in the preceding theorem, G denoting the larger of the two sets, and G_1, say, the component belonging to the class. As before, $I = I_1 + I_2$. We have to prove that G_2 belongs to the class. Suppose this is not the case; then the (inner) content of $G_2 + H$ must be greater than $I_2 + J$, say $I_2 + J + k$. But, by hypothesis, G_1 belongs to the class; hence the (inner) content of $G_1 + (G_2 + H)$ is $I_2 + J + k + I_1$, i.e., it is $I + k + J$. But $G_1 + G_2$ is G, and G belongs to the class; therefore the (inner) content of $G_1 + G_2 + H$ is $I + J$; therefore k must be zero. Therefore, &c.   Q. E. D.[36]

THEOREM 26. *If a set belonging to the (inner) additive class be divided into two components the sum of whose inner contents is equal to that of the original set, each of the components belongs to the class.*

Let G be the set, G_1 and G_2 the components, H any other set whatever. Denote the corresponding (inner) contents by I, I_1, I_2 and J. Suppose, if possible, that the (inner) content of $G_1 + H$ be not $I_1 + J$; then it must be greater than $I_1 + J$. Add the set G_2 to the set $G_1 + H$. Then the (inner) content of the set $(G_1 + H) + G_2$ would be greater than $I_1 + I_2 + J$. But, by hypothesis, $I_1 + I_2 = I$; therefore the (inner) content of the set $G_1 + H + G_2$ would be greater than $I + J$; that is, the (inner) content of the set $G + H$ would be greater than $I + J$. But G belongs to the class in question; therefore the (inner) content of $G + H$ is equal to $I + J$. Therefore, &c. Q. E. D.

THEOREM 27. ADDITION THEOREM FOR THE (INNER) CONTENTS.

If G_1 and G_2 be two sets of the (inner) additive class, of (inner) content I_1 and I_2, (a) the set consisting of all the points common to G_1 and G_2 is a set of this class, say G', of (inner) content I'; and (b) the set consisting of all the points belonging to one or both of G_1 and G_2 is a set of this class, say G, of (inner) content I; further, (c) $I_1 + I_2 = I + I'$.

For suppose the (inner) contents of the parts of G_1 and G_2 which are not common to be $I_1 - x$ and $I_2 - y$ respectively. Then, since the (inner) addition theorem holds for G_2, $I_2 + (I_1 - x) = I$. Similarly, since the (inner) addition theorem holds for G_1, $I_1 + (I_2 - y) = I$; whence

$$x = y = I_1 + I_2 - I.$$

Also $I' + (I_1 - x) \leqslant I_1$; therefore $I' \leqslant x$, that is,

$$I' \leqslant I_1 + I_2 - I \quad \dots\dots\dots\dots\dots(1).$$

Again, take in each component a closed set of content greater than $I_1 - e$, $I_2 - e$ respectively. Then the common part of these closed sets lies in G', and has therefore content $\leqslant I'$. The set of points belonging to one or both of these closed sets lies in G, and has therefore content $\leqslant I$. Then, by Lemma 3,

$$(I_1 - e) + (I_2 - e) < I + I',$$

however small e may be, that is,

$$I' \geqslant I_1 + I_2 - I \quad \dots\dots\dots\dots\dots(2).$$

Comparing (1) and (2), we have

$$I_1 + I_2 = I + I'. \quad \text{Q. E. D.}$$

Again, *the (inner) contents of the parts of G_1 and G_2 which are not common are $I_1 - I'$ and $I_2 - I'$.* In fact, from the result just obtained, we have $x = I'$. It at once follows, by Theorem 25, *that the sets G, G', $G_1 - G'$, $G_2 - G'$, all belong to the class in question.*

Q. E. D.

The theorems which we have obtained enable us, starting from closed sets, to build up a very extended class of open sets, possessing the property that the (inner) addition theorem holds for any set of the class in combination with any set whatever. The great generality of the class obtained suggests the possibility that the (inner) addition theorem holds for all sets without exception. We must be careful, however, not to jump to this conclusion. We have, at most, shewn that all known open sets belong to the class in question. All the known operations employed on members of the class lead to members of the same class ; in modern phraseology, they form a *corpus*. If we could assert that there were no other open sets than those formed from closed sets by these processes, we should have settled, once for all, the difficult question of the classification of open sets.

In connection with the class of operations made use of in this section, the theorems of § 34, Ch. IV, which bear also on the classification of open sets, will be of interest, and are needed in what follows.

56. In Art. 55 I shewed that, in the discussion of the question whether, or no, the (inner) addition theorem holds always, we might confine our attention to sets of zero (inner) content. We may remark that *the general problem of classifying open sets may be reduced to the corresponding problem for sets of zero (inner) content.*

In fact, if we take any open set of (inner) content a, two cases at most can present themselves : either it contains a closed set of content a or it contains closed sets of content as near a as we please. In the former case the given set is the sum of a closed set of content a and an open set of (inner) content zero ; in the latter case we may first subtract a closed set of content $a - e$, and so obtain an open set of content e ; in this latter set we may subtract a closed set of content e', where e' is as small as we please ; and so on. We thus get, by successive subtraction of closed sets, a series of open sets, each lying inside the preceding and having zero for the lower limit of their contents; their deduced set is therefore either altogether absent or has content zero. In the former case the given open set is an ordinary outer limiting set; in the latter case it is the sum of an ordinary outer limiting set and a set of zero (inner) content. In other words, we have the following theorem :—

Every set of (inner) content a is either a closed set or an ordinary outer limiting set, or is equal to the sum of one or other of these and of a set of zero (inner) content.

As the properties of an ordinary outer limiting set may be regarded as known, this theorem confirms the statement made above as to the classification of open sets.

57. The (Outer) Content. The definition adopted makes the (inner) content of an open set depend on that of a closed set; moreover, we get as the (inner) content the content [for we shall see that we can here suppress the term (inner)] of a certain ordinary outer limiting set contained in it. If we attempt to give a definition of content equally applicable to all sets of points, we are met at once by difficulties which might seem to be insuperable.

The ordinary definition of the content of a closed set is, as we saw, equivalent to the following: Describe little intervals of constant length *e* round the points of the set: these fill up a finite set of intervals the content of which is, in the limit, when *e* is indefinitely diminished, the content of the closed set.

If this definition be applied to an open set, it gives us, as we saw, the same content as that of the set got by closing it, and thus fails to distinguish between the set and its component.

In the definition given of the content of a closed set it is, however, unnecessary to take the intervals all of the same length : not only so; it is not necessary to specify that they have a positive lower limit. In fact, if round every point of a closed set we describe a little interval, say < *e*, according to any law, it is clear that the content of such a set of intervals is not greater than it would be if each interval were (if necessary) extended, until the corresponding point of the closed set is its middle point and its length is 2*e*. When *e* describes any sequence with zero as limit, the content of these extended intervals approaches as limit the content of the closed set. Thus the limit of the content of the original intervals is not greater than the content of the given closed set. But by Theorem 4, neither is it less than that content. Thus the limit of the content of the set of intervals constructed, when e is indefinitely diminished, will be the same quantity as before.

If we try to apply this modified form of the definition of the content of a closed set to open sets in general, we are met by a similar difficulty to that which occurred before. Whereas in the case of a closed set no other points are left in ultimately, when e is indefinitely diminished, this is not true of open* sets, unless they belong to the class of what we have called "ordinary inner limiting sets." Thus, if it be legitimate to ascribe content to an ordinary inner limiting set and to define it in this manner, the process in question, when applied to an open set in general, would give us the content of an ordinary inner limiting set of which it is a component. The content, defined in this manner, is called *the outer measure of the content*, or, briefly, the "*(outer) content*."

DEF. Round every point of the set G describe an interval; find the content of the set of intervals so formed; this content has a lower limit for the various possible modes of construction; this lower limit is called the *(outer)*[37] content of the set of points.

It is easily seen that Theorem 18 holds if for (inner) we substitute (outer). Corresponding to Theorem 19 we have the following :—

THEOREM 28. *The (outer) content of a generalised outer limiting set is the limit of the (outer) content of the defining sets* $G_1, G_2 ..., G_n ....$

Let J_n be the (outer) content of G_n and J of the outer limiting set G, and let us denote the limit of J_n when n is indefinitely increased by j. It is evident that, as each G_n is contained in the following G_{n+1}, the quantities J_n never decrease, and j is their upper limit.

Let us commence at such a set G_1 that, e being any small positive quantity, $j - e \leqslant J_n \leqslant j$, for all values of n, and let $e_1 + e_2 + ... < e$. Enclose G_n in a set of intervals of content less than $J_n + e_n$, for all values of n.

Then the parts common to the $(n-1)$th and nth sets of intervals contain G_{n-1}, and must therefore have content $\geqslant J_{n-1}$. Thus, if we take all the intervals together which we have constructed, we have a set of overlapping intervals containing every point of G, and their content is less than or equal to

$$(J_1 + e_1) + (J_2 - J_1 + e_2) + ... + (J_n - J_{n-1} + e_n) + ...,$$

that is, less than $j + e$. Thus $J < j + e$. But J cannot be less than j; for otherwise we could enclose G in a set of intervals of content less than j, which is evidently impossible. Thus $J = j$.

Q. E. D.

It will often be convenient to denote the inner and outer measures of the content by prefixing a subscript i or o; thus $_iI$ and $_oI$ (inner I and outer I) will be used for the inner and outer measures of the content of a set denoted by G.

THEOREM 29. *If G be the set of all the points of two sets G_1 and G_2 without common points,*

$$_iI \leqslant {_oI_1} + {_iI_2} \leqslant {_oI}.$$

Enclose G in a set of intervals of content $_oI + \theta e$, and in G_2 take a closed set of content $_iI_2 - \theta'e$, where θ and θ' lie between 0 and 1.

By the generalised Heine-Borel Theorem (Theorem 6, Chap. IV, p. 41) the number of these intervals which contain points of this closed set may be taken to be finite. Inside each of these intervals the points of the closed set form a closed set, and the sum of the contents of these partial closed sets is, by Theorem 12, the content of the whole closed set. Thus those points of the intervals which are not points of the closed set form a set of intervals of content $_oI + \theta e - {_iI_2} + \theta'e$, containing G_1. Since this is true for all values of e,

$$_oI - {_iI_2} \geqslant {_oI_1},$$

or
$$_oI_1 + {_iI_2} \leqslant {_oI} \quad\quad\quad\quad\quad\quad\quad\quad (1).$$

Again, in G take a closed set of content $_iI - \theta e$, enclose G_1 in a set of intervals of content $_oI_1 + \theta'e$.

The points of this closed set not internal to this set of intervals form a closed component of G_2 of content $x \leqslant {_iI_2}$. By Theorem 13, therefore, the closed set contains closed components of content as near as we please to $_iI - \theta e - x$, internal to the set of intervals constructed. Thus

$$_iI - \theta e - x \leqslant {_oI_1} + \theta'e.$$

Since $x \leqslant {_iI_2}$, and e may be made as small as we please,

$$_iI - {_iI_2} \leqslant {_oI_1},$$

that is
$$_oI_1 + {_iI_2} \geqslant {_iI} \quad\quad\quad\quad\quad\quad\quad\quad (2).$$

By (1) and (2) the result follows.

58. Measurable Sets. For closed sets we know that (inner) and (outer) content are merely different aspects of the same thing, the content of the closed set. A set for which the (inner) and (outer) contents coincide is called *a measurable set*; for such a set we may, without scruple, use the term " content."

It is evident that any definition of the content which agrees in the least with our fundamental ideas must make the content of a set greater than, or at least equal to, that of any of its components ; so that, if the (outer) and (inner) contents ever do not coincide, the former gives us an upper limit and the latter a lower limit for the content. Thus, in the case of measurable sets no other defini- tion of the content is possible.

Theorem 28 now gives us the following important theorem *.

THEOREM 30. *If G be a measurable set, and G_1 and G_2 complementary components with respect to G (that is G is the sum of G_1 and G_2)*

$$_0I_1 + {}_iI_2 = I.$$

COR. *If G_1 is measurable, so is G_2, and the content of G_2 is the difference of the contents of G and G_1, viz. $I_2 = I - I_1$.*

THEOREM 31. *The sum of any countable number of measurable sets without common points is a measurable set, and has for content the sum of their contents.*

Let the sets be G_1, G_2, ..., the number of them being finite or countably infinite.

Enclose G_n in a set of intervals of content less than $I_n + \dfrac{e}{2^n}$, for all values of n.

In this way we can enclose the set G, which is the sum of G_1, G_2, ..., in a set of intervals of content less than $e + I_1 + I_2 + \dots$.

Also we can find a closed component of each set G_n of content greater than $I_n - \dfrac{e}{2^n}$. The sum of these is a closed set, or an ordinary outer, limiting set, the inner measure of whose content is greater than $- e + I_1 + I_2 + \dots$.

Since e may be taken as small as we please, this shews that the outer measure of the content of G is not greater than the inner measure, so that G is measurable.

It also shews that
$$I = I_1 + I_2 + \dots,$$
which proves the theorem.

---

* This property was taken by Lebesgue as the definition of the inner measure of the content of a set in terms of the outer measure of the content of the comple- mentary set with respect to the fundamental segment, supposed finite.

    Cp. Lebesgue, "Intégrale, Longueur, Aire," § 6, *Ann. di Mat.* (1902).

THEOREM 32. ADDITION THEOREM FOR THE CONTENT OF MEASURABLE SETS.

If G_1 and G_2 are measurable sets of points of content I_1 and I_2, the set consisting of all the points common to G_1 and G_2 is a measurable set, say G' of content I', and the set consisting of all points belonging to at least one of G_1 and G_2 is a measurable set, say G of content I ; further $I_1 + I_2 = I + I'$.

We can find closed components of G_1 and G_2 of contents greater than $I_1 - e$ and $I_2 - e$ respectively. By the Addition Theorem for Closed Sets (p. 84), the sum of these contents is equal to the sum of the content of the sets consisting of all their points and of their common points respectively. The set of all their points is a component of G, and the set of their common points a component of G', so that the sum of the contents of these two sets is not greater than $_iI + _iI'$. Thus

$$I_1 + I_2 - 2e < {}_iI + {}_iI'.$$

Similarly, enclosing G_1 and G_2 in intervals, and using the Addition Theorem for Sets of Intervals (p. 81),

$$I_1 + I_2 + 2e > {}_oI + {}_oI'.$$

Since e is at our disposal, it follows that

$$_oI + {}_oI' \leqslant I_1 + I_2 \leqslant {}_iI + {}_iI'.$$

But, unless G and G' are measurable, $_oI$ is greater than $_iI$, and $_oI'$ is greater than $_iI'$. Thus the preceding relation is only possible when the signs of equality are taken, and G and G' are measurable, which proves the theorem.

COR. *The points of a measurable set which do not belong to another measurable set form a measurable set.*

This follows from the above and the Cor. to Theorem 30.

59. From the point of view of an exhaustive classification of open sets, it must be shewn whether sets other than measurable sets exist. This point is still open to question. If there are other sets, then, as will be shewn, all the ordinary sets are included in a class which is included in the class of measurable sets, but may consist of only a part of it: this class has itself the potency of all sets in any segment finite or infinite, and, from the point of view of content, possesses most important characteristics; this is none other than the class of those sets which in combination with *any other set whatever* are such that the sum of the (inner) contents is the (inner) content of the sum, and the sum of the (outer) contents is the (outer) content of the sum *.   As we already have, for definiteness, called the class of sets for which the (inner) addition theorem holds the (*inner*) *additive class*, we shall call that for which the (outer) addition theorem holds the (*outer*) *additive class*; the class above referred to will then belong to both these classes, and will be termed *the additive class.*

Theorem 19 of this chapter shews that for *an ordinary inner limiting set the* (*outer*) *content coincides with the* (*inner*) *content*; it shews, moreover, that, in the case of an ordinary inner limiting set, however we construct the intervals round the points of that set, the content of those intervals always approaches the same limit when the intervals are decreased without limit, viz., the content of the ordinary inner limiting set, provided ultimately no points are left in except those of the given inner limiting set.

In the case of a set which is not an inner limiting set we cannot so construct the intervals that no other points are left in, and there might seem to be a certain degree of arbitrariness in the selection of those points which are to be admitted.

According to the law of construction adopted, we may, as the length of the separate intervals is indefinitely decreased, approach the actual lower limit, that is the (outer) content, or some other quantity lying between this and the content of the set got by closing the given set.

If I be the (inner) content of a set, it is evident that the set cannot be enclosed in a set of intervals of content less than I; thus the defining property of measurable sets may be expressed by saying that *a set of* (*inner*) *content I is measurable if, and only if, it can be enclosed in a set of intervals of content $I + e$, where e is as small as we please.* This property is, as we saw, possessed *par excellence* by ordinary inner limiting sets.   It is remarkable that, as easily follows from Theorem 28, it is also possessed by ordinary outer limiting sets, though, except in particular cases, an ordinary outer limiting set cannot be defined as the inner limiting set of a sequence of sets of intervals.

Thus we have the following theorem which is a special case of Theorem 35, proved below :—

An ordinary outer or inner limiting set is measurable, that is, if its content be I, it can be shut up in an infinite set of intervals whose content lies between I and $I + e$, and it contains closed components of content lying between $I - e$ and I, where e is as small as we please.

* This is, by Theorem 31, true of measurable sets in combination with measurable sets, but, perhaps, not with *any set whatever.*

60. The following theorem proves that all sets belonging to what I have called the (inner) additive class are measurable.

THEOREM 33.[38] *If a set is such that when added to any other set whatever which has no points in common with it the sum of the (inner) contents is the (inner) content of the sum, the set in question is measurable.*

Let I_1 be the (inner) content of the set, and I_2 be that of the set of points required to close it, and I that of the whole set so obtained; then, by hypothesis, $I = I_1 + I_2$. As usual, let the sets whose contents are I, I_1, and I_2 respectively be denoted by G, G_1, and G_2.

Take a closed component G_2' of content $> I_2 - \frac{1}{2}e$ in G_2. The set G_1 lies, of course, in the black intervals of this set. Next shut up the set G in a finite number of intervals $d_1, d_2, \ldots d_n$, of content $< I + \frac{1}{2}e$.

In any one of these intervals d_r, the points of G_2' form a closed set, of content I_r' say, where $\overset{r=n}{\underset{r=1}{\Sigma}} I_r' > I_2 - \frac{1}{2}e$.

The points of G_1 which lie in d_r lie in the black intervals of this closed component of G_2', that is, in intervals whose sum is $d_r - I_r'$. Thus *all* the points of G_1 are enclosed in a set of intervals whose sum is

$$\overset{n}{\underset{1}{\Sigma}} \{d_r - I_r'\} < I + \tfrac{1}{2}e - I_2 + \tfrac{1}{2}e < I_1 + e.$$

This, therefore, proves the theorem.

It is easy to see that, if a set does not belong to the (inner) additive class, we can no longer assert that it is measurable. Take, for example, a closed set of content a, and suppose it, if possible, divided into two components which do not belong to the (inner) additive class, so that the sum of their (inner) contents is less than a. Then, if both these components are measurable, we could enclose the closed set in an infinite set of intervals whose sum is less than a, and therefore in a finite number of these intervals; which is impossible. Thus at least one of the components and therefore by Theorem 30, Cor., both the components, cannot have the property in question.

We have not, however, proved that, if there are sets which do not belong to the (inner) additive class, they may not be further sub-divided into those which are and those which are not measurable.

61. The properties which we have found for the (inner) content have their exact counterparts for the (outer) content; so that we cannot say that either concept seems more fundamental than the other.

A set of (outer) content J is evidently measurable if, and only if, it contains closed components of content J − e, where e is as small as we please.

That this is the case when the set belongs to the (outer) additive class is shewn as follows; the theorem is the counterpart to Theorem 33.

THEOREM 34. *If a set be such that, when added to any set whatever having no point common with it, the sum of the (outer) contents is the (outer) content of the sum, the set in question is measurable.*

As before, let G_1 be the set, G_2 the set required to close it, and G the sum of G_1 and G_2, and let the corresponding (outer) contents be J_1, J_2, and J.

Let us enclose G in a finite number of intervals of content lying between J and $J + e$, and G_2 in a set of intervals of content lying between J_2 and $J_2 + e$.

The points of the former intervals which are not internal to the latter intervals form a closed set of content lying between $J − J_2 − e$ and $J − J_2$; that is, between $J_1 − e$ and J_1, by the hypothesis. The points of this closed set which also belong to the closed set G form a closed component of G, which, since it has no point common with G_2, is also a closed component of G_1. Let its content be denoted by K; then we can enclose it in a finite number of intervals of content less than $K + e$, and these, together with the intervals constructed round G_2 contain all the points of G; hence $K + J_2 + 2e \geqslant J_1 + J_2$, that is $K \geqslant J_1 − 2e$, which proves the theorem.

COR. *The sets of the additive class are all measurable.*

62. From Theorems 19 and 28 we have immediately the following theorem :—

THEOREM 35. *An outer or inner limiting set of measurable sets is measurable and has for content the limit of the contents of the defining sets.*

In particular we have the following special cases :—

COR. 1. *An ordinary inner limiting set is measurable and*

its content is the limit of the contents of sets of normal intervals defining it.

Cor. 2. *An ordinary outer limiting set is measurable and its content is the limit of the contents of the defining closed sets.*

In the special case when the ordinary outer limiting set is a closed set dense nowhere, this is Osgood's theorem [*]. Another important special case gives rise to the following corollary :—

Cor. 3. *If G be the ordinary outer limiting set of a series of closed sets of zero content, every closed component of G has zero content*[†].[39]

It may be noted that this indicates the existence of unclosed sets, such that, though every closed set contained in them has content as small as we please, every closed set containing them has content as large as we please.

The simplest example of the kind is that of the rational points. Let G_n here stand for all the proper fractions with n as denominator. Then G consists of all the rational numbers between 0 and 1. The content of G_n and therefore of G is zero: the set Γ got by closing G is however the continuum from 0 to 1, and its content is therefore unity. This set is dense everywhere. By using the (1, 1)-correspondence given in § 23, p. 46 however we may deduce a similar set which is dense nowhere. The following is an example of the same type obtained directly.

Ex. 5. Consider the following sequence of sets :—

G_1 is H. J. S. Smith's ternary closed set[‡] of the first kind in the segment (0, 1).

By means of repetitions of the processes by which G_1 was constructed in the segment (0, 1), we propose to construct a series of closed sets whose limit G, when closed by the addition of those limiting points not already included in it, is identical with H. J. S. Smith's ternary closed set of the second

* *Amer. Journ.* xix.   † *Quart. Journ.* No. 138, 1903.
‡ p. 79, footnote. The black intervals of this set are got by dividing a segment into three equal parts, blackening the right-hand part ; then repeating this process in each of the two white parts and so on. The construction for the closed set of the second kind is similar, only instead of dividing at each stage into three, we divide successively into 3, 3^2, 3^3, ... parts. The expression for the content of the latter set is the same as that of Ex. 1, p. 78, viz.

$$I_P = \underset{n=\infty}{\mathrm{Lt}} \left(1 - \frac{1}{3}\right)\left(1 - \frac{1}{3^2}\right)\left(1 - \frac{1}{3^3}\right)...\left(1 - \frac{1}{3^n}\right).$$

The former set is of zero content. *Proc. Lond. Math. Soc.* Vol. vi. p. 948 ; cf. *Proc. Lond. Math. Soc.* Vol. xxxiv. p. 286, footnote, for the reason of the insertion of the term "closed," *i.e.* the set in question is that got by adding to H. J. S. Smith's ternary set of the first kind its limiting points.

kind. This latter set we denote by Γ. For this purpose the following should be noticed :—

(1) If we divide the segment in which an H. J. S. Smith's set of the first kind is given into 3^n equal parts, certain of them will be entirely black for the set, and in each of the others there is an H. J. S. Smith's set of the first kind.

(2) If we divide the segment in which an H. J. S. Smith's set of the second kind is given into $3^{\frac{1}{2}n(n+1)}$ parts, certain of them will be entirely black for the set, and in each of the others the given set has precisely the same form, though this form is not an H. J. S. Smith's set of the second kind, because the largest black interval in each part is not $\frac{1}{3}$ of that part.

Having premised this, we proceed to the construction. (See Fig. 11.) The largest black interval of G_1 is the same as the largest black interval of Γ, and

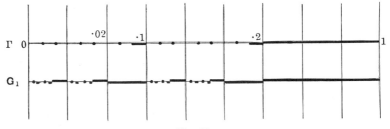

Fig. 11.

in each of the two remaining segments $(0, \cdot 1)$ and $(\cdot 1 \text{ to } \cdot 2)$ in the ternary notation, Γ has the same form, and G_1 consists of an H. J. S. Smith's set of the first kind. We need, therefore, only consider what modification is necessary in the segment $(0, \cdot 1)$, the same modification being supposed made in the segment $(\cdot 1, \cdot 2)$, and the segment $(\cdot 2, 1)$ being left unaltered. In the segment $(0, \cdot 1)$ Γ has its largest black interval of length $\dfrac{1}{3^{1+2}}$ on the extreme right ; in all the other segments of the same length, $(0, \cdot 0^2 1)$, &c., it has the

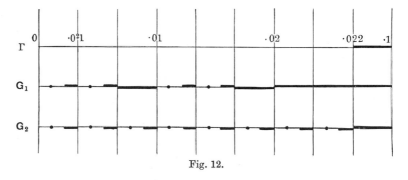

Fig. 12.

same form, whereas G_1 has the form of an H. J. S. Smith's set of the first kind in only some of these segments, $(\cdot 0^2 2, \cdot 01)$, $(\cdot 012, \cdot 02)$, and $(\cdot 02, \cdot 1)$ being entirely black. If, however, in each of these three largest black intervals of

G_1 we insert an H. J. S. Smith's set of the first kind, then the extreme right-hand segment ($\cdot022$, $\cdot1$) will be entirely black for the new set G_2, and in each of the other eight segments G_2 will consist of an H. J. S. Smith's set of the first kind.

We can, therefore, as before, consider what modifications are necessary in the segment (0, $\cdot0^2 1$) only, the same modification being supposed made in the other seven segments, and the segment ($\cdot022$, $\cdot01$) being left unaltered.

By a precisely similar argument as before, it is easily seen that, if we form G_3 by inserting in each of the three largest intervals of G_2 an H. J. S. Smith's set of the first kind, and G_4 by inserting in each of the 3^2 largest intervals of G_3 an H. J. S. Smith's set of the first kind, the extreme right-hand segment of

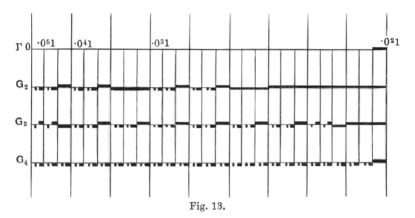

Fig. 13.

length $\dfrac{1}{3^{1+2+3}}$, ($\cdot0^3 222$, $\cdot0^2 1$), will be entirely black, as it is for Γ, and, in all the other segments of length $\dfrac{1}{3^{1+2+3}}$, G_4 will consist of an H. J. S. Smith's set of the first kind.

The general law is now obvious. We shall only have to consider the modifications necessary in $G_{1+1+2+3+\ldots+n}$ in the segment (0, $\cdot0^{1+2+\ldots+n} 1$), in which it consists of an H. J. S. Smith's set of the first kind, the same modification being made in all the other segments of the same length except those which are by our construction already black for Γ.

The modification will consist in inserting H. J. S. Smith's sets of the first kind in the 3, 3^2, 3^3, ... 3^{n+1} largest intervals in turn to form $G_{\frac{1}{2}[n\,(n+1)]+2}$, $G_{\frac{1}{2}[n(n+1)]+3}$, up to $G_{\frac{1}{2}[(n+1)(n+2)]+1}$.

Comparing this series of sets with H. J. S. Smith's ternary closed set of the second kind, we see that, given any small quantity ϵ, we can assign a stage m in the series such that for all values of $n \geqslant m$ all the black intervals of G_n which are $\geqslant \epsilon$ are identical with those $\geqslant \epsilon$ of H. J. S. Smith's ternary closed set of the second kind. But the content of G_n *is always zero*; while that of H. J. S. Smith's ternary closed set of the second kind lies between $\frac{1}{2}$ and $\frac{2}{3}$.

The set G obviously consists of all the isolated points and limiting points on one side only of H. J. S. Smith's ternary set of the second kind, together with some of its limiting points on both sides. The set got by closing G *is* therefore H. J. S. Smith's ternary closed set of the second kind ; that is, Γ.

We will now prove that G does not contain all the points of Γ, and is therefore unclosed. To do this it is sufficient to prove that the point

$$P = \cdot 1212212221\ldots$$

(where the right-hand side of the symbolic equation represents a ternary fraction, the number of 2's between consecutive 1's increasing each time by one), which is a limiting point on both sides of Γ (and a limiting point on one side only of Γ'), is an *internal* point of a definite black interval of G_n for *every* value n, and is therefore *not* a point of G. This interval is $(\cdot 12, \cdot 2)$ of G_1, $(\cdot 1212, \cdot 122)$ of G_2, $(\cdot 12122, \cdot 122)$ of G_3, $(\cdot 1212212, \cdot 121222)$ of G_4, $(\cdot 12122122, \cdot 121222)$ of G_5, $(\cdot 121221222, \cdot 121222)$ of G_6, $(\cdot 12122122212, \cdot 1212212222)$ of G_7, and so on. The general law is now evident, and hence the assertion is proved. Similarly it is evident that any ternary fraction of Γ which involves an infinite number of 2's* cannot belong to any G_n, since, apart from a finite number of figures at the beginning, the numbers of G_n involve only the figures 0 and 1. Each such point is a limiting point on both sides of Γ and is interior to a black interval of G_n whose length diminishes indefinitely as n increases indefinitely.

Hence the set G which we have constructed as the limiting set of a sequence of closed sets, nowhere dense, is open, and has zero content while the content of any closed set containing it is greater than $\frac{1}{2}$.

63. Corresponding to Theorem 20 we have now the following :—

THEOREM 36. *Given an infinite number of sets of points, components of a set of finite (outer) content L, the (outer) contents of these sets having a positive upper limit g, then an infinite number of these sets exists, which can all be enclosed simultaneously in a set of intervals of content $< g + e$, where e is as small as we please.*

If more than a finite number of the sets have zero (outer) content, the theorem is obviously true ; we assume therefore that this is not the case; then there is certainly at least one proper upper limit $g' \leqslant g$ such that, for all values of e, there are a more than finite number of the sets whose (outer) contents lie between $g' - e$ and g', both inclusive.

This being so, let us replace the sets by ordinary inner limiting sets containing them, having the same (outer) content and contained in an outer limiting set of content $L\dagger$, and let G_1, G_2, G_3, ... be a countable set of these ordinary inner limiting sets such that, if the content of G_n be denoted by I_n, $\quad g' \geqslant I_n > g' - \dfrac{e^n}{2^{n+1}}$.

---

* Other than $\dot{2}$, of course.

$\dagger$ It is easy to see how to do this ; we can enclose each of the sets in a set of intervals of content within e of its content, and the whole set in a set of intervals

Then, since an ordinary inner limiting set has the same (inner) and (outer) content, we can, since they are all contained in a set of content L, and have content $> g' - \dfrac{e}{2^2}$, apply to these sets the result of Theorem 20, that is, there must be a countable number of them, say, in order, G_1', G_2', G_3', ..., having in common a set of (inner) content $\geqslant g' - \dfrac{e}{2^2}$, and therefore containing an ordinary outer limiting set of content $\geqslant g' - \dfrac{e}{2^2}$. Let us denote this latter by C_1.

Similarly, there must be a countable number of the sets G_1', G_2', ..., whose contents are greater than $g' - \dfrac{e}{2^3}$, and among these we can find a countable infinite set G_2'', G_3'', ..., having in common a set of (inner) content $\geqslant g' - \dfrac{e}{2^3}$, and therefore containing an ordinary outer limiting set of content $\geqslant g' - \dfrac{e}{2^3}$. Let us call this C_2. In this way we obtain a series of the sets G_1', G_2'', G_3''', ..., and a corresponding series of ordinary outer limiting sets C_1, C_2, C_3, ..., such that C_1 is contained in all the sets G_1', G_2'', ..., C_2 in all but the first, C_3 in all but the first two, and so on.

By Theorem 22 the outer limiting set of C_1, C_2, ... is an ordinary outer limiting set—let us call it C—and its content is the limit of the content of C_n, that is g'.

Now, since G_1' and C_1 are both additive sets, their difference has content $\leqslant \dfrac{e}{2^2}$. Similarly, the difference between G_2'' and C_2 has content $\leqslant \dfrac{e}{2^3}$, and so on. Thus, if we enclose C in a set of intervals of content $< g' + \tfrac{1}{4}e$, we shall be able to enclose the remaining points of G_1' in a set of intervals of content $< \dfrac{e}{2^2} + \dfrac{e}{2^3}$, and the remaining points of G_2'' in a set of intervals of content $< \dfrac{e}{2^3} + \dfrac{e}{2^4}$, and so on. In this way we enclose simultaneously G_1', G_2'', G_3''', ... in a set of intervals of content $< g' + e$. These intervals, of course, contain the original sets from which we obtained G_1', G_2'', G_3''', ... ; so that this proves the theorem.

of content lying between L and $L + e$; if we now omit any parts of the former intervals external to the latter intervals, and let e describe a sequence having zero as limit, we get the sets above referred to.

64. It was pointed out in § 52 that the inner measure of the content possesses the same connection with the potency as the content itself; this cannot be asserted for the outer measure. On the other hand the outer and not the inner measure possesses the properties of the content enunciated in Theorems 9 and 10. The proofs there given, depending as they do on the property that the content of a closed set is identical with its outer measure, are valid in the general case, and need not be repeated; the enunciations of these generalisations are as follows :—

The outer measure of the content of any set is the same as that of any of its coherences.

A set which is more than countable has the same (outer) content as its ultimate coherence.

The (outer) contents of two sets are equal, if the points of each set which do not belong to the other are countable.

65. The (Outer) Additive Class. It is not difficult to shew that all closed sets belong to the outer additive class. That the (outer) content of the sum G of two non-overlapping sets G_1 and G_2 is the sum of their (outer) contents, provided both G and G_1 are closed, has already been pointed out as the correlative to Theorem 5; that this is still the case if G is open can be shewn as follows.

Let G' be an ordinary inner limiting set containing G and having as content the (outer) content of G, that is I. G' contains G_1 (the closed set), and the other component (which contains G_2) is, by Theorem 17, an ordinary inner limiting set, and has therefore, by what has been proved for the (inner) content, content $I - I_1$; therefore $I_2 \leqslant I - I_1$; but, since G can certainly be enclosed in a set of intervals of content as near as we please to $I_1 + I_2$, we cannot have $I_1 + I_2 < I$; therefore $I_1 + I_2 = I$.

Thus we have the theorem :—

The (outer) content of the sum of two sets which do not overlap is the sum of their (outer) contents, provided one of the component sets is closed.

It does not follow that, if the (outer) addition theorem holds when the sum is closed, it holds generally. Instead of this, however, if we could assume that it holds when the sum consists of all the points of an interval, we could, as in § 54, shew that the theorem would be true generally.

The sum of the (outer) contents of two non-overlapping sets is evidently not less than the (outer) content of the sum ; thus the question corresponding to that asked in § 55 is the following :—

Can a segment of length a be divided into two sets of points the sum of whose (outer) contents is greater than a ?

By applying Theorem 27, we can, precisely as in the corresponding discussion of the (inner) additive class, prove the following :—

THEOREM 37. *The (outer) addition theorem holds for an inner limiting set of sets of the (outer) additive class.*

COR. *The (outer) additive class includes all ordinary inner limiting sets.*

THEOREM 38. *The (outer) additive class includes all the outer limiting sets of sets of that class.*

COR. *This class includes all ordinary outer limiting sets.*

The proof given of Theorem 24 serves, with the mere alteration of the word "(inner)" into "(outer)" to prove the corresponding theorem, viz. :—

THEOREM 39. *If each of two sets which do not overlap belong to the (outer) additive class, their sum also belongs to that class.*

Similarly, with the same alteration, and writing "less than" for "greater than" and $-k$ for k, the next proof can be applied, and we get the following :—

THEOREM 40. *If each of two sets one of which is a component of the other belong to the (outer) additive class, so does their difference.*

Similarly we have the following theorem :—

THEOREM 41. *If a set belonging to the (outer) additive class be divided into two components the sum of whose (outer) contents is equal to that of the original set, each of the components belongs to that class.*

The next theorem may be immediately deduced from Theorems 32 and 41.

THEOREM 42. *If G_1 and G_2 be two sets of the (outer) additive class of (outer) content I_1 and I_2, (a) the set consisting of all the points common to G_1 and G_2 is a set of this class, say G', of outer content I', and (b) the set consisting of all the points belonging to one or both of G_1 and G_2 is a set of the class, say G, of (outer) content I; further (c) $I_1 + I_2 = I + I'$.*

Cor. The sets G, G', $G_1 - G'$, $G_2 - G'$ all belong to the (outer) additive class.

66. The Additive Class. The theorems proved enable us without further proof to sum up the chief properties of the additive class.

Def. *The additive class* consists of all sets which have the property that, if one of them be added to any other set, having no point common with it, the sum of the contents, whether (inner) or (outer), is the corresponding content of the sum.

(1) The additive class consists entirely of measurable sets, that is, the (inner) and (outer) contents are the same ; so that we may properly speak of the content of any additive set.

(2) The additive class includes all closed sets, and ordinary inner and outer limiting sets.

(3) The additive class includes all inner and outer limiting sets of additive sets.

(4) The additive class includes the sum and difference of any two additive sets.

(5) If G_1 and G_2 be two sets of the additive class, their common component G' and the set G, consisting of all the points belonging to one or both of them, both belong to the additive class, and the sum of the contents of the two former sets is the same as the sum of the contents of the two latter sets.

(6) The additive class includes all sets of (outer) content zero or (inner) content infinity, and has therefore in any portion of the straight line the potency of all possible sets.

This last property requires proof.

If E be a set of infinite (inner) content, it is evident that the outer content will also be infinite, and that the sum of E and any other set will contain closed components of content as large as we please, and cannot be enclosed in a set of intervals of finite content ; thus E belongs to the additive class.

Next, let E be a set of (outer) content zero ; then the (inner) content of E must also be zero ; so that E is measurable*. Let G be any set of (inner) content a and (outer) content b, having no point common with E. Then $G + E$ can be enclosed in a set of intervals of content as near as we please to b, but not in a set of content less than b ; thus b is the (outer) content of the sum.

---

* Cp. Lebesgue, "Intégrale, Longueur, Aire," § 6, *Ann. di Mat.* (1902).

Again, $E + G$ contains closed sets of content as near as we please to a. Suppose it contains a closed set K of content a' greater than a. Let E' be an ordinary inner limiting set containing E and having zero content. Then, since K and E' are both additive sets, their common part K' is additive and has content zero. Therefore $(K - K')$ is additive and has content a'. But $(K - K')$ is a component of G, and G contains no components of content higher than a; so that this is impossible; therefore $E + G$ does not contain any components of content higher than a; so that a is the content of $E + G$. Thus E is additive. Q. E. D.

Now, if F be a perfect set of content zero, any component E of F has (outer) content zero, and belongs therefore to the additive class; but the potency of the components of F is evidently the same as that of all possible sets. This proves the whole of (6).

It is unnecessary to say more to shew the importance of this class of sets; it includes all the familiar sets and while it consists entirely of measurable sets, we have at present no information shewing that there are other measurable sets than those belonging to this class. If, on the other hand, there be other measurable sets, it possesses distinctive peculiarities distinguishing it from the class of measurable sets *in toto*. The fundamental property of additive sets embodied in the definition enables us to extend the theory of content to all sets of the additive class without any scruple. The extent to which that theory can be still further extended, on the one hand to the (inner), and on the other to the (outer), additive class, and a step further to all measurable sets has been now fully discussed. The only point which remains uncertain is whether or no sets other than these exist.

67. It will be noticed that the additive class includes all countable sets, and that *the content of every countable set is zero*.

Again, *the content of the set of irrational numbers in any segment of a straight line is that of the segment itself*.

By making use of the theorems already proved, we obtain a proof of this theorem, which may subsequently be applied to prove the more general one for space of any number of dimensions. For the sake of variety, and also because it throws fresh light on the subject, an independent proof of the theorem for one dimension is given below.

As in Ex. 2, § 40, divide the segment $(0, 1)$ into m parts, where m is any odd number except unity. Blacken the central part.

Divide each of the $(m-1)$ remaining parts into m^2 parts, and blacken each central part.

Then divide each of the $(m-1)(m^2-1)$ remaining parts into m^3 parts, and blacken each central one; and so on.

The set consisting of the end-points and external points of the set of intervals constructed thus is a perfect set, nowhere dense, whose content* lies between 1 and $1-1/(m-1)$.

Thus, by suitably choosing m, we can get a perfect set, nowhere dense, in the segment $(0, 1)$, whose content is as near as we please to unity. The points of this perfect set are not all irrational, but we will now shew how to obtain from it a similar set in which every point is irrational.

Scheeffer's theorem (Ch. IV, Theorem 18) asserts that, *given two sets, one closed and nowhere dense, and the other countable, and any two quantities a and b, we can find a quantity c, $a < c < b$, such that, if one of the sets of points be pushed a distance c along the straight line, all the points of the countable set lie inside the black intervals of the closed set.*

Choose as the countable set all the rational numbers between 0 and 1, and as the closed set the perfect set just constructed, so that its content is greater than $1-\frac{1}{2}e$, where e is as small as we please. Then we can find a positive quantity $c < \frac{1}{2}e$, such that, shifting the perfect set to the left a distance c, all its points which remain in the segment $(0, 1)$ become irrational. Since these points form a perfect set nowhere dense, of content greater than $1-e$, *we have in this way constructed a perfect set of irrational numbers in the segment $(0, 1)$ of content as near as we please to unity.* Q. E. F.

* For $m=3$ this is the most convenient example of a perfect set of positive content (p. 78). A similar calculation shews that Cantor's typical ternary set (p. 20) is a perfect set of zero content, since, in this case,

$$I_\delta = 1 - \operatorname*{Lt}_{n=\infty}\left(1-\frac{1}{3}\right)^n.$$

CHAPTER VI.

ORDER.

68. In dealing with the potencies of sets, we regard the individual elements (points) of the sets as indistinguishable, or more properly as not distinguished from one another, so that potency enables us to compare sets, regarded as troops in uniform, with one another. The idea embodied in content is totally different: here the individual points are no longer to be regarded as indistinguishable, indeed certain of the points, viz. the semi-external points of the black intervals, seem to play a different *rôle* from the others. The distinguishing property, however,—viz. the relative position of the points—was dependent for its very definition on the existence of the underlying straight line as fundamental region. This will all become still more evident when we come to deal with sets in the plane and higher space. Content is not, like potency, a property of the set *per se*, but a property of the set with respect to the fundamental region.

Order is another property of the set *per se*, but in the determination of the order each individual again bears its own share. In dealing with order we come first to consider the mutual relations of the individuals as such among themselves, and the question arises how are these mutual relations to be characterised, what can we adopt as a measure of order? As before the measurement of order will be made to depend on (1, 1)-correspondence between a given set and a set of known standard form, the characteristic property being maintained, the orders of these standard sets are called *the ordinal types*. Sets of the same ordinal type are said to be *similar*, and if A and B are similar, we write $A \sim B$, or $B \sim A$. *Similar sets clearly have the same potency.*

A set given in order will be called an *ordered set*, and throughout the discussion, unless the contrary is stated, the given order will be supposed to be maintained, not only in the set itself, but also in its components, which, for definiteness, may be distinguished as *ordered components*.

69. The characteristic of order. A set of points, or other elements, containing at least two elements, is said to be *in* *simple order**, if it has two characteristics. The first is that a and b being any two of the elements, distinct from one another, there is some characteristic property, such that we can say without any ambiguity that a possesses it in a superior degree to b, or b in a superior degree to a, this will be indicated in symbols by

$$a < b \text{ in the former case,}$$

and $\qquad\qquad a > b$ in the latter case,

and we shall use the expression " a comes before b " in the former case and " a comes after b " in the latter case to denote the fact in question. If a and b are two elements satisfying neither $a < b$, nor $a > b$, they must possess the characteristic property in the same degree, it follows that they cannot be regarded as distinct, but are only defined in formally different modes. This will be denoted by

$$a \equiv b$$

and we shall say " *a and b are identical†*."

The second characteristic is that if $a < b < c$, then $a < c$. In this case b is said to lie *between a and c*.

If there is an element which comes *before* (*after*) all the others, it is called the *first* (*last*) element.

70. Finite ordinal types. To take first a finite set of, say, n elements. However arranged, these elements are in (1, 1)-correspondence with the first n integers. Such an ordered set, if it consists of more than one element, has a first and a last element: the same is true of every ordered component. Conversely given a set with these properties, it is a finite set. For let a_1 be the first. If a_1 is not the only element, the remaining elements form an ordered component; let a_2 be the first of these. Similarly we determine a_3, unless a_2 is the last. We must arrive at the last after a finite number of such determinations, since the existence of an infinite ordered component a_1, a_2, ... a_m, ... without a last element, is expressly excluded. Thus the above are the characteristics of the *finite ordinal types*.

71. Order of the natural numbers. We have already had to deal with the idea of an infinite set in order, in relation to the potency of a countably infinite set. Here the characteristic property of the set, when arranged in countable order, is that we can say of any element a whether, or no, it corresponds to a smaller integer than b or *vice versa*; if a and b correspond to the same

---

* Cantor, *Math. Ann.*, xlvi. (1895) p. 496.

† If a and b really are identical, the order is said to be *pure*, otherwise *mixed*. Unless stated to the contrary, *order* is always understood to mean *pure order*.

integer they are identical. Thus, *when arranged in countable order, the order of any countably infinite set is that of the natural numbers.* The characteristic peculiarities of this order are that it, as well as each of its ordered components, has a first element, and that every element, with the exception of the first element, has another element immediately preceding it, while there is no last element.

That these peculiarities suffice to define this ordinal type may be shewn as in § 70. For a similar argument proves the existence of an infinite ordered component a_1, a_2, ..., while the existence of a component formed by adding another element b_1 is excluded, since b_1 would have no element immediately preceding it.

To this ordinal type belongs an open sequence lying entirely on one side of its limiting point, the points being taken in order leading up to the limiting point. When taken in this order the sequence is sometimes called a *progression,* and in the reverse order a *regression**. The negative integers in descending order, -1, -2, ..., $-n$, ... form a regression.

72. Orders of closed sequences, etc. A countable set may however be considered as arranged in some other order. When the set is defined in a manner giving the elements in order, this order is called the *natural order.* We have already had numerous examples illustrating this. For instance the sequence given in Ex. 1, Ch. III, p. 17 is given in order, the points being assigned from left to right, in order of position, or the corresponding numbers in ascending order of magnitude. Here again the set, as well as each component, has a first element. There is a last element, viz. the point 1, but it is such that it has no definite element preceding it, while every element except the first 0 and the last 1 has a definite element immediately preceding it and another immediately following it: thus the natural order here is different from the finite orders and from the order of the natural numbers. As before it may be shewn that these peculiarities suffice to characterise the ordinal type. All closed sequences lying only on one side of their limiting point, considered in order of position approaching the limiting point, evidently belong to this ordinal type. We may call this order *the progressive order of a one-sided closed sequence.* To the same ordinal type belongs every open sequence on one side only of its limiting point, with one point added beyond the limiting point.

---

* Russell, *Principles of Math.* p. 199.

A two-sided open sequence, for instance

$$-1, -\cdot01, -\cdot0^21, \ldots\ldots, \cdot0^21, \cdot01, 1,$$

(using the binary notation), is an example of another countable set given in natural order of position, or ascending magnitude. It has all the characteristics of a finite set, except that which postulates that what is true of the set is true of its components. It has indeed ordered components having the order of the natural numbers, or of a closed one-sided sequence *.

The negative and positive integers in the following order,

$$-1, -2, -3, \ldots\ldots, -n, \ldots\ldots, n, \ldots\ldots, 3, 2, 1,$$

belong to this ordinal type.

A two-sided closed sequence, for instance

$$-1, -\cdot01, -\cdot0^21, \ldots\ldots, 0, \ldots\ldots, \cdot0^21, \cdot01, 1,$$

has proper components of all the above types. The whole set however is such that every element except the first -1, the last 1, and one other element, the origin 0, has a definite element immediately preceding it, and another immediately following it. The origin however has no element either immediately preceding or following it.

Ex. 3, Ch. III, § 12, p. 23 gives us another countable set in a more complicated order; and the method indicated at the conclusion of that section enables us to build up sets, whose orders become more and more difficult to describe. These sets will occupy us more closely in the next chapter (cp. also Ex. 5, p. 28).

73. Graphical and numerical representation. The idea of the orders of countably infinite sets which are simply ordered is nothing more than a discontinuous function of two variables, and may be graphically represented by means of the diagram of a rectangular trellis, so familiar in the theory of numbers†.

If we take any countably infinite set, whether it be of points on the straight line, or anything else, and arrange them in countable order, say $E_1, E_2, \ldots, E_n, \ldots$ (this may generally be done

---

* Finite sets, progressions, regressions and open two-sided sequences are classed together as *discrete series*, and a set of postulates given for them by Huntingdon, "The Continuum as a Type of Order," *Annals of Math.*, Series 2, Vol. VI. (1905) p. 164. The interest of this class of sets lies in the fact that mathematical induction may be applied to discrete sets, and to these sets only.

† F. Bernstein, *Inaug. Diss.*, Gött. 1901; *Math. Ann.* LXI. p 118 (1905). Bernstein uses the complete trellis, instead of the wedge-shaped diagram of this article, and obtains the necessary and sufficient condition that the diagram should represent order in a corresponding form. The wedge-shaped diagram has the practical advantage of greater simplicity.

in a variety of ways, but we choose out one particular arrangement), the idea of the *natural order* of these E's is completely embodied by giving a law by which we can say whether, or no, E_i came after E_j originally, i being the less of the two integers i and j. Unless we can give such a law, we cannot speak of a "natural order" at all; *vice versa*, given such a law, we can determine the position of any element E_i with respect to any other one E_j, and so can always say whether, or no, it lies between any two assigned elements : this is what we mean by saying we know the natural order, or the ordinal type.

The diagrammatic representation of the natural order depends on the customary representation of the pair of integers (j, i) by means of the cross points of a rectangular trellis, so that the point (j, i) is the point whose coordinates are j and i. Each such point (j, i) in the wedge-shaped diagram in which $i < j$, we mark with a black spot if E_j comes before E_i (in symbols, if $E_j < E_i$). This wedge-shaped diagram gives us all the information we require. The order-diagrams determined in this way by the various ordinal types of countably infinite sets, do not exhaust all the possible patterns formed by placing spots at the trellis-points of the wedge. Indeed the first characteristic of simple order without the second would suffice to determine such a pattern. *The necessary and sufficient condition that such a pattern should serve as an order diagram is as follows: — If i, j, k be any three integers in order $i < j < k$, the right-angled triangle whose vertices are (j, i), (k, i), and (k, j) cannot be such that both ends of the hypotenuse are spotted (or unspotted) and the opposite vertex unspotted (or spotted).* In fact there are six ways of arranging E_i, E_j, E_k in ascending order of magnitude, and therefore of the eight possible patterns two have to be rejected. Plotting down the six cases, it will be found that the above statement is true, so that the condition is necessary. On the other hand, if it is satisfied, each such triangle in the diagram will correspond to an order of E_i, E_j, E_k so that the condition is sufficient.

We may transform such a diagram into a numerical representation by interpreting each black spot by a 1, and each unspotted trellis-point by a 0, and reading the columns from bottom to top in order from left to right. Prefixing a point and interpreting in the binary scale, we get a binary fraction corresponding to each diagram, and *vice versa*.

These numbers, which we may call *the binary order-fractions*, *form a perfect set nowhere dense, and have therefore the potency c.*

In fact the condition that a diagram should be an order-diagram does not permit any order-fraction to begin with 00 without having 0 as the third digit, and generally, given n digits, imposes a restriction on the $(n + 1)$th digit which is equivalent to cutting out from the continuum $(0, 1)$ a certain set of intervals, one of which is $(\cdot 001, \cdot 01)$. It is easily proved that the end-points of these intervals are limiting points of the set; for instance, in the case of the given interval, which may also be written $(\cdot 000\dot{1}, \cdot 01)$, the end-points may be approached by the points $\cdot 0001^4 1^5 \ldots 1^n$ and $\cdot 0100^3 0^4 \ldots 0^n \dot{1}$ for all integers n; these points correspond to diagrams clearly satisfying the condition for order-diagrams. Thus by Theorem 15, Ch. IV, p. 48 the result follows.

A given ordinal type will in general be represented by a variety of diagrams. In this way the terminating binary fractions, and zero, or, which is the same thing, the diagrams with a finite number of spots, and with no spots, all represent the ascending order of the natural numbers*, and correspond to different ways in which it can be arranged in countable order. In like manner all the simple circulators represent the descending order of the negative integers†. In particular the binary number $\cdot 1\ 11\ 111\ \ldots$, or $\dot{1}$, or 1, represents the order of the negative integers in the countable order $-1, -2, \ldots -n, \ldots$, the corresponding diagram having dots at all the trellis-points.

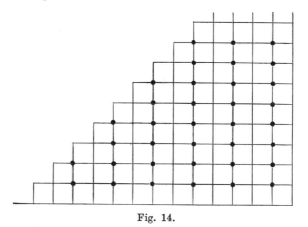

Fig. 14.

---

* Cantor's number ω. These are not the only binary order-fractions representing progressions; for instance $\cdot 0\ 11\ 0^3\ 1010\ 0^5\ 1010^3\ 0^7\ 1010^n\ 0^{n+4}\ \ldots$ represents the progression 2, 4, 1, 6, 3, 8, 5, $\ldots$ in which each odd integer has been moved on, and each even integer moved back, two places.

† Cantor's number $^*\omega$.

Generally given a binary number representing the order of a certain simply ordered set, it may be shewn that, if we can* alter the first n figures of the binary number in such a way as to obtain a new number representing order, the new binary number will represent the same order as before. For instance the diagram whose columns are alternately plain and spotted (Fig. 14), or the corresponding number

$$\cdot 0 \ 11 \ 0^3 \ 1^4 \ldots 0^{2n-1} \ 1^{2n} \ 0^{2n+1} \ldots$$

represents the natural order of the negative and positive integers including zero, and corresponds to the order

$$0, \ 1, \ -1, \ 2, \ -2, \ 3, \ldots$$

while $\cdot 1 \ 11 \ 0^3 \ 1^4 \ldots 0^{2n-1} \ 1^{2n} \ 0^{2n+1} \ldots$

corresponds to the countable order

$$1, \ 0, \ -1, \ 2, \ -2, \ 3, \ldots$$

and $\cdot 1 \ 10 \ 0^3 \ 1^4 \ldots 0^{2n-1} \ 1^{2n} \ 0^{2n+1}, \ldots$

to the countable order

$$1, \ -1, \ 0, \ 2, \ -2, \ 3, \ldots.$$

74. The rational numbers. Close order. We have seen that the rational points are countable, their natural order of position, or ascending magnitude, is such that between any pair of elements there is another element. Any simply-ordered set which has this property will be said to be *in close order*.

It is to be noticed that while every set which is dense everywhere is in close order, the converse is not true†.[3] Every set which is dense in itself on both sides (Ch. III, § 10, p. 22), or on one side only, is in close order (cp. Ex. 1, Ch. IV, p. 45).

It is evident that the property of being in close order is invariant for (1, 1)-transformation maintaining the order. Thus

* In general this is not possible. It is only possible when, apart from a finite number of columns, each column of the order-diagram is either wholly plain or wholly spotted.

† H. J. S. Smith uses the term as equivalent to dense everywhere. The term has been here adopted but in a slightly different sense, for obvious reasons. *Proc. L. M. S.* VI. p. 145. Unfortunately Cantor, who had previously only used the expression *dense everywhere* in a strictly relative sense (cp. p. 21 footnote), uses in *Math. Ann.* XLVI. p. 504, the expression *dense everywhere* in an absolute sense as equivalent to the expression *in close order*. The same remark applies to the terms "limiting element," p. 509, "closed set," "perfect set," p. 510. This has been frequently copied without comment, thereby introducing quite unnecessary confusion into the subject. It is to be recommended that in using these words in an absolute sense, the word *ordinally* should be prefixed, see § 77.

if one set of a certain ordinal type is in close order, every set of the
type is in close order.

As typical of a countable set in close order it is convenient to
take the set consisting of all the rational numbers between 0 and
1 whose denominators are powers of 2, viz.

$$\tfrac{1}{2},$$
$$\tfrac{1}{4}, \qquad\qquad \tfrac{3}{4},$$
$$\tfrac{1}{8}, \quad \tfrac{3}{8}, \quad \tfrac{5}{8}, \quad \tfrac{7}{8} \;\ldots\ldots\ldots\ldots(A),$$
$$\text{etc.}$$

or, expressed in the binary scale,

$$\cdot 1,$$
$$\cdot 01, \qquad\qquad \cdot 11,$$
$$\cdot 001, \quad \cdot 011, \quad \cdot 101, \quad \cdot 111 \;\ldots\ldots\ldots(B),$$
$$\text{etc.}$$

that is all the finite binary fractions. Considering, as is often
convenient, every number as expressed by an infinite number of
figures, and not ending in a series of zeros, by which means every
number is expressed in one and only one way, the numbers in
question are those ending in $\dot{1}$, viz. as follows :—

$$\cdot 0\dot{1},$$
$$\cdot 00\dot{1}, \qquad\qquad \cdot 10\dot{1},$$
$$\cdot 00\dot{1}, \quad \cdot 010\dot{1}, \quad \cdot 100\dot{1}, \quad \cdot 110\dot{1} \;\ldots\ldots(C),$$
$$\text{etc.}$$

Let N denote n zeros and ones in any order, $\cdot N1$ or $\cdot N0\dot{1}$
is any one of this set of numbers, and it is the limit on the right
of the sequence

$$\cdot N01, \quad \cdot N011, \quad \cdot N0111, \ldots\ldots$$

and on the left of the sequence

$$\cdot N\,101, \quad \cdot N\,1001, \quad \cdot N\,10001, \ldots\ldots$$

the set being dense in itself on both sides.

The limits which are not included in the set consist of all
the remaining binary fractions, whether non-periodic, or periodic
but not simple circulators.

Given any number between 0 and 1, we can at once, by
expressing it in the binary scale, (1) determine a sequence belonging to our set, and defining the number, and (2) ascertain whether,
or no, the number belongs to our set.

For instance, the rational number $\tfrac{1}{3}$ has the following defining
sequence,

$$0, \quad \cdot 01, \quad \cdot 0101, \quad \cdot 010101, \quad \text{etc.} \,;$$

it does not belong to the set, and is a periodic binary fraction, viz. $\cdot 0\dot{1}$.

The most general countable set in close order, without first or last element, can be brought into $(1, 1)$-correspondence maintaining the order* with the above typical set in the following manner:—

Let the length of the fundamental segment (A, B) be denoted by s_1. The points A and B may, without loss of generality, be assumed to be limiting points of the set.

Arrange the points of the set in (AB) in countable order

$$P_1, P_2, P_3, \ldots\ldots\ldots$$

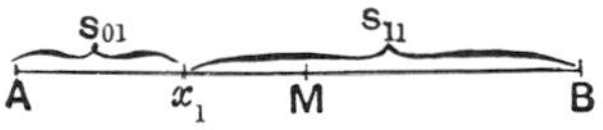

Take the point P_1 which comes first in the series

Fig. 15.

$$P_1, P_2, P_3, \ldots, P_k, \ldots\ldots$$

and denote it by x_1, and denote the lengths of (Ax_1) and (x_1B) by s_{01} and s_{11} respectively.

Since A is a limiting point of the set, there are an infinite number of the points in s_{01}, and we denote the first of the points in the countable order $P_1, P_2, \ldots$ which lies in s_{01} by x_{01}. Similarly the first in s_{11} by x_{11}. We notice that one of the points x_{01} or x_{11} is P_2, and that the points x_{01}, x_1, x_{11} are in the same order as the binary points $\cdot 01, \cdot 1, \cdot 11$, viz. $x_{01} < x_1 < x_{11}$.

We continue in this way, denoting the segments on the left and right of x_{N1} by s_{N01} and s_{N11} respectively, and the first of the points $P_1, P_2, \ldots$ which lies in s_{N01} by x_{N01} and in s_{N11} by x_{N11}. Since between any two points of the set there is another point of the set, and A, B are limiting points of the set, these points will exist for every combination of n digits N.

In this way we attach to each binary number $\cdot N1$ a point x_{N1} of the set, and to each point P_k of the set a definite number $\cdot N1$, where $n \leqslant k$. The $(1, 1)$-correspondence between the points of our set and the typical set in close order thus obtained, is then such that the order is maintained, that is to say if

$$\cdot M1 < \cdot N1 < \cdot R1,$$

then

$$x_{M1} < x_{N1} < x_{R1}.$$

We have thus set up a $(1, 1)$-correspondence maintaining the order between our set in close order and the typical set of the same ordinal type.

---

* Cantor, *loc. cit.*, §9; Brodén, *J. f. Math.* Vol. cxviii. This ordinal type is denoted by η.

75. The same notation being used as in the preceding section, let us write

$$s_{01} = \tfrac{1}{2} s_1 (1 + j_1),$$

$$s_{11} = \tfrac{1}{2} s_1 (1 - j_1),$$

and generally

$$s_{N01} = \tfrac{1}{2} s_{N1} (1 + j_{N1}),$$

$$s_{N11} = \tfrac{1}{2} s_{N1} (1 - j_{N1}),$$

where j_{N1}, and in particular j_1, is any number greater than -1 and less than 1.

By proper choice of the quantities j_{N1} we can construct any set of the ordinal type considered. *The necessary and sufficient condition that the set should be dense everywhere is that the length of the maximum segment after n divisions may be zero:* this can be expressed as a condition to be satisfied by the quantities j_{N1}. For the above formula shews that, when calculated out, $s_{N1} : s_1$ is expressed as a product of n factors of the type $\tfrac{1}{2} (1 + e j_{i1})$ where e is 0 corresponding to each figure 1 in N, and e is ± 1 corresponding to each figure 0 in N, the index i in each case being that approximation to N got by stopping at the 0 or 1 in question. An example will make this clearer than any explanation; in practice the rule is excessively simple :—

$$s_{1011101} = s_1 \tfrac{1}{2} (1 - j_1) \tfrac{1}{2} (1 + j_{11}) \tfrac{1}{2} (1 - j_{101})$$

$$\tfrac{1}{2} (1 - j_{1011}) \tfrac{1}{2} (1 - j_{10111}) \tfrac{1}{2} (1 + j_{101111}).$$

Let $j_n{}'$ denote the maximum value of $|j|$ used at the nth division, and let

$$J_n{}' = \tfrac{1}{2} (1 + j_1{}') \tfrac{1}{2} (1 + j_2{}') \dots \dots \tfrac{1}{2} (1 + j_n{}');$$

also let $s_1 J_n$ be the length of the maximum segment after n divisions, so that

$$J_n \lessgtr J_n{}'.$$

The necessary and sufficient condition that the set should be dense everywhere is then

$$J \equiv \operatorname*{Lt}_{n=\infty} J_n = 0.$$

If, which is more easily ascertained when the j's are given,

$$J' \equiv \operatorname*{Lt}_{n=\infty} J_n{}' = 0,$$

it is clear that J is also zero, and the set is dense everywhere;

this latter condition is sufficient but not necessary. An equivalent form of this condition is that $\sum\limits_{n=1}^{\infty} \frac{1}{2}(1-j_n')$ must diverge*.

Ex. 1. In the typical case (§ 74) all the j's are zero. Hence

$$J_n = J_n' = 1\,(\tfrac{1}{2})^n,$$
$$\therefore J = J' = 0.$$

Ex. 2. Divide the segments always in the ratio 2 : 3. All the j's are equal to $\frac{1}{5}$.

$$J_n = J_n' = s_1\,(\tfrac{2}{3})^n,$$
$$J = J' = 0.$$

The set is dense everywhere.

Ex. 3. Take the set of all the rational fractions between 0 and 1, whose denominators are powers of 3, and arrange them thus :—

We divide the segment $(0, 1)$ or s_1 at the point $\frac{2}{3}$ or $\cdot 2$ in the ternary scale. The segment s_{11}, of length $\frac{2}{3}$, we then divide in the middle at the point $\cdot 1$ in the ternary scale. The segment s_{01}, of length $\frac{1}{3}$, we divide in the ratio 2 : 1 at the point $\cdot 22$ in the ternary scale. Generally, any segment whose length is $\dfrac{2}{3^\mu}$ we divide in half, and any segment whose length is $\dfrac{1}{3^\mu}$ in the ratio 2 : 1.

We then get the following scheme, where on the left-hand side the indices of the points x_{N1} are given, and at the corresponding places on the right-hand side the ternary fractions expressing the corresponding points x_{N1}.

		$\cdot 1,$			$\cdot 2,$	
	$\cdot 01,$		$\cdot 11,$	$\cdot 1,$		$\cdot 22,$
$\cdot 001,$	$\cdot 011,$	$\cdot 101,$	$\cdot 111,$	$\cdot 02,$	$\cdot 12,$	$\cdot 21,$ $\cdot 222$
$\cdot 0001,$	$\cdot 0011, \ldots\ldots\ldots\ldots\ldots\ldots\ldots\ldots\ldots\ldots$			$\cdot 01,$	$\cdot 11, \ldots\ldots\ldots\ldots\ldots\ldots$	

It is clear that at every division, the extreme left-hand segment has the maximum, and the extreme right-hand segment the minimum value. The maximum segment is alternately bisected and divided in the ratio 2 : 1. The j corresponding to bisection is 0, and that to division in the ratio 2 : 1 is $\frac{1}{3}$†.

* "Density of Linear Sets of Points," *Proc. L. M. S.* Vol. xxxiv. p. 285 *seq.*, where the above condition is given in the special case considered.

† We notice that all the j's are either 0 or $\frac{1}{3}$, and if j_{N1} is $\frac{1}{3}$, j_{N01} is 0 and j_{N11} is $\frac{1}{3}$, while if j_{N1} is 0, j_{N01} and j_{N11} are both $\frac{1}{3}$.

Hence if ϕ_n be the number of j's at the nth division which are equal to $\frac{1}{3}$, and $f(n)$ the number equal to zero,

$$f(n) = \phi(n-1),$$
$$\phi(n) = \phi(n-1) + 2f(n-1) = \phi(n-1) + 2\phi(n-2),$$

the general solution of which is

$$\phi(n) = 2^n A + (-1)^n B.$$

Since $\phi(1) = 1$, $\phi(2) = 1$, this gives

$$\phi(n) = \frac{2^n + (-1)^{n-1}}{3}, \qquad f(n) = \frac{2^{n-1} + (-1)^{n-2}}{3} = 2^{n-1} - \phi_n.$$

Thus at the 1st division there was $1\,k$ which was $\frac{1}{3}$, and $0\,k$'s zero.

2nd	,,	,, 1	,,	,,	1	,,
3rd	,,	were 3 k's which were ,,	,,	,,	1	,,
4th	,,	,, 5	,,	,,	3	,,
5th	,,	,, 11	,,	,,	5	,,

Hence $\qquad J_n = \tfrac{2}{3} \cdot \tfrac{1}{2} \cdot \tfrac{2}{3} \cdot \tfrac{1}{2} \cdot \tfrac{2}{3} \ldots\ldots\ldots,$

the last factor being $\tfrac{1}{2}$ if n is even, and $\tfrac{2}{3}$ if n is odd.

Thus $\qquad J_{2m} = (\tfrac{1}{3})^m, \qquad\qquad J_{2m+1} = \tfrac{2}{3} (\tfrac{1}{3})^m,$

but $\qquad\qquad\qquad J_n' = (\tfrac{2}{3})^n > J_n.$

Both J and J' however vanish, the condition for density is therefore satisfied.

76. Suppose we have a sequence of the typical countable set in close order, lying entirely on one side of its limiting point L, say on the left, and let L be a point of the set : then, since the order is maintained, there are only a finite number of the corresponding points of the general countable set G in close order lying between the point corresponding to the first point of the sequence and that corresponding to any other point of the sequence, but an infinite number between the former point and the point corresponding to the limiting point L of the sequence : thus these corresponding points also form a sequence lying entirely on one side of the point L' corresponding to L; L' may be the limiting point of the sequence, or may lie beyond it, in the latter case the limiting point L'' of the sequence will not belong to the set G and between L' and L'' there will be no point of G. *Vice versa*, to every such sequence of the set G corresponds such a sequence of the typical countable set.

If we take any other sequence of the typical countable set on the same side of L as before, and having L as limiting point, it forms with the former sequence another such sequence, and there corresponds to it a sequence of G, whose limiting point must be the same as before. If, on the other hand, we approach L by means of a sequence of the typical set on the other side, say the right, the limiting point of the corresponding sequence must again be such that between it and L' there is no point of G; thus either it is L', or it is some point L''' on the other side of L' from L''; in this case between L'' and L''' there is no point of the set G except L'.

Similarly if the limiting point L be not a point of the typical set in close order, we get two points L'' and L''' (or one point, if these coincide), not belonging to G, but limiting points of G, such that between them there is no point of G; these being the limiting points of sequences in G whose correspondents have L as limiting point.

Thus we see that

(1)   No countable set which is in close order is closed, since this is true of the typical set.

(2)   There are only two kinds of such sets; either such a set is dense everywhere in a segment, or in a closed set of potency c.

(3) Extending the correspondence between the points of two such sets to their limiting points, in such a way that to each limiting point of a sequence corresponds the limiting point of the corresponding sequence, each non-included limiting point has at most two correspondents, between which lies no point of the set in question.

Examples of both kinds of sets in close order have been given in Ch. IV, p. 45. A countable set in close order may*, however, be dense in itself on one side only. For instance, the terminating ternary fractions not involving the figure 1 (cp. Ex. 2, p. 20). Nor does this exhaust the possibilities; we might, for instance, replace any of the terminating ternary fractions just mentioned (which are limits on the right only) by the corresponding circulators (which are limits on the left only) ending in $0\dot{2}$ instead of 2.

On the other hand any two sets, each consisting of all the end-points of the black intervals of a perfect set, are in (1, 1)-correspondence maintaining the order, and constitute an ordinal type different from any yet discussed.

77. By reasoning precisely similar to that used in the particular case of the preceding section, it appears that, if two simply ordered sets on the straight line are in (1, 1)-correspondence maintaining the order, their limiting points are in a special kind of (2, 2)-involution (which in special cases may become a (1, 2) or a (1, 1)-involution), such that only the end-points P_1 and P_2 of a black interval can have the same correspondent P and this can only take place if P be not an end-point of a black interval; if, on the other hand, P_1 corresponds to an end-point Q_1 of a black interval (Q_1, Q_2), P_1 has no other correspondent and P_2 corresponds uniquely to Q_2.

If any limiting point P has two correspondents P_1 and P_2 in the involution, both these points cannot be points of the set of which they are limiting points, since this set is in (1, 1)-correspondence with the other set. Thus if a limiting point P_1 does not belong to the one set, and the corresponding point P does belong to the other set, P_1 must be an end-point of a black interval, and the other end-point P_2 must belong to the set and be the correspondent of P.

From all this it follows that *if there be one set of an ordinal type which is closed, all sets of that type are either closed or can be*

---

* See second footnote, p. 127.

closed by adding to them a finite or countably infinite number of limiting points on one side only.

Such sets are said to be *ordinally closed,* and their type is said to be a *closed ordinal type.* Given any progression (or regression) of the elements of such a set, there is a definite element of the set which comes immediately after the progression (or before the regression). This element is called the *ordinal limiting element* of the progression (or regression), and is said to be a *principal element* of the set. A simply-ordered set all of whose elements are principal elements is said to be *ordinally dense in itself,* and if also ordinally closed is said to be *ordinally perfect*.* All these properties are invariant for (1, 1)-correspondences which maintain the order.

78. Order of the continuum. The next ordinal type†
to be discussed is that of the closed continuum itself. In the introductory discussion of the linear continuum in Chapters I and II, we assumed the straight line to consist of the rational points with all their limiting points, that is to say *the continuum is* (1) *ordinally closed, and* (2) *has a component which is dense everywhere in it, and is countable and in close order.* These two characteristics completely determine the ordinal type of the continuum‡. It follows therefore from the preceding section that *any linear set which has the order of the continuum can only be itself a segment, or a closed set of potency c omitting one endpoint of each black interval.*

That a (1, 1)-correspondence between the segment (0, 1) and such a set exists can be easily shewn. When the set is perfect and dense nowhere, the following mode is convenient.

---

* See second footnote, p. 127.

† Denoted after Cantor by the symbol θ.

‡ The problem of determining under what circumstances a simply-ordered set in any fundamental region whatever has the ordinal type of the continuum, was proposed and solved as above by Cantor in *Math. Ann.* XLVI. The problem has frequently been rediscussed. References will be found in Russell, Huntingdon, *loc. cit.,* Veblen, *Trans. of the Amer. Math. Soc.* Vol. VI. pp. 165—171. After what has been said it is unnecessary to remark again on the use of terms; attention may however be directed to the fact that in all these discussions the word continuum must be understood to apply to the *ordinal continuum,* not to the *actual continuum,* still less to the *actual linear continuum* for which (§ 95) a still further specification is necessary.

In this connection notice may be directed to Veblen's use of the Heine-Borel theorem as an axiom equivalent to the Cantor-Dedekind axiom; *Bull. of the Amer. Math. Soc.* X. p. 436 (1904).

Let (A, B) be the fundamental segment, A and B being points of the set. Then the set in question, say G, consists of all the external points of a set of non-overlapping and non-abutting intervals, together with one end-point from each interval.

Let the intervals arranged in any way in countable order, *e.g.* in order of magnitude, be denoted by

$$d_1, d_2, \ldots \ldots$$

Let us take the first in this series which lies between A and d_1 and denote it by d_{01}, and the first between d_1 and B by d_{11}. Similarly the first between A and d_{01} by d_{001}, the first between d'_{01} and d_1 by d_{011}, the first between d_1 and d_{11} by d_{101}, and the first between d_{11} and B by d_{111}; and so on. In this way we get all the black intervals arranged in the order of the typical binary set,

$$d_1, d_{01}, d_{11}, d_{001}, d_{011}, d_{101}, d_{111}, d_{0001}, \ldots \text{ etc.}$$

That end-point of any interval d_{N1} which belongs to the set we denote by P_{N1}, then it is clear that the countable set

$$P_1, P_{01}, P_{11}, P_{001}, P_{011}, P_{101}, P_{111}, P_{0001}, \ldots \text{ etc.}$$

is in close order, and in $(1, 1)$-correspondence maintaining the order with the typical set in close order. Those of the limiting points of the above countable set which do not belong to it are limits on both sides, no two of them therefore by § 76 can have the same correspondent; further no one of them can have two correspondents because the typical set in close order has no limiting points on one side only. Thus the involution between the limiting points (§ 77) is a $(1, 1)$-involution, so that the whole set G is in $(1, 1)$-correspondence maintaining the order with the continuum.

79. The next set whose order is to be considered is the set of all the derived and deduced sets of a set whose first derived set is more than countable.

In this investigation we have outstepped the limits within which we had hitherto confined ourselves; the fundamental region F is no longer the straight line.

Form first the fundamental region F_1 whose elements are all the closed sets of points on the straight line. It is shewn in § 97 that the position of any element G of F_1 may be determined in the same way as that of a point on a straight line, in other words F_1 may be taken to be a straight line. F_1 is part of F and we then construct the whole fundamental region F in the following manner: starting with any closed set G (which is an element of F_1 and therefore of F), form all the sets from G which can be formed by the processes of derivation and deduction in order; these, although they are elements of F_1 which, as closed sets, are the same as these derived and deduced sets, are to be regarded as new elements of F, and the position in F of any one of them G' is determined by two criteria, firstly by the position of G in F_1 and secondly by the position of G' in the series of derived and deduced sets of G. The fundamental region F consists of F_1 and these new elements. F is not simply ordered; but given any element G of F_1, the sets which can be obtained from G by the processes of derivation and deduction clearly form a simply ordered set in F, the position of any element G' in the set being determined by the series of processes by which G' is obtained from G, which, as is shewn in § 97, is equivalent to assigning a certain linear set of points of potency c.

If G is countable, the whole set so obtained is countable, and, when arranged in its natural order, it has a last element, which is a set consisting of a finite number of points of the straight line.

If G is not countable, the set when arranged in its natural order has no last element, since although only a countable number of elements can be passed over, which as sets of points on the straight line are distinguishable from one another (and are therefore in (1, 1)-correspondence with a certain ordered set in F_1), yet after we get to the nucleus we still have elements of F which as such are distinguishable from one another, namely, by the second criterion of position in F.

If G and Γ be two closed sets whose points are more than countable, we can then clearly set up a (1, 1)-correspondence maintaining the order between the elements of F obtained from G and those obtained from Γ, by making two elements correspond which are obtained from G and from Γ respectively by the same series of the processes of derivation and deduction. Thus there will be in the theory of sets of elements in F a definite potency $\aleph_1$ (Aleph-eins, Aleph-one) and a definite ordinal type Ω corresponding to a series of elements obtained from any element G of F_1 by the processes of derivation and deduction.

It has been necessary to go fully into this point, since the step which we have taken into a new fundamental region F, is one of great importance, and the way is beset with new and still unsurmounted difficulties. It must be emphasized that we have not yet shewn that there is a set of points on the straight line of potency $\aleph_1$, or of ordinal type Ω: we cannot infer the existence of such a set from the fact that such a set exists in a totally different fundamental region.

THEOREM 1. *The whole set of all the derived and deduced sets of a perfect set is not countable.*

For suppose it were countable, and let the sets arranged in countable order be

$$E_1, E_2, E_3 \dots\dots\dots\dots\dots\dots\dots(1).$$

Then E_2 may be subsequent to E_1, in the natural order, or it may precede E_1; in the latter case however there must be a *first* set E_i which is subsequent to E_1 in the natural order. In the same way let E_j be the *first* set after E_i in the order (1), subsequent to E_i in the natural order ; and so on.

In this way we obtain a new series,

$$E_1, E_i, E_j, E_k, \dots E_m, E_n \dots\dots\dots\dots(2),$$

which is never-ending, since any set has sets after it in the natural order ; here these sets are in their natural order and also

$$1 < i < j < k < \dots < m < n \dots\dots\dots\dots(3).$$

These sets will have a deduced set, which will be one of the sets of the series (1); let its index be λ, and determine between which two numbers of the series (3) λ lies, suppose

$$m < \lambda \leqslant n.$$

Then E_λ is subsequent to E_m; but E_n is the *first* set in the order (1) subsequent to E_m, therefore λ cannot be less than n, and E_λ must be identical with E_n. But this cannot be true, for by our determination of E_λ (although as a set of points it is the same as E_n), it is distinguishable from E_n by its place in the natural order, and is subsequent to E_n. Thus it is impossible that the sets can be arranged in countable order.

80. Well-ordered sets. A set of ordinal type Ω in F has the following important characteristic, *the set itself, as well as every one of its components, has a first element.* Such a set is said to be *well-ordered.*[4]

The set of all the negative integers in order of ascending magnitude $\dots -n, -(n-1), \dots -3, -2, -1$, is an example of a set which is not well-ordered : the ordinal type of this set is denoted by $*\omega$.

It is clear that the set of all the derived and deduced sets of a set whose first derived is countable, also has this characteristic ; such a set is *a countable well-ordered set.*

There will be a variety of ordinal types corresponding to such series. These ordinal types belong to the theory of linear sets of points, since, as we saw (Ch. III, §12 *seq.*), we can construct sets of points on the straight line whose natural order from left to right is precisely that of any given series of such sets.

The following theorems with respect to well-ordered sets, have therefore certainly an application to countable sets on the straight line, and to countable sets and sets of potency $\aleph_1$ in the fundamental region F of the preceding article. How far they are of general application is, because of our ignorance as to the properties of different fundamental regions, indeed, even of the straight line itself, at present uncertain. We shall now suppose that we are working in some fundamental region R, at present unspecified, in which there exist well-ordered sets of various potencies.

THEOREM 2. *Every component of a well-ordered set is well-orderable.*

That is to say, when we take the elements of such a component in the order in which they occur in the original set, the component is well-ordered, for it has a first element, and so has every one of its components, since these are components of the original set.

THEOREM 3. *Every component set of a well-ordered set determines an element which is the first of the remaining elements to follow every element of the component, except when there are no elements following every element of the component. In particular every element except the last, if there is one, has one immediately following it.*

For the component in question determines another component consisting of all the remaining elements which do not precede any element of the former component; the first element of this second component is the one referred to in the enunciation.

It is to be remarked that the series of all the derived and deduced sets has no last element, unless the first derived set is countable. But, for instance, the series of these sets which in any particular case are distinct has a last element, viz. the nucleus, if the set first derived be more than countable, and otherwise a finite set of points.

The component of a well-ordered set consisting of all the elements preceding any particular element, say a, is called *a segment* (Abschnitt), and will be said to be cut off by a; we shall use A to denote the segment cut off by a. It will be shewn that of two well-ordered sets which are not similar, one is always similar to a segment of the other, so that *all well-ordered sets in any particular fundamental region can be considered as segments of one particular well-ordered set*; it is not however certain that in that fundamental region the latter well-ordered set, as such exists at all.

It is clear from the definition that *if A is a segment of E and A′ of A, A′ is a segment of E.*

THEOREM 4. *A well-ordered set is not similar to any of its segments.*

For consider the component of the well-ordered set E consisting of all those elements which cut off segments similar to E ; if this exists it will have a first element, let this be a. Then the segment A, cut off by a, is contained in every segment similar to E, and may be said to be the smallest segment similar to E.

Now since A is similar to E, every element of A, considered as an element of E, has a correspondent in A ; but, since A is a segment of E, there is at least one element of A whose correspondent is not a point of A. Thus A, considered as a component of E, corresponds to a proper component of A. But the element a is the first of the elements of E to follow all the elements of A ; hence, since E is similar to A, there is an element $a′$ of A, which, of those elements of A not belonging to the proper component in question, is the first to follow all the elements of that proper component. Thus the proper component is the segment $A′$ cut off by $a′$. Now $E \sim A$, and $A \sim A′$, therefore $E \sim A′$. But $A′$ is a segment of A, contrary to the hypothesis that A was the smallest segment of E similar to E. Therefore there is no such smallest segment, and therefore no such first element a, so that the hypothetical component does not exist, and there are no segments of E similar to E. Q. E. D.

THEOREM 5. *If every segment of a well-ordered set E is similar to a segment of a well-ordered set F, E is similar to F or to a part of F.*

For, by Theorem 3, there will be only one segment of F similar to any particular segment of E, since otherwise there would be at least two such segments, and they would be similar to one another; but, since, of the two elements cutting off these segments, if not identical, one must precede the other, all the elements preceding that one would form a segment of the segment cut off by the other, thus the one segment would be a segment of the other, and be at the same time similar to it, contrary to Theorem 4.

Thus there is a (1, 1)-correspondence between all the segments of E and all or some of the segments of F. If, now, we make any element a of E, cutting off a segment A, correspond to the element a' of F, cutting off the corresponding segment A' of F, this sets up a (1, 1)-correspondence between all the elements of E and all or some of the elements of F.

In this correspondence the order will be maintained; for, if a and b be any two elements of E and a' and b' the corresponding elements of F, then, if a and b are distinct, one of them precedes the other, let a denote the one which precedes the other b. Then the segment B' being similar to B, there will be a segment of B' similar to A ; and, since there is only one segment of F similar to A, it will be none other than this segment of B' ; that is, A' is a segment of B', and therefore a' precedes b' ; thus the order is maintained.

So far we have proved that E is similar to F, or to a proper component of F. If this proper component were not a segment, there would be at least one element a' of F which came before some element b' of that component, although a' would not belong to that component. But the correspondence was such that the whole segment cut off by b' was similar to a certain segment B of E, and therefore the segment cut off by a', being a segment of B', is similar to a certain segment A of E, thus a' will have a correspondent a in E, and will therefore belong to the proper component in question, contrary to the hypothesis. Thus the proper component can only be a segment of F, which proves the theorem.

COR. *The necessary and sufficient condition that a well-ordered set E should be similar to another well-ordered set F is that every segment of E is similar to a segment of F, and every segment of F to a segment of E.*

For in this case E cannot be similar to a part of F, since that part of B would be similar to a part of A, and therefore A would be similar to a part of itself, contrary to Theorem 4.

THEOREM 6. *Of two well-ordered sets E and F, which are not similar, one is always similar to a segment of the other.*

Consider the component of E consisting of all those elements of E which cut off segments similar to no segments of F. If this component exists at all, it has a first element a and the segment A, cut off by a, is evidently contained in every segment of E which is similar to no piece of F; so that it may be said to be the smallest such segment. Hence any segment of A is similar to some segment of F, so that, by Theorem 5, A is similar to F, since, by hypothesis, it is not similar to a segment of F.

Thus, either F is similar to a segment of E, or there is no segment of E which is not similar to a segment of F. In the latter case, however, by Theorem 5, E is similar to a segment of F, since by hypothesis, E is not similar to F. This proves the theorem.[5]

COR.   *Every component of a well-ordered set E, is either similar to E or to a segment of E.*

In Ch. III it was shewn that, given any particular operation of the series of derivations and deductions, we can construct a countable set of points on which every one of the operations up to that particular operation can be carried out, and which is reduced to a single point by that operation.   If therefore we take any perfect set of points not everywhere dense, and in one of its black intervals introduce such a countable set, having a point of the perfect set as limiting point of highest order, we have in this way constructed a set of points whose first derived is more than countable, such that all the derived and deduced sets, corresponding to the operations originally given, are distinct and form a segment of the whole series of derived and deduced sets ; all the remaining sets of the series are identical with the nucleus.   Each of these segments is therefore finite or countably infinite.

Thus we have the following theorem :—

THEOREM 7.   *Every well-ordered set E whose potency is less than $\aleph_1$, is finite or countably infinite, and is similar to a segment of the set of all the derived and deduced sets of a set E' whose first derived is more than countable.*

The set E' can be so chosen that each derived and deduced set of the segment in question is distinct from any other derived and deduced set, while all the derived and deduced sets not belonging to the segment are identical with the nucleus of E'.

Every such set E is also similar to the series of all the countably infinite sets which can be obtained by derivation and deduction from a certain countable set E''.

We have now got a complete grasp of the well-ordered sets of potency less than $\aleph_1$, and we see that they arise in order.   The earliest are the finite ordered sets, and as typical we may take the finite sets of ordinal numbers,

1st,

1st, 2nd,

1st, 2nd, 3rd,

and so on.   Or using the integers in their ordinal capacity,

1,

1, 2,

1, 2, 3,

and so on.

The first transfinite well-ordered set is of type ω (see p. 126) and as typical we may take the set of all the ordinal numbers,

1, 2, 3, 4, ... *ad inf.*

Then comes a set of the type of a closed sequence; and then a closed sequence followed by one isolated point.   Then in turn, a closed sequence followed by any finite number of isolated points; then a closed sequence followed by an open sequence.   After these comes a closed sequence followed by another closed sequence, and so on.

From Theorem 6, Cor. the following theorem at once follows:

THEOREM 8.   *Every transfinite well-ordered set has at least one component of type ω, and no component of type $*\omega$.*   (See p. 126.)

THEOREM 9.   *Every well-ordered set of well-ordered sets is a well-ordered set.*

For if we take all the elements of any component A of the set E so formed, they determine a certain set of the well-ordered sets, viz. those of the sets to which the elements in question belong. This set of well-ordered sets is itself well-ordered, since it is a component of a well-ordered set.   Thus it has a distinct *first* set, and the elements which belong to this first set have themselves a first, since that first set is well-ordered.   This first element evidently comes before every other element of the original component A.   Thus every component set of the elements has a first element; that is, the set is well-ordered.

THEOREM 10.   *If A and B be two well-ordered sets, and we form the set whose elements are pairs of elements, one from A and one from B, and we arrange this set in such an order that of two elements which contain the same element of A, that one comes first which contains the earlier element of B, while of two elements not containing the same element of A, that one comes first which contains the earlier element of A, this set is well-ordered.*

For it certainly has a first term, viz. that one which consists of the first term of A with the first term of B.   Further, if K be any component of the set, the elements of A which occur in K, forming a component of A, have a first term, say α, and the elements of B which occur in K in combination with α form a component of B and have therefore a first term β. The term (α, β) then is an element of K and, by definition, comes first in order in K, so that K has a first term: thus the set in question is well-ordered. Q.E.D.

COR. *If we combine any finite number of sets A, B, C, ... in the manner indicated in order, the resulting set is well-ordered.*

This theorem can not be extended to a more than finite number of sets. Suppose we have a sequence of well-ordered sets each of order 2, for instance each consisting of the two figures (0, 1). The set formed will consist of zero and all the binary fractions in their ascending order. Take the component consisting of all the binary fractions, the first element of A which occurs is 0, and so it is of B, and so of C, and so on; the element

$$\cdot 0000 \ldots \ldots$$

however does not occur in the component in question, so that the argument used in the proof of the preceding theorem breaks down.

81. Multiple Order. A set is said to be in *multiple order* when there are various characteristics, each of which enables the set to be arranged in mixed simple order, grouping together all elements which possess any one of these properties in the same degree. When this is the case, and conversely these characteristics determine each element uniquely, the order is said to be *pure*, otherwise it is said to be *mixed*. The characteristics may, or may not, themselves have an order. Thus, for instance, the set of all closed intervals on the straight line may be arranged in double order, characterising them by means of the position of their middle points and their length; these characteristics are not given in order. If, however, we determine an interval first by the position of its left-hand end-point and then by that of its right-hand end-point, the characteristics are given in order.

The set of all sets of closed intervals on the straight line may be arranged in ω-order, the characteristics being the positions of the left and right hand end-points of the intervals taken in countable order.

The only multiply ordered sets which have been seriously investigated are finite sets. Here the connection of the subject of n-ple order with the theory of sets of points in n-dimensional space is well brought out[*]. Suppose for simplicity we have a

---

[*] Vivanti, *Ann. di Mat.* Series 2, Vol. XVII.

doubly ordered set of m elements. Take a rectangular trellis (§ 73) and mark the trellis-point (i_1, i_2) corresponding to any element which stands in the i_1th and i_2th places respectively, according as we order the set in the first or second of the given ways. We thus get a set of m trellis-points having the same ordinal type as the set given.

In general, taking an n-dimensional trellis, we can assign a finite set of trellis-points to characterise any finite ordinal type of an n-ply ordered set.

If s_1, s_2, ... s_n be the number of trellis lines in the various directions on which points of the representative set lie, s_1, s_2, ... s_n are called the *dimensions* of the set. The *dimensions* of a multiply ordered set are as characteristic as its potency: these characteristics however do not determine the ordinal type, as in the case of a simply ordered set. The number $\phi(m, n)$ of n-ply ordered types of potency m is clearly connected with the number $V^{(m)}_{s_1, s_2, \dots s_n}$ of types of potency m and dimensions s_1, s_2, ... s_n by the formula

$$\phi(m, n) = \sum_{s_1=1}^{s_1=m} \sum_{s_2=1}^{s_2=m} \dots \sum_{s_n=1}^{s_n=m} V^{(m)}_{s_1, s_2, \dots s_n}.$$

This formula has been used[*] to determine $\phi(m, n)$ after obtaining formulae for calculating $V^{(m)}_{s_1, s_2, \dots s_n}$. On the other hand Cantor[†], who was the first to investigate the subject, calculated $\phi(m, n)$ directly, by means of recurring formulae.

    * Hermann Schwarz, *Ein Beitrag zur Th. d. Ordnungstypen*, Halle a. S. 1888.
    † Cantor, *Zeitschr. f. Phil. Kritik*, Vol. XCII. (1888).

CHAPTER VII.

CANTOR'S NUMBERS.

82. Cardinal numbers. In Ch. IV we discussed the various potencies which could occur on the straight line, and found that, as far as the present state of knowledge extends, these may be characterised by the positive integers, together with two new symbols, for which we took the letters a and c, a denoting the potency of a countably infinite set, and c that of the continuum.

In the preceding chapter we have seen that there is a fundamental region F, in which there is a potency $\aleph_1$ which is different from any of those mentioned, unless it is possibly the same as c; this latter point has not yet been settled[*].

These are particular cases of a great theory of potencies and of cardinal numbers in general, due to Cantor, and which has occupied numbers of mathematicians ever since: the theory, at first apparently simple, is by no means in a satisfactory state, and it will only be possible in this chapter to give a short account of it, avoiding controversial points, and giving references as to the literature.

83. The word *set* has come to be used in general for what Cantor called a *well-defined* set. The following is the definition originally given by Cantor:—

"Ein Inbegriff von Elementen, die irgend welcher Begriffssphäre angehören, nenne ich wohldefinirt, wenn auf Grund ihrer Definition, und in Folge des logischen Princips des ausgeschlossenen Dritten es als intern bestimmt angesehen werden muss, sowohl ob irgend ein derselben Begriffssphäre angehöriges Object zu der gedachten

---

[*] Bernstein, *Jahresb. d. d. Mathvgg.* 1905, p. 449, gives a sketch of the method by which he hopes to obtain a proof that $\aleph_1 = c$.

Mannichfaltigkeit als Element gehört oder nicht, wie auch ob zwei zur Menge gehörige Objecte, trotz formaler Unterschiede in der Art des Gegebenseins einander gleich sind oder nicht*."

Here Cantor emphasizes the fact that in any logical and more especially mathematical thinking, we must confine our ideas to some particular field; the collection of all the distinct objects of thought which belong to that field we shall call *the fundamental region*. The objects in question are called the *elements* (points) of the fundamental region. It will no longer be assumed that the fundamental region is the straight line.

If a law be given, of such a nature that it enables us to divide all the elements of the fundamental region into two distinct classes, those which do, and those which do not, satisfy that law (the latter of those classes possibly containing no elements at all, the so-called *null-set*), the former of these classes is called by Cantor *a set*.

There are reasons for modifying this definition; it may be contended that it should be postulated that the second of the above classes must also exist, that is to say, the fundamental region itself is not to be regarded as a set unless there is another fundamental region containing the first fundamental region together with other elements. For most practical purposes this distinction is immaterial, since such an extended fundamental region can be found, but it is perhaps a proper law of thought that we cannot regard a lot of objects in their totality unless we can get beyond them.

Cantor adds a very important gloss to his definition. The law need not be such that, given any element, we can *actually* determine whether or no that element belongs to the set, but it must be such that there can be no doubt that the element either does, or does not, belong to the set. Thus, at the time Cantor wrote, it was uncertain whether the number π was algebraic or not; it was however logically certain that every number was either algebraic or transcendental, so that the algebraic numbers certainly form a set in the fundamental region of all numbers, and the remaining numbers, viz. the transcendental numbers, also form a set. It has now been proved that π is transcendental, but this does not affect the case at all; the algebraic numbers form a set, although no general calculus is known by which we can differentiate the algebraic numbers from the transcendental.

* *Math. Ann.* xx. p. 14.

A great deal of the difficulty with respect to the development of the theory of sets in general turns however on the exact meaning to be applied to the term "logically determinate*."

84. If two sets A and B in any fundamental region are such that they can be brought into (1, 1)-correspondence with one another, they are said to be *equivalent†*, or to have the same *potency‡*. (Cp. Ch. IV.)[6]

We write this symbolically $A \sim B$.

It is completely in accordance with this definition that we say that in the theory of potencies we *regard* the elements of the set as indistinguishable, or, more properly, as undistinguished from one another§. It must not be supposed that the elements of a set can

---

* Cp. Hobson, "The General Theory of Transfinite Numbers," *Proc. L. M. S.*, Series 2, Vol. III. Part 3, p. 182.[40]

† In this connection it must be emphasized that there are difficulties in accepting the statement that every set has a potency. Different explanations of this are given by Hobson, *loc. cit.*, and Jourdain, *Phil. Mag.* VII. Ser. 6, p. 66 (1904). While this question is unsettled, Cantor's proof that there is no greatest potency (*Jahresb. d. d. Mathvgy.* I. p. 77 (1890); see § 22, p. 43, and Russell, § 346, p. 365) cannot be regarded as convincing. The proof turns on the fact that, if E be any fundamental region of potency $\bar{\bar{E}}$, it is impossible to set up a (1, 1)-correspondence between all its elements and all its sets of elements. Now each element of E constitutes a set of potency 1, thus, *if all the sets of elements of E can be considered as defining a potency*, that potency is (§ 85) greater than $\bar{\bar{E}}$. A discussion of the difficulties which arise from this proposition will be found in Russell, see also footnote †, p. 156.

‡ Cantor, *Borchardt's J.* LXXVII. p. 257, LXXXIV. p. 242. If A and B are so defined that there is a particular (1, 1)-correspondence connecting them, they are said to be *simply equivalent*. It may be, however, that, although the existence of (1, 1)-correspondences connecting A and B is implicitly involved in the definitions, these correspondences form a set of potency μ greater than 1 which is so defined that it is only possible by using the law of arbitrary choice to pick out one of the correspondences. In this case A and B are said to be *multiply* (μ-*ply*) *equivalent*.[7]

Logically multiple equivalence means less than simple equivalence, and, whenever possible, proofs should be made to depend on simple equivalence, or at least the multiplicity of the equivalence should be kept in view. For instance, proofs depending on the choice of a single unit involve c-ple equivalence.

It is easily proved that if
$$A \sim B \text{ (Mult. } a\text{)},$$
and
$$B \sim C \text{ (Mult. } \beta\text{)},$$
then
$$A \sim C \text{ (Mult. } a\beta\text{)}.$$

In the proof of the Cantor-Bernstein-Schröder Theorem given in the text, if we bring out explicitly the multiplicities of the equivalences, it is easily seen that if A is a-ply equivalent to B_1, and B is β-ply equivalent to A_1, then A is $a\beta$-ply equivalent to B. Bernstein, "Bemerkung zur Mengenlehre," *Gött. Nachr.* 1904, Heft 6.

§ See pp. 76 and 121. Cantor, *Math. Ann.* XLVI. § 1.

be actually indistinguishable, which is inconsistent with the logical use of the plural at all*. What we mean is that those characteristics which enable us to perceive the plurality of the elements, though involved in the definition of the set, are in themselves irrelevant, and may be disregarded.

THEOREM 1. (THE CANTOR-BERNSTEIN-SCHRÖDER THEOREM.) *If a set A is such that it has a proper component A_1 which is of the same potency as a set B, while A is itself of the same potency as a proper component B_1 of B, then A and B have the same potency†.*

For since there is a $(1, 1)$-correspondence between A and B_1, in this correspondence there will be a proper component B_2 of B_1 which corresponds to A_1.

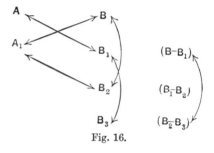

Fig. 16.

And since A_1 is in $(1, 1)$-correspondence with B, this gives us a $(1, 1)$-correspondence between B and B_2. In this there will be a proper component B_3 of B_2 in $(1, 1)$-correspondence with B_1. In this there will be a proper component B_4 of B_3 in $(1, 1)$-correspondence with B_2, and so on. (Cf. Fig. 16.)

This series of sets $B_1, B_2, \ldots$ cannot come to an end after a finite number of stages, because at any stage having found B_n, there must be a next set B_{n+1} since there is a preceding set B_{n-1}, and these have the same potency, viz. that of B or of B_1.

Let D denote the inner limiting set of $B, B_1, B_2, \ldots$; it may be that D contains no points, otherwise D is a proper component of each B_i. Then

$$B = D + (B - B_1) + (B_1 - B_2) + (B_2 - B_3) + \ldots,$$
$$B_1 = D + \qquad\quad (B_1 - B_2) + (B_2 - B_3) + \ldots.$$

Here each set in brackets really contains points, since each B_i is a proper component of B_{i-1}.

Writing E for the sum of all the sets $(B_{2n-1} - B_{2n})$ and F for the sum of all the sets $(B_{2n} - B_{2n+1})$, we have

$$B = D + (B - B_1) + E + F,$$
$$B_1 = D \qquad\qquad + E + F.$$

* Russell, § 345, p. 363.

† Bernstein's proof. Borel, *Théorie des Fonctions*, p. 105. G. Cantor, *Ztschr. f. Philosophie*, Bd. XCI. E. Schröder, *Jahresb. d. d. Mathvgg.* Bd. V. (p. 81). E. Zermelo, *Gött. Nachr.* 1901, pp. 1–5.

Now in the $(1, 1)$-correspondence of B_2 with B, $(B_2 - B_3)$ was in $(1, 1)$-correspondence with $(B - B_1)$. Similarly $(B_4 - B_5)$ was in $(1, 1)$-correspondence with $(B_2 - B_3)$, and therefore with $(B - B_1)$, and generally $(B_{2n} - B_{2n+1})$ with $(B_{2n-2} - B_{2n-1})$ and therefore with $(B - B_1)$.

Thus to any point P of $(B - B_1)$ there is a point P_1 of $(B_2 - B_3)$, a point P_2 of $(B_4 - B_5)$, ... a point P_n of $(B_{2n} - B_{2n+1})$, ... and all these points are distinct from one another. That is to each point P of $(B - B_1)$ there are a points of F.

Since we can set up a $(1, 1)$-correspondence between the points

$$P, P_1, P_2, \ldots \text{ and } P_1, P_2, P_3, \ldots$$

it follows that $(B - B_1) + F$ is in $(1, 1)$-correspondence with F, and therefore B is in $(1, 1)$-correspondence with B_1, and therefore with A. Q.E.D.

Ex. 1. By this theorem we can, for instance, prove, as follows, that *the straight line with an interval as element is of potency c.*

It is clearly only necessary to prove the theorem in the segment $(0, 1)$.

Consider any interval (X, Y) in the segment $(0, 1)$, where, using the binary notation,

$$X = \cdot x_1 x_2 x_3 \ldots ,$$

$$Y = \cdot y_1 y_2 y_3 \ldots ,$$

and X is less than Y, and let

$$Z = \cdot x_1 y_1 x_2 y_2 x_3 y_3 \ldots .$$

This gives us a $(1, 1)$-correspondence between all the intervals (X, Y) of the segment $(0, 1)$ and all the Z-points. These latter form a proper component of the segment $(0, 1)$, since, if x_1 is 1, y_1 cannot be 0, so that the interval $(\cdot 1, \cdot 11)$ contains no Z-point.

But, if x_1 is 0 and y_1 is 1, the remaining figures may be any we please, so that every point of the segment $(\cdot 01, \cdot 0\dot{1})$, or $(\cdot 01, \cdot 1)$, is a Z-point. Thus there is a proper component of the Z-points in $(1, 1)$-correspondence with the segment $(0, 1)$. Thus, by the preceding theorem, the Z-points* have the potency c, and therefore the same is true of all the intervals in the segment $(0, 1)$. Q.E.D.

85. If we have two sets A and B, then there are four mutually exclusive logical possibilities :—

---

* The Z-set and its complementary set are two interesting sets of intervals. Each interval is bounded at one end by an isolated end-point of each set, and at the other by a semi-external point of each set. These end-points must, however, be considered as belonging to the complementary set, since they correspond to values $X \geqslant Y$. Disregarding this, the two sets of intervals are reflexions of one another in the point $\cdot 1$. Amalgamating the pairs of abutting intervals of the two sets, we get *an infinite set of intervals having no isolated end-points and no external points, but only semi-external points*, in the segment $(0, 1)$.

(1)   There may be a component* A_1 of A in (1, 1)-correspondence with B, and also a component B_1 of B in (1, 1)-correspondence with A ;

(2)   there may be such an A_1, but no such B_1;

(3)   there may be such a B_1, but no such A_1;

(4)   there may be no such A_1, and no such B_1.

It is clear that there are no other possibilities.

In the first case, by definition, or by the Cantor-Bernstein-Schröder Theorem, A and B have the same potency and we write $A \sim B$ or $\overline{\overline{A}} = \overline{\overline{B}}$, $\overline{\overline{A}}$ being the symbol used for the potency of the set A. In the second case A is said to have *a greater potency* than B, $\overline{\overline{A}} > \overline{\overline{B}}$; in the third case B has a greater potency than A, and A is said to have *a lesser potency* than B, $\overline{\overline{A}} < \overline{\overline{B}}$†. The fourth case alone presents any difficulty ; it is clear that, if there are such sets A and B, they have not the same potency, nor is the potency of either greater than that of the other ; but, whereas it is easy to give sets A and B illustrating any of the first three cases, no sets are known illustrating the fourth case‡, and it is possible that the case does not really exist at all ; at present however this has not been proved. If the fourth case could be proved not to exist, the potencies of sets would be comparable in respect of magnitude, and we could properly speak of potencies as *numbers* and set up a calculus of such numbers. As it is such a calculus has been worked out, but it labours under the difficulty above explained.

86.   The addition and multiplication of potencies.

Addition.   Let A and B be two sets, in the same fundamental region, and not having any common element, and let their potencies

* Proper or not.   This point constitutes the difference between this and Borel, *Théorie des Fonctions* (1898), p. 102.

† Here it should be remarked that when we have the two relations $A \leqslant B$ and $B \leqslant A$, we may at once deduce the logical conclusion $A = B$. Any reference to the Cantor-Bernstein-Schröder Theorem at this point is an error.

‡ It has been suggested by Beppo Levi, "Intorno alla teoria degli aggregati" (*Lomb. Ist. Rend.* II. 35, p. 863), and Hobson, *loc. cit.*, that this case may occur when a set G is in (1, γ)-correspondence with a set Γ, that is, when to each element of G there corresponds a non-countable set of elements of Γ of constant potency γ. We had a case of this in § 73, where G is the set of all simply-ordered sets, and Γ the set of all the binary order-fractions; γ is here the potency of all the different arrangements of the natural numbers in countable order. This difficulty only occurs if we restrict the word equivalent to mean simply equivalent. Such a set G is multiply equivalent to Γ. (See footnote ‡, p. 147.)

be a and b, then the *sum* $(a + b)$ of the two potencies is defined as the potency of the set $(A + B)$ consisting of all the elements of the two sets together.

More generally if k be any potency, and we replace each element of a set K of potency k, by a set of elements, the sum of the k potencies of these sets is defined to be the potency of the whole set of elements so obtained. It can easily be shewn that the sum of any number of potencies obeys the commutative and also the associative law.

Multiplication.[41] Let A and B be two sets as before of potencies a and b. Form a set C of pairs of elements, one from each of the sets A and B; the potency of this set is defined to be the product of ab.

More generally if K be any set of potency k, and we have a set Ⅱ of potency p corresponding to each element P of K, then if we form a set each of whose elements is a set of potency k of elements chosen one from each of the above sets Ⅱ, the potency of this whole set is said to be the product of the k factors p*.

The product obeys the commutative, associative and distributive laws.

Ex. 2. We may use this definition to prove that the product of all the integers

$$1 . 2 . 3 \dots = c.$$

For to find $1 . 2 . 3 \dots$, we must take the natural numbers, and, in place of each, say n, insert n elements, say $(n, 1)$, $(n, 2) \dots$. We may do this conveniently in the form of a wedge,

$$(1, 1)\ (2, 1)\ (3, 1)\ (4, 1), \dots\dots$$
$$(2, 2)\ (3, 2)\ (4, 2), \dots\dots$$
$$(3, 3)\ (4, 3), \dots\dots$$
$$(4, 4), \dots\dots$$

We must now form a set G whose element is determined by choosing one from each of the columns of this wedge ; by definition the potency of such a set of elements will be $1 . 2 . 3 \dots$.

We can form such a set of binary fractions by replacing all the chosen brackets in the wedge (1) by ones, and all the non-chosen brackets by zeros so as to form a new wedge (2), the figures of which we read off in the usual way, from top to bottom of each column in succession from left to right.

The points of the segment $(0, 1)$ which are not obtained in this way are then easily seen to fill up a set of black intervals which may be obtained as follows. Bisect the segment $(0, 1)$, and blacken the left-hand interval. Then divide the white segment into 2^2 parts and blacken the extreme intervals on the left and right. Then divide each of the 2 white segments into 2^3 parts and blacken the extreme intervals in each segment, and so on.

* Schoenflies, *Ber.* p. 9 (1900). Whitehead, *Am. J. of Math.*, Vol. xxiv. (1902).

These abutting intervals correspond to a set of non-abutting intervals, the largest of which reaches from the origin to the point ·0 01 001 0001 .... Thus our set is a perfect set nowhere dense* and has therefore the potency c, which proves the result stated.

A particular case of multiplication is the process of raising to a certain power; here the sets Π are to be taken all of the same potency p.

Ex. 3.   The method used in Ex. 2 may be used to prove that $2^a = c$.

Here, instead of a wedge, we have two rows of brackets,

$$(1, 1)\ (1, 2)\ (1, 3) \ldots\ldots$$
$$(2, 1)\ (2, 2)\ (2, 3) \ldots\ldots\ .$$

Substituting 1 for each chosen bracket and 0 for every other, and reading off the columns from top to bottom in succession from left to right, we again get a perfect set, whose black intervals are obtained by dividing the segment (0, 1) into four parts and blackening the extreme parts, and repeating this construction in the two intermediate segments, and so on.

Ex. 4.   Similarly it may be shewn that $n^a = c$, for every positive integer n. Or the following method may be employed, assuming the results of Exs. 2 and 3.

If instead of two rows as in Ex. 3, we take three rows, and proceed as in that example, we get a set of binary fractions whose potency is, by definition, 3^a. Confining our choice to the first two rows, we get the set of Ex. 3 as a component of our set.   Thus

$$3^a \geqslant c.$$

On the other hand, if in the wedge of Ex. 2 we confine our choice to the first three rows, we get a component of the set of Ex. 2 of potency 3^a.   Thus

$$3^a \leqslant c.$$

It follows that

$$3^a = c.$$

Ex. 5.   Similarly, taking a countably infinite number of rows,

$$a^a \geqslant c.$$

On the other hand, corresponding to each choice we can assign a continued fraction

$$\cfrac{1}{n_1 +}\ \cfrac{1}{n_2 +}\ \cfrac{1}{n_3 +}\ \ldots\ldots,$$

which is uniquely determined by the choice of the n_1th bracket in the first column, the n_2th in the second, and so on.   Thus

$$a^a \leqslant c.$$

It follows that

$$a^a = c.$$

Grouping these results together,

$$c = 1 . 2 . 3 \ldots = 2^a = 3^a = \ldots\ldots = n^a = \ldots\ldots = a^a.$$

* The content, being $\underset{n=\infty}{\mathrm{Lt}}\ (\tfrac{1}{2})^n$, is zero.

87. When we restrict ourselves to the potencies of well-ordered sets the difficulty referred to in § 85 disappears. It was shewn in the preceding chapter (p. 140) that, if A and B are well-ordered sets, A is similar to B or to a segment of B, or else B is similar to a segment of A. Thus the potency of A is either equal to, less than, or greater than that of B. The potencies of the earlier well-ordered sets in order of magnitude are:—

$$1, \quad 2, \quad 3, \quad \ldots\ldots n, \quad n+1, \quad \ldots\ldots$$

The next in order is a (§ 18), which is denoted by Cantor by the symbol $\aleph_0$ (Aleph-null, Aleph-zero); then follows $\aleph_1$, since any segment of the set of potency $\aleph_1$ defined in § 79 has, by Theorem 25, Ch. IV, p. 56, one of the above as potency, while $\aleph_1 > \aleph_0$.

These numbers form the beginning of the series of *Cantor's cardinal numbers*, properly so called, because there can be no doubt as to their comparability with respect to magnitude. We have now to show how this series is to be continued. To do this it is best to define Cantor's ordinal numbers.

88. *Cantor's ordinal numbers* are none other than the ordinal types of well-ordered sets[*], or, if we prefer to say so, they may be taken to characterise those types.

As in the case of the Cantor cardinal numbers, we have a right here to speak of ordinal *numbers* instead of ordinal *types*, since, given any two of the numbers, say a and b, they correspond to the ordinal types of sets A and B such that either A and B are similar, or A is similar to a segment of B, or B to a segment of A; in the first case we say that $a = b$, in the second $a < b$, in the third $a > b$. Thus the ordinal numbers themselves are naturally arranged in simple order, it will be shewn (§ 89) that they form a well-ordered set.

By what has been said in the preceding chapter the series of ordinal numbers begins, as the cardinal numbers do, with

$$1, \quad 2, \quad 3, \quad \ldots\ldots n, \, n+1, \ldots\ldots$$

Instead of a however we now have ω (§ 73 and footnote[*], p. 126).

In any fundamental set, *e.g.* the straight line, in which there is a well-ordered set of type ω, we only have to alter the order of the set, removing the first element and placing it at the end

---

[*] Cantor, "Grundlagen," § 3, *Math. Ann.* XLIX. § 12. Russell, Chap. XXXVIII.

and we get a set whose ordinal type is the next in magnitude after ω, this we denote * by $\omega + 1$.

In this way the series of ordinals may be continued,

$$\omega + 1, \quad \omega + 2, \quad \ldots\ldots \omega + n, \quad \omega + n + 1, \quad \ldots\ldots$$

All the corresponding sets are countable; without rearrangement such sets exist on the ordered straight line, for instance in Fig. 4, p. 24 we may take as representative of any one of these the sequence in the segment $(0, \cdot1)$ followed by a finite number of points in the segment $(\cdot1, \cdot1^2)$. This figure gives us convenient sets representing each of the succeeding numbers

$$\omega \cdot 2, \quad \omega \cdot 2 + 1, \quad \ldots \omega \cdot 2 + n, \quad \omega \cdot 2 + n + 1, \ldots$$
$$\omega \cdot 3, \quad \omega \cdot 3 + 1, \quad \ldots \omega \cdot 3 + n, \quad \omega \cdot 3 + n + 1, \ldots$$
$$\ldots\ldots\ldots$$
$$\omega \cdot m, \quad \omega \cdot m + 1, \quad \ldots \omega \cdot m + n, \quad \omega \cdot m + n + 1, \ldots$$

the set $\omega \cdot m + n$ consisting of all the points to the left of the point $\cdot1^m$ followed by the next n points. The ordinal type of the set T_2 itself is denoted by ω^2. Fig. 5, p. 24, gives us convenient sets representing the succeeding numbers whose types are of the form $\omega^k + a\omega^{k-1} + b\omega^{k-2} + \ldots$, while the ordinal type of the set E in that figure is denoted by ω^ω.

The method, as indicated at the end of § 12, enables us to find always higher numbers, but we shall require more and more diagrams. On the other hand, using these ordinal numbers as indices for the derived and deduced sets of a set whose first derived is more than countable, we have corresponding sets of elements in the fundamental set F, described in § 79. This fundamental set may serve us most conveniently for characterising the numbers which next follow in succession.

The derived sets of E are denoted by the indices $1, 2, \ldots, n, \ldots$, the deduced set of these by ω; the derived sets of E_ω by $\omega + 1$, $\omega + 2, \ldots, \omega + n, \ldots$, and the deduced set of these by $\omega + \omega$ or $\omega \cdot 2$; and so on. The set deduced from $E_\omega, E_{\omega \cdot 2}, \ldots, E_{\omega \cdot n}$ has the index ω^2, that deduced from $E_{\omega^2}, E_{\omega^3}, \ldots, E_{\omega^n}$ the index ω^ω, that deduced from the E's with indices $\omega^\omega, \omega^{\omega \cdot 2}, \ldots, \omega^{\omega \cdot n}$ the index ω^{ω^2}, and that deduced from the E's with indices $\omega^\omega, \omega^{\omega^2}, \omega^{\omega^3}, \ldots$ the index ω^{ω^ω}.

For the set deduced from the sets with indices $\omega^\omega, \omega^{\omega^\omega}, \omega^{\omega^{\omega^\omega}}, \ldots$

* This notation, as well as that used for the succeeding numbers, enables us to dispense with the introduction of new symbols till a much later stage. It agrees with the theory of ordinal addition (§ 91) and multiplication (§ 92); $a + \beta$ is the ordinal type of a set of type a followed by one of type β, and $a\beta$ of a set consisting of a set of type β of sets of type a.

the notation breaks down, but the principle can be carried on *ad infinitum*. This principle, or more properly these *two* principles, are—(1) to every number α there is a *next* number*, which shall be denoted by $\alpha + 1$; (2) to every infinite set of numbers without a greatest number there is a *next* number†, which shall be called a " limiting number " (*Limeszahl*).

The ideal symbols defined in this way are called *Cantor's transfinite ordinal numbers of the first potency* $\aleph_0$, *or of the second class*. The sets corresponding to them are all countable, and sets of points of each of these ordinal types exist on the straight line.

It was pointed out in the preceding chapter that the ordinal types of such sets could be represented by diagrams, or by corresponding binary numbers.

Fig. 17 is one of the diagrams representing the natural order of the derived and deduced sets up to E_{ω^2} inclusive, the countable order being obtained from the diagram

ω^2	ω	1	2	3	4	5	...
	2ω	$\omega + 1$	$\omega + 2$	$\omega + 3$	$\omega + 4$	...	
		3ω	$2\omega + 1$	$2\omega + 2$	$2\omega + 3$	...	
			4ω	$3\omega + 1$	$3\omega + 2$	...	
				5ω	$4\omega + 1$	...	
					6ω	...	

the columns being read in order, thus

ω^2, ω, 1, 2ω, 2, $\omega + 1$, 3ω, 3, $\omega + 2$, $2\omega + 1$, 4ω, 4, $\omega + 3$, ...

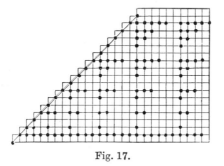

Fig. 17.

The number equivalent to the diagram is

·1 11 100 1101 10010 100000 1101011 10010010 ....

If we start with any generic binary order-fraction, or, which is the same thing, with any generic order-diagram, we shall not in general

---

* Corresponding to derivation.
† Corresponding to deduction.

have a corresponding Cantor number. The only sets whose ordinal types are Cantor numbers are well-ordered sets ; a set which corresponds to a generic binary order-fraction is only simply ordered.

89. The next Cantor ordinal after all those of the first potency is Ω, the ordinal type of the set of all the derived and deduced sets of a set whose first derived is more than countable. Sets representative of this number (having it as ordinal type) exist as we know in the fundamental set F, described in § 79.

In the same way in which we formed the ordinals of the second class, by rearrangement of a sequence in the straight line, we can now form a set of ordinal type $\Omega + 1$ in F, and generally in F we can form all possible well-ordered sets of potency $\aleph_1$. The corresponding ordinal numbers are said to form *the third class of transfinite ordinals.*

It is to be noted that these numbers of the third class have not been as yet used to any great extent, and that many people have scruples as to using them at all. The same objection which is made to these numbers has been made to the higher numbers, viz. their existence as ordinal types of sets in any practical fundamental region, *e.g.* in the straight line, has been doubted*.[7] As to the doubts thrown on the existence and utility of these numbers, and of the numbers of the fourth and higher classes, the reader must be referred to the literature on the subject, which is too controversial to find a place here†.

It remains only to add that Cantor‡ continues his series of ordinal numbers beyond those which correspond to the potency $\aleph_1$, justifying the procedure in two different ways. The first way is by the hypothesis that the two principles mentioned may themselves be used to set up an ideal set of numbers, and that such an ideal set of numbers is a collection of proper objects of thought and form of themselves a proper fundamental region, provided only the system so formed is in accordance with the laws of logical thought. The second way is based on the hypothesis that well-ordered sets of higher and higher potencies do exist,

---

* The proofs of the theorem $\aleph_1 \leqslant c$ supplied by Bernstein in his dissertation (§ 8 (1900), see also *Proc. L. M. S.* Ser. 2, Vol. I. p. 243 and §§ 73 and 88 *supra*) and Hardy (*Quart. J. of Math.* xxxv. p. 88, 1903), depending as they do on multiple equivalence, do not serve to set up a definite linear set of points of potency $\aleph_1$, in the sense in which Fig. 2, p. 17, does for a set of ordinal type ω.

† Russell, Hobson *loc. cit.*, Zermelo, Jourdain, Schoenflies, Bernstein and König, in *Math. Ann.* LIX. and LX. (1904–5), where other references will be found.

‡ Cantor, *Math. Ann.* XLIX. (1897) ; Schoenflies, Chap. VII ; Jourdain, *Phil. Mag.* (1904–5).

and that, from the properties proved for well-ordered sets in the preceding chapter, they may all be considered as segments of one great well-ordered set, which itself may be taken as fundamental region.

Whether or no the series of ordinal numbers be assumed to have the unlimited extent ascribed to it by Cantor, we can easily prove that it is a well-ordered set. For any component of it either has 1 for its *first* element, or it defines a second component, consisting of all the ordinals which precede every element of the given component. If β be any number of the given component, there is a fundamental set in which there is a well-ordered set of the type β, and this set will have, as proper components, segments of all the ordinal types of the second component, and will therefore also have a component, proper or not, whose type is the next in order after these, viz. $\alpha + 1$ if they have a last one α, or the next limiting number α if they have no last number. This number $\overline{\alpha + 1}$, or α, will then belong to the given component, and between it and the numbers of the first component there can be no other number, thus it is the *first* number of the given component. Thus, by definition, *the ordinal numbers form a well-ordered set.*

90. Assuming the existence of Cantor's scheme of ordinal numbers to be accepted, it gives us a means of determining the system of cardinal numbers already referred to, the so-called system of Aleph-numbers or Cantor's cardinals. Just as all the ordinals of the second class correspond to sets of potency a or $\aleph_0$, while the potency of these numbers themselves is $\aleph_1$ and is greater than $\aleph_0$, so the ordinals of the third class correspond to sets of potency $\aleph_1$; and these numbers have a greater potency $\aleph_2$ and so on. The indices of the Aleph-numbers may themselves be Cantor ordinals as great as we please.

91. The theory of ordinal addition*. The general theorem given as Theorem 9, of the preceding chapter—*Every well-ordered set of well-ordered sets of elements, regarded as an ordered set of elements, is well-ordered*—must be taken into account in forming any system of transfinite ordinal numbers; in Cantor's system it is the basis of the theory of addition. The order of such a set of elements must be considered known in terms of the

---

* Cantor, *Math. Ann.* xlvi. § 8 (1895), "Grundlagen," § 3. Beside addition and multiplication, Cantor considers in the Grundlagen *subtraction, division,* and *prime numbers.* See also Russell, § 294.

orders of the separate sets; in the special case when all the sets are finite, this is embodied in the fact that the ordinal number of the final set is the sum of the ordinal numbers of the components in order.

In general *the sum of the ordinal numbers of the components in the order given by the first well-ordered set is defined to be the ordinal number corresponding to the final set.*

For instance, (1, 2, 3), (4, 5), (6, 7, 8) is a finite ordered set of ordinal type 3 of finite ordered sets; regarded as an ordered set of elements, in which each element of any component set retains its place with respect to the others of that set and precedes every element of any subsequent set, viz. (1, 2, 3, 4, 5, 6, 7, 8), it is a finite ordered set of ordinal type $8 = 3 + 2 + 3$.

It is a mere accident, depending on the fact that the sets are finite, that the sets could be added in any order, that is to say that cardinal and ordinal addition are the same; this will be found not to be true of transfinite ordinal numbers: *the commutative law only holds for finite ordinal numbers.*

The associative law however will be found to be perfectly general in its application. In the case of transfinite as well as finite ordinal numbers it is in fact an immediate consequence of the definition of ordinal addition, that *if a set be divided up in two different well-ordered manners into well-ordered components, the two sums so obtained are equal.* The set whose ordinal type is represented by $(a + b + c)$ will have the same elements in the same order as one whose ordinal type is represented by

$$(a + b) + c \quad \text{or} \quad a + (b + c).$$

This proves the associative law,

$$a + (b + c) = (a + b) + c.$$

It follows that *any addition may be regarded as a certain ordinal number of repetitions of the process of adding two ordinals;* since however the commutative law does not hold, this point of view does not afford such a simplification as it does in the case of finite integers.

The order in which two numbers are to be added is conveniently determined by speaking of the first of the two sets as the *augendum* and the second as the *addendum*. Suppose, for instance, a set of type ω is the augendum and a single point is the addendum. The final set will be of the ordinal type of a closed sequence, which, as in § 88, may be denoted by $\omega + 1$, we do not need to introduce a new symbol.

On the other hand, if the augendum is a single point and the addendum a set of ordinal type ω, we have a set whose ordinal number is $1 + \omega$. This gives us a new theorem connecting the finite integers with ω, for the final set is of the type of an open sequence; hence

$$1 + \omega = \omega \neq \omega + 1.$$

Hence, generally,

$$n + \omega = \omega \neq \omega + n.$$

This shews that the commutative law cannot be assumed to hold.

When we are dealing with finite numbers the process of addition always leads us to new numbers, but we see from the above that when we are dealing with transfinite ordinal numbers the process of addition may only lead us back to one of the original numbers.

92. The law of ordinal multiplication. The law of multiplication of ordinal numbers in order has not as yet been defined except when the number of factors is finite*. If we could define in a satisfactory manner the ordinal product of factors, we should be able to arrange the continuum in the form of a well-ordered set, which has not yet been done (cp. Ch. VI, concluding remarks). The law of multiplication of a finite number of factors depends on Theorem 10 of the preceding chapter, and it was pointed out at that place that the process could not be extended to an infinite number of sets.

DEF. *Given any finite ordinal number n of well-ordered sets, form the product set as in cardinal multiplication, by replacing each set by any one of its elements, and taking the set of all such possible different combinations. This set can be arranged in such an order that given two terms having the same element in each of the first m places, but different elements in the $\overline{m + 1}$th place, that term whose $\overline{m + 1}$th element comes before the other in the $\overline{m + 1}$th well-ordered set, is taken to precede the other term. By Theorem 10, Chap. VI the set is then a well-ordered set; its ordinal number is defined to be the product of the n ordinal factors in order.*

In the case of two factors, the first of which we may call the

---

* Exponentiation however can be satisfactorily defined, as exemplified by the number ω^ω, the ordinal type of the set of points in Fig. 5, p. 24. Cantor, *Math. Ann.* XLIX. §§ 18—20. Schoenflies, pp. 47, 48. Hausdorff, *Jahresb. d. d. Mathvgg.* XIII. (1904), p. 569.

multiplier and the second the *multiplicand*, the rule is, take the first term of the multiplier and combine it in turn with each term of the multiplicand in order, and so with each term of the multiplier in order; the set of all such pairs in order is well-ordered and its ordinal number is the product of the two factors.

Another form of the definition of multiplication of two factors is the following:—Take each term of the multiplier in order and replace it by a set similar to the multiplicand, the well-ordered set so obtained has for ordinal type the product of the two ordinal types in order.

Symbolically we place the number corresponding to the multiplier on the right and that corresponding to the multiplicand on the left*, that is, we read the product symbol from right to left. Thus $\omega \cdot 2$ is the ordinal type of a set got by placing instead of each of the elements of a set of ordinal type 2 a set of ordinal type ω, and this is the same therefore as $\omega + \omega$. But 2ω is the ordinal type of a set got by putting in the place of each element of an open sequence a pair of elements, so that $2 \cdot \omega$ is the same as ω.

$$\omega \cdot 2 = \omega + \omega > \omega,$$
$$2 \cdot \omega = \omega \neq \omega \cdot 2.$$

---

* This notation is so far adopted that it does not seem advisable to revert to Cantor's original notation in which the product symbol is read from left to right.

CHAPTER VIII.

PRELIMINARY NOTIONS OF PLANE SETS.

93. When we come to deal with points which do not lie in a straight line, the fundamental region will now be taken to be a plane, or a flat space of three or more dimensions, just as in Chs. I—V it was the straight line. The full discussion will in all cases be given for the plane, in general it will only need small verbal alterations for higher space. Later on the fundamental region may be taken to be a set of points contained in space of a finite number of dimensions, this will include the special case of ordinary curved space. The theory must not be considered to be applicable without fresh investigation to a fundamental space of an infinite number of dimensions.

94. Just as the straight line was to be considered as the geometrical representative of the arithmetic continuum, so the plane is to be regarded as the geometrical representative of the two-fold arithmetic continuum, each point of the plane corresponding uniquely to two numbers in order, (x_1, x_2), called its coordinates, and conversely each pair of coordinates determining uniquely a point of the plane; the order of the coordinates is formally material, the points (a, b) and (b, a) being different. It will generally be assumed that the coordinates are ordinary rectangular Cartesian coordinates, giving the distances of the point from two perpendicular straight lines, but this is by no means essential, and the idea of coordinates in the plane, or in n-dimensional space, is as independent of the idea of measurement as it was in the straight line. Often it is convenient to use ordinary plane polar coordinates, r and θ, r being the distance of the point P from a given point or origin O, and θ the angle made by the radius vector OP with a fixed direction.

95. The question of the *dimensions* of the fundamental region is one of interest. The plane is commonly said to be of two dimensions, since it is the geometrical representative of the two-fold arithmetic continuum ; it is however shewn in § 96 that the points of the plane (or numbers of the two-fold arithmetic continuum) can be arranged in simple order, and the question arises, are we justified in speaking of the plane as two-dimensional, and what do we mean by the dimensions of the fundamental region ?

Riemann, in his *Habilitationsschrift* (1854), assumed explicitly that the essential characteristic of a variety or manifold of n dimensions was that the position of a point in the region required n numerical determinations or coordinates. According to this conception the plane would have the same right to be denominated a one-dimensional variety as the straight line. Riemann had however implicitly assumed that the coordinates were themselves *continuous* functions of the position of the corresponding point, and moreover his mode of generating a variety of n dimensions is by a continuous process performed on a variety of $\overline{n-1}$ dimensions. Continuity was in those days habitually assumed as something which did not need special reference or discussion, and it was not till the investigations of Cantor, Dedekind and others into the meaning of continuity and discontinuity that a proper concept of a continuous set of points, a variety, manifold, or region was possible. The idea of the dimensions of a variety, like the idea of potency, is one which depends on $(1, 1)$-correspondence, but the idea of continuity enters in the former and not in the latter idea.   It will be seen from § 96 that the points of the plane have the same *potency* as those of the straight line, but, since the correspondence there given is not continuous, it does not shew that they have the same *dimensions*. Subsequent theorems* however will shew that they have not the same dimensions, so that *the dimensions of the plane are none other than two.* The discussion of these theorems will however be deferred until some of the characteristic properties of plane sets of points and plane regions have been developed.

96. Cantor's (1, 1)-correspondence between the points of the plane, or of n-dimensional space and those of the straight line†.

* Chap. IX, p. 216.
† Cantor, J. für Math. LXXXIV. p. 245 (1877).

Let (x_1, x_2) be any point of the area of a unit square, and let x_1 and x_2 be expressed as binary fractions, with an infinite number of figures, zeros and ones,

$$x_1 = 0 . e_1 e_2 e_3 \ldots\ldots,$$
$$x_2 = 0 . e_1' e_2' e_3' \ldots\ldots;$$

this can always be done in one and only one way. Corresponding to the point (x_1, x_2) of the plane we take the point x of the straight line, where

$$x = 0 . e_1 e_1' e_2 e_2' e_3 e_3' \ldots\ldots;$$

the point x is then uniquely determined by the point (x_1, x_2). Conversely any point x of the straight line, lying between 0 and 1, determines uniquely a point (x_1, x_2) of the unit square, whose coordinates are got from x by taking the figures of x alternately, x_1 beginning with the first figure after the point, and x_2 with the second. This gives us a $(1, 1)$-correspondence between the points of the unit square and those of the unit segment. The same principle may be used to set up a $(1, 1)$-correspondence of the plane and the straight line.

In exactly the same way it may be proved that the points of n-dimensional space can be put in $(1, 1)$-correspondence with those of the straight line; in this case, instead of taking the figures of x alternately, we must take every nth figure, and so form the n coordinates of the corresponding point, x_1 beginning with the first figure after the point, x_2 with the second, x_3 with the third, and so on.

97.   Cantor's $(1, 1)$-correspondence between the points of space of a countably infinite number of dimensions and those of the straight line*.

Here the same principle may be used as before; we have only to take

$$x_1 = 0 . e_{11} e_{12} e_{13} \ldots\ldots,$$
$$x_2 = 0 . e_{21} e_{22} e_{23} \ldots\ldots,$$
$$x_3 = 0 . e_{31} e_{32} e_{33} \ldots\ldots,$$
$$\ldots\ldots\ldots\ldots\ldots\ldots$$

and so on, and, for the corresponding point,

$$x = 0 . e_{11} e_{12} e_{21} e_{13} e_{22} e_{31} \ldots\ldots,$$

the indices being obtained as in Theorem 3, p. 35, by writing down the figures of $x_1, x_2 \ldots$ in the form of a wedge and reading them in columns.

* Cantor, *J. für Math.* LXXXIV. p. 245 (1877).

COR. 1. *All countable sets of points on the straight line form a set of potency c.*

COR. 2. *All countable sets of intervals on the straight line form a set of potency c.*

This follows from Cor. 1, using Ex. 1, p. 149. Hence by Theorem 4, p. 19, we have the following :—

COR. 3. *All closed sets on the straight line* form a set of potency c.*

COR. 4. *All ordinary inner and outer limiting sets form a set of potency c.*

It follows from these theorems that the potency of the points of space of any finite or countably infinite number of dimensions is the same as that of the linear continuum, that is c. It also follows that any plane set G is in (1, 1)-correspondence with a certain set on the straight line. It does not however follow that the characteristic properties of plane sets of points can be deduced from this correspondence. The chief reason why this is not the case is that *such a correspondence is never continuous.*

98. Continuous Representation. Suppose we have a correspondence, not necessarily (1, 1), between a plane set G and a set g on the straight line, in which to each point of g there is one and only one corresponding point of G; such a correspondence is said to be a representation of the plane set on the linear set. Let p be a point of g and P a corresponding point of G.

Describe a circle C_P with P as centre ; then, in general, we cannot find any interval d_p with p as centre such that every point of g in d_p has its correspondent inside C_P; when, however, this is the case, the representation is said to be *continuous* at the point p with respect to the set g.

---

* The set F_1 of § 79. Borel, *Leçons sur la Théorie des Fonctions*, 1898, p. 50. Hence it follows that *to assign a certain series of the processes of derivation and deduction is equivalent to assigning a certain set of points of potency c on the straight line.* For if E be any closed set on which the series of processes can be performed, without reducing it to a null-set or to its nucleus, and E' the first of the derived and deduced sets not obtained by the processes of the series, (E, E') may be regarded as a point in the plane, and therefore as a point on a straight line. Since, however, there are evidently c such sets E, we get c such points determining and determined by, the series of processes in question. This proves the theorem, which may be expressed in symbols by the equation

$$\aleph_1 c = c,$$

or $\aleph_1 \leqslant c$ (Mult. c^c),

(see footnote *, p. 156, and Appendix).

This definition is the same as saying that *the coordinates x_1 and x_2 of the point P are single-valued functions of the coordinate x of the point p, which are continuous at the point p with respect to the set of points g.*

If the set g is a segment of the straight line, the words " with respect to the set of points g " may be omitted. If again the representation is continuous at every point of g, it is said to be *continuous with respect to g,* and, if g is a segment, to be *a continuous representation.*

Even a continuous representation, however, does not enable us to deduce properties of a plane set from those of a linear set, without special precautions. The whole plane can be represented continuously on the straight line, and indeed in such a simple manner that a point of the plane has in general only one correspondent on the straight line, and has at most only four; yet the characteristics of the plane, in particular its form, are in no sense deducible from this correspondence. On the other hand a (1, 1)-correspondence which is also continuous is of immense value in enabling us to deduce properties of plane sets of points from sets on the straight line.

99. Peano's continuous representation of the points of the unit square on those of the unit segment.[6]

Take a unit square and divide it into nine (or m^2 where m is any odd integer) equal squares. Take also the unit segment and divide it into nine (or m^2) equal segments. Let the nine squares correspond to the nine segments, the order of the segments being their natural order from left to right: and the order of the squares being obtained as follows:—begin at the bottom left-hand corner, take the squares straight up the first column, down the second and up the third, till we arrive at the top right-hand corner. The diagonals of the squares in order, which abut end to end as in Fig. 18, form a polygonal line of nine (m^2) stretches, with nodes at two (or $\overline{m-1}$) points. This polygonal line may be used to denote graphically the order of the

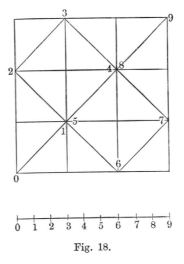

Fig. 18.

squares as they correspond to the segments on the straight line in their natural order.

In repeating this construction in each small square, it is best for definiteness so to turn the figure that the diagonal of that small square which formed part of the polygonal line just constructed, should run as before from the bottom left-hand corner to the top right-hand corner.

Repeating this construction *ad infinitum*, the polygonal line becomes at every stage more and more crinkly, so that the order of the small squares becomes more and more complicated; it is however always determinate, and is such that adjacent segments on the straight line correspond to adjacent squares. Any point p of the straight line which is not a point of division, determines uniquely a series of segments, one from each stage, lying one inside the other and having p as their common internal point; these determine uniquely a corresponding series of squares, lying inside one another and having only one common internal point (Theorem 1, p. 17), this point P we take to correspond to p. If, on the other hand, p is a point of division, it determines two series of segments, one lying on its right and the other on its left, and each having p as the common end point; corresponding to these we have two series of squares, but since the square in one series at any stage is always adjacent to the square in the other series at the same stage, both series define the same point P, which is a common boundary point. Thus again we have to p a single correspondent P.

It is clear that this correspondence is continuous; for suppose

$$\frac{1}{3^{2k+2}} \leqslant d < \frac{1}{3^{2k}}, \quad \left[\text{or } \frac{1}{m^{2k+2}} \leqslant d \leqslant \frac{1}{m^{2k}} \right].$$

An interval of length not greater than d with the point p as centre will then overlap at most into two adjacent intervals at the kth stage; therefore two at most of the corresponding squares, and these adjacent to one another, will suffice to contain a correspondent for each internal point of the interval d_p. Thus a circle with centre P and radius $e > \dfrac{\sqrt{5}}{m^k}$ will certainly enclose a correspondent for each internal point of d_p. Now whatever circle C_P be described with P as centre, we can determine k so that $\dfrac{\sqrt{5}}{m^k}$ is less than its radius; taking the length of d_p to lie between $\dfrac{1}{m^{2k+2}}$ and $\dfrac{1}{m^{2k}}$ it follows that the correspondents of every point of d_p lie inside C_P. Thus the correspondence is a continuous one.

On the other hand it is not a (1, 1)-correspondence. For though every point of the square which does not lie on a dividing line clearly has only one correspondent, yet a point Q lying on a dividing line but not at a corner of a square at any stage will have two different correspondents on the straight line, one for each series of squares lying one inside the other and having Q as common boundary point. Again any corner of the squares, unless it lie on the polygonal line at some stage, will have four distinct correspondents : for instance the point where the 5th, 6th, 7th and 8th squares in Fig. 18 meet, will have four distinct correspondents, one in each of those squares.

100. This construction of Peano's was given in answer to a query of Jordan's as to whether it was possible for a curve to fill up completely a space. The answer is that a curve, or a "continuous line," as Jordan tentatively defined it, may do so, and hence the expression "space-filling curve" has crept into mathematical literature. Jordan assumed that any plane set which can be brought into continuous correspondence with the points of a closed segment of a straight line ought to be called a *continuous line*, and the term Jordan curve has, since Jordan wrote, been often used* for such a set of points. Peano's correspondence, however, shows that such a set of points may, without further restriction, constitute a region, and there are obvious objections to allowing the word curve to be applied to a region. It does not seem desirable that the word curve should be applied to any set of points which is not *dense nowhere* in the plane, or in the space in which it lies. The term "space-filling curve" seems to be one which should be dismissed, but it is necessary here to mention and explain it.

The polygonal lines constructed at each stage in Peano's construction, are curves, which become more and more crinkly at each stage, and will eventually, as far as the eye can distinguish, fill up the whole square. At no stage, however, will they actually exhaust the points of the square. The outer limiting set of these curves (that is the set consisting of all the points which at any stage belong to one of these polygonal lines) will be an open set, dense everywhere in the square. This set, *together with all its limiting points*, has been called the *limit* of the polygonal lines; it

* Some writers however only use Jordan curve for a set in continuous (1, 1)-correspondence with a closed segment, *i.e.* a Jordan curve without double points, *e.g.* Osgood, *Trans. Am. Math. Soc.* IV. (1903) p. 107 footnote. See below.[42]

must however not be supposed that, using the word in this sense, the form of the " limit " of a series of sets of points is in essentials reproduced by the form of the sets of points themselves.

The points of this limit can in our case however be described in order continuously, and we can imagine a point passing from point to point, while the length x on the straight line represents the time taken in passing. This idea, which seemed to Jordan the essential characteristic of a curve, has occasioned the use of the term " space-filling curve " for this limit regarded as arranged in this order.

It hardly seems that with our present advanced knowledge of the theory of sets of points this concept of Jordan's, which was at the time a great advance on the current concept of a curve, can be retained. Jordan's definition of a continuous line marks a stage in the development of the theory of curves, and leads to important consequences in the case when the correspondence is (1, 1), but the concept of a curve, as such, regarded as a set of points, must surely be recognised as a conglomeration of ideas of which that of order is only one, and is, from many points of view, a subservient one. A curve, like any other plane set of points, has a form, and it is in many respects this form which is its most interesting characteristic; this form is a property independent of order, and whereas it is a unique property, the order is one which is at least determinable in two ways.

In the present volume therefore the expression Jordan curve will never be used except with the implicit limitation "without double points," in which case the points of the curve do not fill up a region, the correspondence being (1, 1)*.

101. The mode of construction given by Peano has been modified in various ways since. In Peano's construction the points of one polygonal line in which it intersects the next correspond to certain fixed points on the straight line; in other words certain points of the "limit" are constructed at each stage and the polygonal line drawn through them. It is however only essential that the polygonal lines should at each stage represent the order of the squares, and they may be drawn as we please, provided they pass in order through the squares, it is not necessary that they should have any fixed or base points at all. Again, the order of the squares can be chosen in a variety of ways. Hilbert† gave a

* See p. 216.
† Hilbert, *Math. Ann.* xxxviii. p. 459.

correspondence in which there were no base points, the principle of which was afterwards used by Moore* from whose work Fig. 19 is reproduced. Moore's polygons have the interesting property that at every stage they enclose a simply connected region.

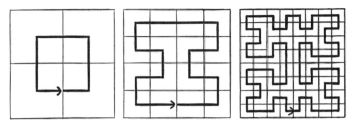

Fig. 19.

Moore's polygons, unlike Peano's, have no nodes; in spite of this however the so-called limit, or space-filling curve, derived from them, must be considered to have multiple points, since, however we bring the points of the square into continuous single-valued correspondence with the points of a segment, the correspondence cannot be (1, 1), and when we describe the points of the square in order we must from time to time return to the same point, which will for this reason be called a multiple point. We saw that this was the case with Peano's construction, and the fact that the approximating polygons have fixed nodes as base points rendered this fact clearer. In Moore's case, where there are no nodes and no fixed, or base points, this fact is only concealed, hence Osgood (who accords the term Jordan curve only to a plane set in (1, 1)-correspondence with a closed segment) denies the right to term the locus of the point moving in the plane according to the law of Hilbert or Moore a Jordan curve at all (the ground given above for refusing to use the term curve in this connection was a different one).

102. Continuous (1, 1)-correspondence between the points of the whole plane† and those of the interior of a circle of radius unity.

Let every point inside or on a concentric circle of radius $\frac{1}{2}$ be

---

* Moore, *Trans. of the Am. Math. Soc.* I. 1, p. 77.

† Regarded as unbounded, or disregarding the boundary, or with suitable convention as to the boundary.

transformed into itself, and any other point (r, θ) into a point (r', θ'), where

$$\theta = \theta',$$

$$r' - \tfrac{1}{2} = \frac{2r - 1}{4r - b},$$

b being an arbitrary constant less than 2. This transforms the whole plane in the manner required.

The fundamental region may therefore, as in the case of the straight line, be considered to be either the whole plane, or the interior of a circle, or of any closed curve whose interior can be suitably transformed into that of a circle. In general it will be assumed that the fundamental region is finite, and does not include the points of its boundary; the general discussion of the fundamental region must be deferred until § 123.

103. A set of points in a plane is a collection of points of the plane, defined by such a law that (1) any point of the plane either belongs to the set or not, but not both, nor neither, (2) assuming we know sufficient of the characteristics of our points, the law enables us to determine whether or no a given point belongs to the set, (3) having found any number or collection of points of the set, it is always possible to find more, if there are any more.

It is clear that the definition is precisely equivalent to the following :—

A set of points in the plane is a collection of points (x_1, x_2), such that the coordinates x_1 form a set of numbers, and so do the coordinates x_2.

The section of a set of points by a straight line is then a linear set of points, and so is the projection of a set on to a straight line, in which case, in general, one point of the projection will correspond to more than one point of the original set.

104. Most of the terminology already used is of general application in the plane or n-dimensional space, and requires no separate discussion. The proofs of theorems require usually trifling verbal alterations, as circle or sphere or n-dimensional sphere for interval. A point L is said to be *a limiting point of a set,* if inside every circle with L as centre there is a point of the set other than L, if L is a point of the set. The terms *isolated points of a set,* and *isolated set* then have the same meanings as on p. 18. A *sequence* is a set with one and only one limiting point, a finite set having clearly no limiting point. A set which consists entirely of limiting points is said to be *dense in itself*. A set which contains all its limiting points is said to be *closed*, otherwise *open*. It is then easily proved that *both the projection and the section of a closed set are closed sets*. A set which is both closed and dense in itself is called *perfect*. A set is said to be *dense everywhere* in the fundamental region, if every circle lying in the fundamental region contains points of the set in its interior; a set is said to be *dense nowhere*, if every such circle contains another circle entirely free of points of the set. It will be found that these definitions are perfectly consistent with the extended idea of a region which we shall subsequently develope, and that the circles may be replaced by regions of no such special form.

The rational points of the plane are those points both of whose coordinates are rational. They form an open set which is dense in itself, dense everywhere in the plane, and, by Theorem 3, p. 35, has the potency a.

A perfect set may clearly be dense nowhere, and it is easy to construct such perfect sets, for instance as follows.[9]

Ex. 1. Let the fundamental region be a circle of radius unity. Take a circle, and divide its circumference into three equal parts. Divide each of these parts, as in Cantor's example p. 20, into three equal parts, and blacken the middle part, and then divide each of the two segments in each of the three parts similarly and so on. The set of points on the circle which are not internal to the blackened arcs, constitutes a plane set of points which is perfect and dense nowhere (Fig. 20).

This could, of course, have been got by transformation from the straight line. The following is a more typical plane set.

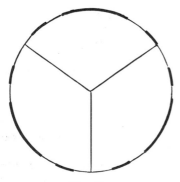

Fig. 20.

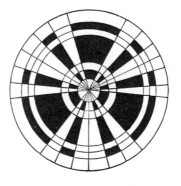

Fig. 21.

Ex. 2.   Let the fundamental region be a circle of radius unity.   Draw a circle of radius ·1 and another of radius ·2, using the ternary notation. Blacken arcs of the circumference and the part of the ring between them as in the preceding example, leaving on the circumference of every concentric circle between them a perfect set dense nowhere.   Then draw circles of radii ·01 and ·02, and blacken the parts between arcs of their circumferences as before, and similarly with circles of radii ·11 and ·12, and so on, using the principle of Cantor's example not only for the radii of the circles but also for the construction of the points of the set lying on the circles.   The set consisting of all the points except the blackened parts is a plane perfect set, dense nowhere (Fig. 21).

Ex. 3.   Instead of subdividing the circumferences of the circles, constructed as in the preceding example, we might blacken only the rings between the circles (Fig. 22), the plane set would still be perfect and dense nowhere in the plane, although on each circle the set would be from the linear or rather circular point of view dense everywhere. Or, we might take on the circles any perfect sets we please.

Fig. 22.

It is clear that we may construct in similar ways sets of points which are dense nowhere in the plane, and which may or may not contain components which are dense everywhere on straight lines or other known curves.   The next example is of a plane set which is not only dense nowhere in the plane, but also on every straight line or curve; it is of historic interest as connected with one of the first attempts to deal practically with some of the less obvious problems of plane sets of points nowhere dense*.

Ex. 4.   Take a unit square, and let e_1, e_2, ... be any set of positive quantities each less than 1.   Cut off from each corner of the unit square a square of side $\frac{1}{2}e_1$, and blacken the cross left in the centre (Fig. 23).   From each corner of each of the four white squares cut off a square of side $\frac{1}{4}e_1e_2$, and so on, ad infinitum.   The set of points inside and on the perimeter of the square, but not internal to any of the blackened crosses, is a plane perfect set, as is easily shewn, dense nowhere in the plane, and whose section on any straight line is also dense nowhere.

Fig. 23 shews the second stage of the construction when each of the e's is equal to $\frac{2}{3}$.   The set so obtained is easily seen to be the same as if we had taken

* Veltmann (*Zeitschrift für Math.* xxvii.) constructed the countable set consisting of all the corners of the crosses in Ex. 4, as a countable plane set which could not be enclosed in a finite number of regions of content as small as we please.

on the axis of x Cantor's perfect set of content zero and on each of the ordinates through points of this set had constructed a similar linear perfect set.

Fig. 23.

In the chapter on Content, we shall see that the content of this perfect set is the difference between the area of the square and that of the blackened region, the latter being calculated as the sum of the areas of all the crosses composing it. It is therefore convenient for reference to calculate here the area of the black region. Denoting this by I_d, we have, since the area of the first black cross is $1 - e_1^2$,

$$I_d = 1 - e_1^2 + e_1^2 (1 - e_2^2) + e_1^2 e_2^2 (1 - e_3^2) + \ldots = 1 - \operatorname*{Lt}_{n = \infty} e_1^2 e_2^2 e_3^2 \ldots e_n^2.$$

In Fig. 23, therefore, the content of the black region is unity, that is, it is the same as that of the unit square, but if we choose the quantities e_1, e_2, ... suitably, it is clear that we can make the area of the black region as small as we please.

Ex. 5. In like manner if we take any linear plane perfect set dense nowhere on the axis of x, and place on each of its ordinates a similar set, we get a perfect plane set dense nowhere in the plane, whose section, by any straight line, is also dense nowhere on that straight line. For instance, if we take the typical ternary perfect set of positive content we get such a set, the construction of which, by means of successive crosses, is given in Fig. 24. The area I_d of the black region in this example is less than that of the unit square, since it is clearly given by the following expression:—

$$I_d = 1 - 4\left(\frac{3-1}{2}\right)^2 \frac{1}{3^2} + 4\left(\frac{3-1}{2}\right)^2 \frac{1}{3^2}\left\{1 - 4\left(\frac{3^2-1}{2}\right)^2 \frac{1}{3^4}\right\} + \ldots$$

$$= 1 - \left[\left(1 - \frac{1}{3}\right)\left(1 - \frac{1}{3^2}\right)\left(1 - \frac{1}{3^3}\right)\cdots\right]^2.$$

Therefore

$$I_d < 1 - (\tfrac{5}{9})^2 < 1 - \tfrac{24}{80} < \tfrac{7}{10},$$

and

$$I_d > 1 - \tfrac{4}{9} \cdot \tfrac{64}{81} > 1 - \tfrac{1}{2} \tfrac{64}{80} > \tfrac{3}{5}.$$

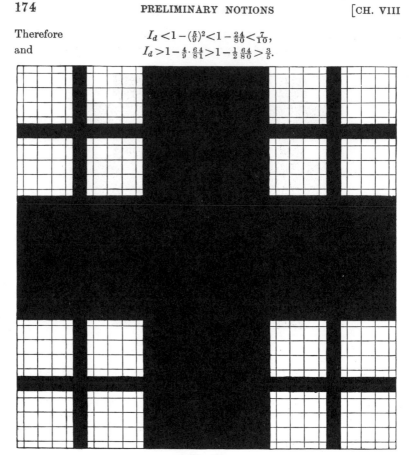

Fig. 24.

The quantity in brackets is the content of the typical ternary set of positive content on the straight line (Ex. 1, p. 78), that is the square of the linear content of the section of the plane perfect set by the axis of x or of y; therefore, its square is the content of the plane perfect set, so that we see that here the content of the plane set has the same relation to its section as the area of the square to its side, the latter being of course a special case of the former. This question will be completely discussed in the chapter on Content.

Ex. 6. Take a circle divided into four quadrants, and in each inscribe a circle. Blacken the parts of the first

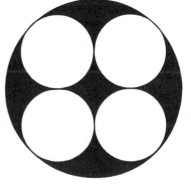

Fig. 25.

circle external to these four circles.   In each of the four small circles perform a similar construction, and so on.   The final set is perfect and dense nowhere in the plane, and on every curve.

105.   THEOREM 1.   *If G be a set of points in the plane and G have a limiting point L, we can find a sequence of points of G having L as sole limiting point.*

For let P_1 be any point of the set.   Describe a circle C_1 with L as centre passing through P_1.   Describe a concentric circle C_1', of half the radius of C_1.   Then since L is a limiting point of the set G, there are certainly points of G inside the circle C_1', and these form a set of points, so that we can assign one of them P_2. Describe a circle C_2, with L as centre, passing through P_2, and a concentric circle C_2' of half the radius of C_2.   As before we can assign a point P_3 of G inside C_2'.   Proceeding thus we get a countably infinite set of points of G,

$$P_1, P_2, P_3, \ldots P_n, \ldots$$

whose distances from L decrease without limit as n is indefinitely increased.   If Q be any point not one of the P's, nor L, there will be a definite value k_1 of k, such that Q lies outside the circle C_{k_1}, but not outside any circle C_k for values of k less than k_1.

Now there are only a finite number of the points P outside the circle C_{k_1}, and we can determine r so that a circle with centre Q and radius r lies entirely outside the circle C_{k_1}, and contains therefore only a finite number of the points P: thus Q cannot be a limiting point of the points P.   Therefore the points P form a sequence having L as sole and only limiting point.

106.   Since the distance between two points is a positive quantity, it follows that, if G_1 and G_2 are two sets, the distances of the points of G_1 from those of G_2 have a lower limit (§ 16), which is positive or zero; this lower limit is called *the minimum distance* of G_1 from G_2†.

THEOREM 2.   *If G_1 and G_2 are closed sets, the minimum distance e between G_1 and G_2 is zero if, and only if, G_1 and G_2 have a common component.   If e is not zero there is at least one pair of points P_1 from G_1 and P_2 from G_2 such that the distance $P_1 P_2$ is e.*

Let $e_1, e_2, \ldots$ be a sequence of continually decreasing positive quantities having e as limit.   Let $G_{1,i}$ be the component of G_1

† Jordan, *Cours d'Anal.* i. § 29.

consisting of all those points of G_1 whose minimum distances of $G_{1,i}$ from G_2 are $\leqslant e_i$; then, since e is the lower limit of the distances of points of G_1 and G_2, $G_{1,i}$ certainly contains points for all values of i. Also, if Q be any point not belonging to $G_{1,i}$, and k be any positive quantity less than the difference between e_i and the minimum distance of Q from G_2, there can be no point of $G_{1,i}$ within a distance k of Q. Therefore Q cannot be a limiting point of $G_{1,i}$, whence, G_1 being closed, it follows that $G_{1,i}$ is a closed set. Further $G_{1,i}$ contains $G_{1,(i-1)}$ for all values of i.

Let us project on to a straight line, then the projections of $G_{1,1}$, $G_{1,2}$, ... are closed sets (§ 104) and each is contained in the preceding set. Therefore, by Cantor's Theorem of Deduction (p. 26), these sets have in common a closed set of points. If L be any one of these points, the projecting line through L contains points of each of the sets $G_{1,i}$, and these sections are closed sets (§ 104), each lying in the preceding one, and have, therefore, again a deduced set. If P be any point of this deduced set, it is a point of $G_{1,i}$ for all values of i, therefore its minimum distance from G_2 is $\geqslant e$ but $\leqslant e_i$ for all values of i, and is therefore e.

If e is zero, it follows that P is a limiting point of G_2, and therefore, G_2 being closed, a common point of G_1 and G_2.

If e is not zero we can in like manner, since the point P is itself a closed set, determine from it a point P' of G_2, whose distance from P (minimum distance) is e, which proves the theorem.

CHAPTER IX.

REGIONS AND SETS OF REGIONS*.

107. Before proceeding to discuss further extensions of theorems proved true for the straight line such as Cantor's Theorem, p. 38, or the Heine-Borel Theorem, it is desirable to explain fully what is meant by *a region* of the plane or higher space.

In the straight line we have only one possible linear element, the small interval or segment. If we lay such elements on one another, side by side, or overlapping, or move such an element about, it generates only one form of region, the larger interval or segment. The common part of two such segments that overlap is a segment, the parts left over in each are again segments. This makes the whole theory of regions on the straight line a comparatively easy one.

In the plane, and still more in higher space, this ceases at once to be the case; there is no single type of plane element which takes precisely the place of the linear segment, to the exclusion of all others, in fact the idea of *form* is one which first occurs in connection with space of more than one dimension, and is of fundamental importance in the classification and recognition of plane sets of points. In the plane the linear element itself exists, but it may be either straight or curved, and if we lay such linear elements end to end, or overlapping, or move such an element about, we generate *stretches* (that is segments of straight lines), or polygons, or curves of various forms, entities which at once have the peculiarities of form already referred to, and do not reproduce on a larger scale the form of the element with which we started. On the other hand, if we start with a small piece of the plane, this may have a great variety of forms, it may be a triangle or a circle, or a figure of more complicated character, and as we move it about, or lay such elements like tiles one overlapping with another, we generate parts of the plane of the

* *Quart. J. of Math.* No. 145 (1905).

most various and peculiar forms, having apparently no special reference to the form of the element with which we started; any such part of the plane will naturally be called *a region*, and it appears as the natural generalisation in two dimensions of the segment in one dimension. It has not, however, the definiteness of the segment, since it permits great varieties of form; in consequence it will be found that it does not possess properties analogous to all those proved for the linear segment. To arrive at these properties, it is necessary to start, for definiteness, with some special form of region, and we shall start with *the triangle**, whose properties, considered as an area, may be supposed known from elementary Geometry, and may be shortly summed up in the following facts:—

(1) A point *inside* a triangle or *internal to a triangular region*, is defined to be such that it lies on the same side of any one of the three straight lines forming the triangle as the intersection of the other two. The set of all the points inside a triangle is called a *triangular domain*, or *the interior of the triangle*, and is a simple case of a region. The periphery of the triangle is said to form *the boundary* of this region. Any point of the plane which is neither internal to the triangle nor on the boundary is said to be an *ordinary external point* of the triangle.

(2) The region defined in (1) is said to have no *edge*; we may however consider this region together with some or all of its boundary points; this also is said to constitute *a triangular region*, which has the same internal points and the same boundary as before; the points of the boundary which are considered as belonging to the triangular region are called its *edge-points*, and constitute its *edge*. If the edge consists of the whole periphery of the triangle, the region is called *a closed triangular region*, its points then constitute a closed set of points, as is easily proved; otherwise the region is said to be *an open triangular region*. An open triangular region may have no edge, it is then said to be a *completely open region*, or a *domain*; but it may have an edge, and the edge is not then necessarily an open set of points, *e.g.* the edge might consist of a finite number of points on the periphery.

The boundary points of an open triangular region which are not edge-points are considered as external to the region, and are called *external boundary points*, they are clearly limiting points of the points of the region. The external points of an open triangular

* Hilbert, *Grundl. d. Geom.* § 4.

region consist therefore of ordinary external points and external boundary points.

(3)   Any internal point of a triangular region can be joined* to any other internal point by a *stretch* (linear segment), whose points are all internal to the region.   Any ordinary external point of the triangle can on the other hand be joined to any other by a simple polygonal path†, whose points are all ordinary external points of the triangle.   Finally if any internal point be joined to an ordinary external point by means of a simple polygonal path, there will be some of the points of the path which are internal and some which are ordinary external points of the triangle, and at least one which is a boundary-point of the triangular region.

108.   A set of intervals on a straight line is in a certain sense a case of a two-dimensional set, since an interval requires two numbers to characterise it, for instance its two end-points, or its middle-point and its length ; regarded however as a set of points on the straight line, a set of intervals possesses the simple property that the intervals over which the points are distributed may always be considered as countable (p. 38); a similar property belongs to a set of triangles, although a triangle, as such, requires six coordinates to determine it, for instance, the coordinates of its three vertices, so that a set of triangles, as such, is a special case of a six-dimensional set ; regarded however as a plane set of points, the triangles over which the points are distributed may always be considered to be countable, as is formally proved below.   The triangles which will most frequently be used for this purpose, and which will be termed *primitive triangles*‡, are those whose vertices are the rational points. Should there be any reason for doing so, we may take, instead of the rational points, the points of any countable set dense everywhere in the plane.   In space of three, or n, dimensions, we may define and use in like manner *primitive tetrahedra*, or *primitive $\overline{n+1}$-hedra*.

Denoting the vertices of any primitive triangle by P_i, P_j, P_k, ($i < j < k$) it follows by Theorem 3, Cor., p. 36 that *the primitive triangles are countable*.

---

* Hilbert, *loc. cit.* § 2.

† That is a finite number of stretches (linear segments) AB, BC, ..., KL, placed end to end, such that the vertices A, B, ..., K, L are all distinct and no internal point of one of the stretches lies on another of them.   If all these conditions are fulfilled except that A and L are identical, the figure is called *a simple polygon* (triangle, quadrangle, pentagon, ...).   Hilbert, *loc. cit.* § 4.

‡ "A Note on Sets of Overlapping Intervals," *Rend. d. Circ. Mat. di Palermo*, Nov. 1905.   See also footnote, p. 40.

THEOREM 1. *Given any set of triangles, overlapping in any way, a set of primitive triangles can be assigned, having the same internal points as the given set.*

Let P be any internal point of one of the triangles ABC. Join AP, BP, CP, and produce them to meet BC, CA, AB in L, M, N. Let the rational points, arranged in countable order, be R_1, R_2, ...; since they are dense everywhere there will be some of them inside any triangle ; let R_i, R_j, R_k be the first lying inside the triangles APM, BPN, CPL respectively (Fig. 26). Then it is easily shewn by elementary geometry that the point P lies inside the triangle $R_iR_jR_k$ which, on the other hand, itself lies inside the triangle ABC. It follows that the set of primitive triangles, each of which has its three vertices internal to some one of the given triangles, contains as internal points every point which is internal to at least one of the triangles, and contains no other internal points, which proves the theorem.

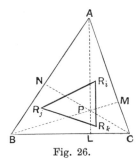

Fig. 26.

COR. *Given any set of triangles, overlapping in any way, a countably infinite set of the triangles can be chosen, having the same internal points as the given set.*

109. Given a set of triangles, whose equivalent primitive triangles are d_1, d_2, ..., it may be that we can find a proper component of this set, d_{i_1}, d_{i_2}, ..., such that no triangle of this component overlaps with any but triangles of this component. If so the given set is said to be *intransitive,* otherwise *transitive.*

A set of triangles is said to *tile over* the part of the plane consisting of all the internal points of the triangles.

DEF. A part of the plane which can be tiled over by a transitive set of triangles is called a *domain,* or a *completely open region* of the plane.[43]

It is shewn below (Theorem 2) that the points of a domain always form an open set. *The most general form of region consists of a domain together with some or all of its non-included limiting points.* If all the limiting points are included, the region is said to be *closed,* otherwise *open.*

It follows now from Theorem 1 that we may, whenever convenient, assume that the triangles generating a region are countable, and, indeed, that they are primitive triangles. Every point of a region is a limiting point of points of that region, thus a region is a special case of a plane set of points dense in itself. From the definition it follows that *a closed region is a closed set of points*, hence *a closed region is a special case of a plane perfect set.*

110. A region divides the points of the plane into three classes :—

(1) *Internal points of the region, i.e.* points internal to at least one generating triangle.

(2) *Boundary points of the region, i.e.* points other than internal points, yet limiting points of the internal points of the region.

(3) *External points of the region* which are neither internal nor boundary points.

The set of all the boundary points is called the *boundary* of the region. The boundary points which are defined as belonging to the region are called *edge points* of the region and the whole set of them is called *the edge*; these may be absent, in which case the region is a domain.

111. It is clear from the known properties of the triangle that an internal point of a region is such that we can describe a circle round it as centre containing only internal points of the region inside it and on its circumference, such a circle is said to *lie entirely in the region.* An external point P, in like manner, is such that corresponding to each generating triangle we can describe a circle with P as centre containing inside it and on its circumference only points which are external to that triangle. Corresponding to each triangle there will be a number of these concentric circular regions, which, taken all together, form a circular domain. Let us take this circular domain corresponding to each triangle; if the radii of these circular domains decrease without limit, it is clear that the centre P and no other point is internal to all the circles; P is then said to be an *external boundary point* of the region. Otherwise the point is said to be an *ordinary external point,* in this case it is easily seen that the common points of the circular domains consist of all the points inside a concentric circle, with possibly some or all of the points of the circumference also *. Thus an ordinary external point is such

---

* This may for instance be proved by taking a straight line through the centre and determining the interval which, with or without its end points, is common to the sections of the circles by the line; rotating this common interval about the centre it generates the circle in question.

that we can describe a circle with the point as centre, every point inside or on the circumference of this circle being an ordinary external point of the region, such a circle is said to *lie entirely outside the region*.

112. In defining a region the edge points must be in some way specified. It is sometimes, but not always, possible to define the region completely by means of generating triangles, even when it is closed, the edge points being defined as those points of the boundaries of the defining triangles which do not become internal points of the region. The region is then said to be *described* by the defining triangles. If however, for instance, the boundary be a curve with a cusp pointing outwards, whether ordinary or rhamphoid, the cusp is external to the defining triangles, in whatever mode the region be described.

A region may often be described in different ways, but the triangles used must be so constructed that in each case they have the same internal and external points, they will then have the same edge points; if this is not true the two sets of triangles will not be considered to describe the same region.

Two regions which have the same internal points are not necessarily identical, for instance the whole of the inside of a circle is one region; it has no edge points; the points of the circumference of the circle, as well as every point outside the circle, are external to this region. Another region is the whole of the inside of the circle with the circumference, the internal points are the same as before, the edge points are the points of the circumference, the remaining points are the external points. These two regions are distinct. The former may be described as follows:—Take three radii of the circle, separated by angles of 120°, divide each of these radii similarly, by continued bisection, at the points distant $\frac{1}{2}, \frac{3}{4}, \frac{7}{8}, \ldots$ of their length from the centre, and join corresponding points to form equilateral triangles; rotate each of these triangles round the centre, until it returns to its original position; the triangles describe the region. The second region can be described by an equilateral triangle inscribed in the circle, rotated round the centre back to its original position.

By Theorem 1, if we are only concerned with the internal points of the region, we can choose out a countable set of the generating triangles which suffice to cover the whole domain of the region; we may in so doing have converted some edge points into external points.

Ex. 1. A region being described by means of an equilateral triangle which revolves round its centre of gravity, the region is a closed circular area. Since the points on the circumference of the circle can be brought by projection into (1, 1)-correspondence with a segment of the straight line, they are of potency c; hence no countable set of the triangles (which are such that one and only one goes through such a point) can include every point of the circumference of the region. When a countable set of the triangles have been chosen, covering the whole domain, their vertices will form a countable set of points on the circumference, dense everywhere in the circle, that is such that every other point of the circle is a limiting point of this set, and therefore an external boundary point of the region described by the chosen triangles.

Ex. 2. Take a square $ABCD$. At every point of a perfect set dense nowhere in AB, erect a perpendicular half as high as the square. These lines, together with the periphery of the square, form the boundary of a region or rather of various regions all having the same internal points. We can describe such a region so as to have no edge, or so that it has edge points ; but however we describe it, we cannot avoid external limiting points. If we describe it so as to leave as few external limiting points as possible, the edge will consist of (1) BC, CD, DA, (2) the black intervals of the perfect set on AB, (3) the whole of the perpendiculars through the extremities of each black interval, and (4) the tops of the remaining perpendiculars*.

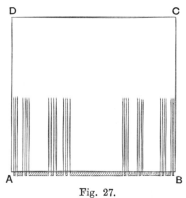

Fig. 27.

113. We can now prove the fundamental property of the internal points of a region.

Two points are said to be *joined* by a set of triangles, when that set is transitive, and has both the points as internal points.

THEOREM 2. *If P and Q are any two internal points of a region, P and Q can be joined by a finite number of the generating triangles.*

Let d_1, d_2, ... be the primitive triangles having as internal points all the internal points of the region. Let P and Q be any two internal points : then either both P and Q are internal points of the same primitive generating triangle or not. In the

* Osgood, *Trans. of the Amer. Math. Soc.* (1900) pp. 310 *seq.*

latter case we take any one of the triangles d_i in which P lies and one d_j in which Q lies.   Let n be any integer which is not less than either i or j.   Construct all the primitive generating triangles of indices less than or equal to n. P and Q are known to be internal points of these triangles, since they lie in d_i and d_j. These n primitive triangles generate one or more non-overlapping regions.   If there is only one region, or if P and Q are internal to the same region, then P and Q are joined by a finite number of triangles.   If, however, for every value of n, P and Q lie in different regions, let D_n denote the transitive set of triangles, whose indices are less than or equal to n, generating the region containing P but not Q. D_n does not then contain d_j; also no triangle of D_n overlaps with a triangle of index less than or equal to n not belonging to D_n. The countable set D of the triangles D_n, for every value of n, would then be a proper component of the original set of primitive triangles, since it would not contain d_j; and no triangle of D would overlap with a triangle not belonging to D.   This is however impossible, since the original set of primitive triangles was transitive.   It follows that at some stage the two regions containing P and Q will have merged into one, so that P and Q can be joined by a finite number of generating triangles.   Q. E. D.

COR.   *Two internal points of a region can be joined by means of a finite number of stretches (linear segments) forming a polygonal path, every point of the stretches being an internal point of the region*.[44]

114.   It is interesting to remark that, although at a finite stage the regions containing P and Q in the preceding proof merge into one, we cannot assign any finite stage at which all the regions formed by the first n triangles must merge into one.   In other words, if round every point of the boundary we describe a circle of radius e, it may happen that, no matter how small e is, the part of the given region external to the circles does not form one single region.   The following is a simple example where this is the case.

Ex. 3.   **The Chow.**   Take a unit square, and divide its lowest side at the points $\frac{1}{2}$, $\frac{1}{4}$, ... $\frac{1}{2^n}$ ......, or in the binary notation, $\cdot1$, $\cdot01$, $\cdot001$, etc.   Mark above each of these points the point whose height is $\frac{1}{4}$ of the distance to the next point, *e.g.* the points ($\cdot1$, $\cdot0001$), ($\cdot001$, $\cdot00001$).   Join these latter points to the points of the top of the square vertically over the middle points of the intervals, as in Fig. 28, forming a zigzag line joining the top right-hand corner of the square to the bottom left-hand corner as limiting point of the zigzag. The part of the square below this zigzag, shaded in the figure, constitutes a region, the "Chow's body," with the bottom and right-hand edges of the square and the zigzag line as edge. Every point of the left-hand side of the square is an external boundary point, and together these form the closely adhering "Chow's tail."

It is easy to see that, if we describe a circle of radius e, as small as we please round every point of the boundary of this region, some of the lowest corners of the zigzag will each give a circle which overlaps with a circle round one of the points of the base of the square ; thus a triangular portion of the region will be isolated from the part of the region lying on the right and left of it, so that the region will always be divided into at least two regions.

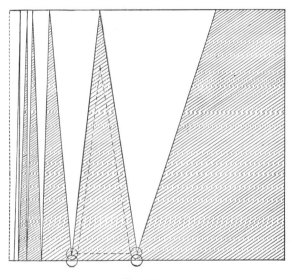

Fig. 28.

115. It is to be noticed that a boundary point is not necessarily such that in any circle with that point as centre there are, apart from the point itself, both internal and external points of the region in the circle. Take, for instance, a triangle revolving round one of its corners ; the region described is a circular area with the centre C as one of its edge-points. But a small circle with C as centre contains no points other than internal points, except C itself.

DEF. A boundary point which has the property that in every circle with the point as centre there are ordinary external points, will be called a *rim point,* and the set of all the rim points will be called the *rim.*

THEOREM 3. *Any region which has ordinary external points has rim points, and there is at least one rim point on every straight line joining an internal point to an ordinary external point.*

(In particular the above is true for any region contained in a finite rectangle.)

For a region has at least one generating triangle, and therefore it always has internal points; let P be an internal point. If Q is an ordinary external point, we can draw a circle with Q as centre, containing only ordinary external points; let this circle cut the straight line PQ, between P and Q, in Q_1; blacken the interval QQ_1. Then every point of the black interval, including its end-points, is an ordinary external point.

Fig. 29.

Similarly every ordinary external point between P and Q is internal to such a black interval. Thus by Theorem 8, p. 41, all the ordinary external points in the stretch PQ fill up a set of non-over-lapping intervals. Let that one of these intervals of which QQ_1 is a part have its other end at L. Then L is not an ordinary external point of the region, since it is not internal to the interval of which QQ_1 forms a part. Neither is it an internal point, since any circle with L as centre contains points of the interval QL, and these are ordinary external points of the region. Thus L, not being either an internal or an ordinary external point, must be a boundary point. Further, since in any circle with L as centre there are ordinary external points, L is a rim point, which proves the theorem.

COR. *There is a rim point on every polygonal line, consisting of a finite number of linear segments placed end to end, and leading from an internal to an ordinary external point of the region.*

The proof of this is of precisely the same character as that of the theorem.

116. THEOREM 4. *Any straight line which meets a linear segment terminated by two internal points of a region cuts from the region at least one interval.*

Given any two points P and Q of a region, they are either internal to one generating triangle, or each is internal to a different triangle. In the latter case, by Theorem 2, the two points can be joined by a finite series of overlapping triangles. If I be any straight line which cuts the linear segment PQ, then I divides the plane into two parts, in one of which P lies and in the other Q; hence I certainly cuts one of the series of triangles above referred to, and therefore cuts from the region at least one segment.

COR. *Any simple polygon* which separates two internal points P and Q of a region from one another contains at least one interval on its perimeter, consisting entirely of internal points of the region.*

The proof is of precisely the same nature as that of Theorem 4.

THEOREM 5.   *A straight line which contains any internal point of a region is such that the internal points of the region on it form a set of non-overlapping intervals.*

This follows at once from the fact that the section of a triangle is an interval, since, by Theorem 8, p. 41, a set of overlapping intervals has one and only one equivalent set of non-overlapping intervals.

COR.   *If the region be a closed region the points of the region on the straight line form a closed set consisting of a set of intervals with their limiting points.*

117.   DEF.[10] If a closed region can be entirely enclosed in a strip of width e, bounded by two straight lines perpendicular to a certain direction L, we shall say that the *span* of the region in the direction L is less than e; the lower limit s_L of these widths is said to be the *span of the region in the direction L.*

The upper limit of the spans in every direction is called *the span of the region.*

By the theory of linear sets a strip can be drawn with sides perpendicular to L of width equal to the span of the region in the direction L, containing inside it every internal point of the region, and having on each of its bounding lines at least one boundary point. Thus if A and B are any two internal points of the region, and x_{AB} the projection of AB on L, it is clear that x_{AB}, though always less than s_L, can, by properly choosing A and B, be made as near as we please to s_L. Thus s_L is the upper limit of x_{AB}. This gives us the following alternative form of the definition of the span: *The span is the upper limit of the length of the distance between any two internal points of the region.*

* See footnote †, p. 179.

118. The expression perpendicular (parallel) to the strip is to be understood as a contraction for perpendicular (parallel) to the straight lines bounding the strip.

THEOREM 6. *If such a strip be drawn, whose width is the span of a region in the direction perpendicular to the strip, any straight line parallel to the strip and inside the strip cuts from the region at least one interval.*

For let P and Q be two points on the bounding lines of the strip which are either edge points or external limiting points of the region. With P and Q as centres describe two circles small enough not to cut the straight line in question. Then inside these circles there are internal points of the region; hence by Theorem 4 the result follows.

DEF. If the section of a region by every straight line that meets it is a single interval, the region is called a *disc*.

The expression disc will be used in the plane or in higher space; if there could be any ambiguity, we might say a *disc of n-dimensions*.

A circle (sphere), triangle ($\overline{n+1}$-hedron), square (cube), rectangle (parallelepiped) are all discs.

119. THEOREM 7. *If $R_1, R_2, \ldots$ be a countably infinite series of closed regions, each lying entirely within* the preceding, and if the span of R_n decrease indefinitely as n increases, there is one and only one point internal to all the regions.*

First let the regions be discs.

Let $e_1, e_2, \ldots$ be the spans of the discs in the direction XX'. Determine strips of widths $e_1, e_2, \ldots$ perpendicular to XX', such that every internal point of R_n is internal to the nth strip. The sections of these strips by XX' are a series of segments, each lying inside the preceding, and their lengths decreasing indefinitely; hence, by Theorem 1, p. 17, there is one and only one point internal to all these segments; by Theorem 6 the perpendicular to XX' through this point meets every one of the discs in a segment, and these segments lie again one inside the other and their lengths decrease indefinitely, since the span of R_n in this direction decreases indefinitely. These segments determine one and only one point L internal to them all, and therefore internal to the

---

* *I.e.* the points of each region, including all its boundary points, are internal points of the preceding region.

discs. It is clear that no second point M could be internal to the discs, for otherwise the projections of the straight line LM in two perpendicular directions could not both be zero, and therefore the spans could not decrease indefinitely. This proves the theorem for discs.

Next let the regions not be discs. Draw the strips perpendicular to XX' as before, and draw similar strips parallel to XX', whose widths are the spans in the direction perpendicular to XX'. Thus we get corresponding to each region a rectangle, the common part of the two corresponding strips, and these rectangles lie one inside the other, and their spans decrease indefinitely. Hence, by what has just been proved, there is one and only one point L internal to all the rectangles. This point L is such that every circle with L as centre contains one of the rectangles, and therefore all subsequent rectangles, since the span decreases indefinitely; therefore any circle with L as centre contains internal points of one of the regions, and of all subsequent regions; thus, the regions being closed, L is not external to any one of these regions, and is therefore internal to all preceding regions, that is to all the regions; which proves the theorem.

COR. 1. *If the regions are not closed, they still determine one and only one point L, which, if not internal to all the regions, is a boundary point of every region after a certain stage.*

COR. 2. *If only the internal points of each region are known to be internal to the preceding region, the result is the same as in Cor.* 1.

COR. 3. *If the span in one direction only diminish without limit, the points common to all the regions form a stretch (linear segment) in the perpendicular direction.*

For, as before, if XX' be the direction in which the span decreases without limit, we determine a single straight line YY' perpendicular to XX' which cuts from each of the rectangles a segment: these segments do not now decrease without limit, and determine therefore a closed stretch LL' consisting of all the points internal to them all (Fig. 30).

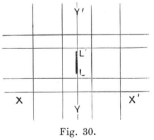

Fig. 30.

If P be any point of LL', the parallel to XX' through P meets each of the regions, by Theorem 6, and therefore contains a point

P_1 of the first region, lying in an interval of length e_1 containing P. Similarly there is a point P_2 of the second region (and therefore of the first region), lying in an interval of length e_2 containing P, and so on.   Since the spans $e_1, e_2, \ldots$ in the direction XX' decrease without limit, this shews that P is a point of the first region since it is closed, similarly it is a point of the second, and therefore an internal point of the first region and similarly of every one of the regions.   Thus the whole stretch LL' is internal to all the regions.

It is clear that no point external to LL' is internal to all the regions.   For if there were such a point it would either lie on YY' or not; it cannot lie on YY' since only the points of LL' are internal to the sections of all the rectangles, and it cannot be off YY' since otherwise the span in the direction XX' would not decrease without limit.   Thus the Corollary is proved.

If the regions are not closed, or if it is only known that the internal points of each are internal to the preceding region, the regions still determine a stretch, every point of which is an internal or boundary point of the regions.

It will be shewn subsequently that if the span in two directions do not decrease without limit the points common to the regions form a region or a curve, or a combination of regions and curves with limiting points.   The following is however a special case which can be treated without any preliminary theorem.

COR. 4.   *If the regions are circles, and the span does not decrease indefinitely, the common points consist of all the points internal to a certain circle, including, or not including, the circumference.*

For, since the span of a circle in every direction is the same and equal to twice the radius, the limiting interval, constructed as in the proof of the preceding theorem, will always be of the same length $2s$, less than the diameters of all the circles, with the possible exception of one, to which it may be equal.   Omitting this latter circle, if it exists, let us diminish the radius of each circle by s, without altering its centre; these new circles will then continue to lie inside one another and will therefore, by Theorem 7, have one and only one common internal point; if this be C it is clearly the centre of the omitted circle and internal to all the remaining circles, and its distance from the nearest point of each of the given circles is greater than or equal to s, so that the whole circle with C as centre and s as radius lies in every one of the given circles. No point outside this circle could be internal to all the given circles, as otherwise the limit of the span in at least one direction would be greater than $2s$. This proves the corollary.

120. THEOREM 8. (WEIERSTRASS'S THEOREM*.) *Any plane set of points, not merely finite in number, has at least one limiting point.*

It is only necessary to prove this theorem for the case when the points all lie in a finite square; the result follows in the general case by correspondence.

Divide the square, in which all the points lie, into four equal squares, by straight lines parallel to the sides of the square, and number them 1, 2, 3, 4, beginning with the top left-hand corner, and proceeding clockwise. Then in at least one of the squares there must be an infinite number of points of the given set. Take the first of the four squares having this property, and proceed with it as with the larger square. This process can be continued *ad infinitum*, since at every stage one of the smaller squares must contain an infinite number of the given points. We thus get a series of squares the internal points of each contained in the preceding and the span decreasing each time by one-half. By Theorem 7, Cor. 2, these determine one point L, internal to all of them, or a common boundary point of them all after a certain stage. L is clearly such that in any circle with L as centre there is one of the defining squares, and therefore there are points of the set in the circle, therefore L is a limiting point of the set, which proves the theorem.

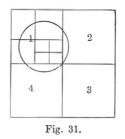

Fig. 31.

This theorem is the basis of the theory of the derived and deduced sets; a theory which is to all intents and purposes the same in linear and in higher space; it enables us to prove the existence of a first derived set for any set which is not a finite set, and it is easily shewn that the first derived set is closed. Cantor's Theorem of Deduction follows, the proof being of the same nature

---

* This name, by which this theorem is known to many mathematicians, is probably due to Weierstrass's use of the theorem in his memoir *Zur Theorie der eindeutigen analytischen Functionen*, *Abh. aus d. Functionentheorie*, 1886, p. 3. The theorem is, however, much older, and seems (at least for countable sets) to have been known to Cauchy. The method of proof is due to Bolzano, *Paradoxien des Unendlichen*, Leipzig, 1851.[45]

as that given for the straight line on p. 26.   From these facts the
whole theory can be developed as in the straight line.

121.   In what follows it will be seen that a close connection
exists between regions, or sets of regions, and closed sets of
points.

THEOREM 9.   *If R_1, R_2, ... be closed regions lying each inside
the preceding, and there be a point L internal to all the regions such
that a circle can be drawn with L as centre containing no other
point common to all the regions, the span of the regions decreases
without limit.*

For let e be any small quantity smaller than the radius of the
circle in question, and describe with L as centre a circle of radius e.
Then inside this circle and on its circumference there is no point
except L common to all the regions.   If we draw a concentric circle
of radius $\frac{1}{2}e$, there is no point common to all the regions in the ring
between these two circles, including their circumferences.   Thus,
if inside the ring there is a point P belonging
to one of the regions, we can assign an integer i,
such that P belongs to R_i, but not to R_{i+1}.
Drawing with P as centre any circle lying entirely inside the ring, there will then (whether,
or no, P is itself an internal point of R_i), be
inside this circle an internal point of R_i not
belonging to R_{i+1}, and therefore a primitive

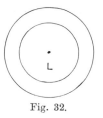

Fig. 32.

triangle consisting entirely of such points of R_i.   Let the primitive
triangles inside the ring be d_1, d_2, ... and take the first of these
which has this property; and let P_i be its centre of gravity.

Then either there is an upper limit m to the possible values of
i, for points inside the ring, or else, corresponding to each integer
i we get a point P_i.   By Weierstrass's Theorem the points P_1, P_2, ...
would then have a limiting point Q, which, since P_i is always
internal to the ring, must be a point inside, or on the boundary of,
the ring, and does not therefore belong to all the regions.   Therefore we can determine an integer k, such that Q is not a point of
R_k.   But this is impossible, since P_i is a point of R_k for all values
of i greater than k.   Thus there must be an upper limit m to the
possible values of i.   *That is, we can assign an integer m, such
that in this ring there is no point of R_m.*   Since however if L
be joined to any point outside the circle of radius e by means of
a finite number of linear segments, forming a polygon, this polygon

must cut the circle, and therefore contain a point not internal to R_m; it follows by Theorem 2, since L is an internal point of R_m, that all the internal points of R_m are inside the circle of radius e, so that the span of the regions decreases below e: since e may be as small as we please, this proves the theorem.

THEOREM 10. *If* R_1, R_2, ... *be regions lying each inside the preceding, and having two or more common points all lying on a straight line, the span in one and only one direction decreases without limit.*

This is at once evident, since, if XX' be perpendicular to the line of common points, the segments on XX' as in Theorem 7, Cor. 3 have only one point in common, so that the span in the direction XX' decreases without limit. On the other hand if P_1 and P_2 be two of the common points, the projection of P_1P_2 on any straight line except XX' is not a point, and therefore the span in no direction except XX' is zero.

THEOREM 11. *If* R_1, R_2, ... *be a countably infinite series of closed regions, each lying entirely within the preceding, and if the span in two directions does not decrease indefinitely as n increases, there are points common to all the regions, which do not all lie on any straight line, and they form a perfect set, such that, if P_1 and P_2 be any two points of the set, any straight line which meets the linear segment terminated by P_1 and P_2 contains at least one point of the set.*

For simplicity of expression we take the two directions to be perpendicular to one another; this is no restriction, only in the contrary case we must work with a parallelogram instead of a rectangle.

As in the proof of Theorem 8, construct rectangles whose sides are parallel to the given directions, the lengths of those sides being the spans in those directions, each rectangle containing all the points of one of the regions inside it and on its sides: then we know that, since the regions are closed, there will be at least one point of the region on each side of the corresponding rectangle. As in the proof of Theorem 7, Cor. 3, these rectangles determine a strip, every straight line in which cuts from each region at least one segment, the bounding lines of the strip being parallel to the first of the given directions. Similarly they determine such a strip whose sides are parallel to the second of the given directions. Let $ABCD$ be the rectangle common to these two strips. Then $ABCD$

is clearly composed of all the points common to all the rectangles; therefore all the common points of R_1, R_2, ... must lie in or on the sides of $ABCD$, and it is easily seen, since there is at least one point of each region on each side of the corresponding rectangle, that there is at least one of the common points on each of the sides of $ABCD$: this is a consequence of Weierstrass's Theorem. Thus there are at least two common points.

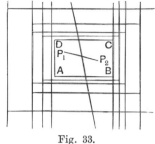

Fig. 33.

Let P_1 and P_2 be any two of the common points. Then, by Theorem 4, any straight line which cuts the linear segment P_1P_2 cuts from the first region R_1 at least one segment, and therefore, by Theorem 5, Cor., a closed set of points. Similarly the straight line cuts from R_2 a closed set of points contained in the former closed set : and so on for all the regions. By Cantor's Theorem of Deduction (p. 26) there is at least one point common to all these sets; this point of course is a common point of R_1, R_2, ..., thus we see that any straight line cutting the linear segment P_1P_2 contains at least one of the common points.

Since there are certainly points on each of two opposite sides of $ABCD$, there must be at least one point common to all the regions on every straight line inside $ABCD$ and parallel to one of its sides, hence there are certainly an infinite number of the common points, and their potency is c. The set has therefore a limiting point; if L be any limiting point, L clearly belongs to each region, and therefore to the set; thus the set is closed.

Further the set is dense in itself, and therefore perfect, by Theorem 9. Thus each of the properties stated in the enunciation of Theorem 11 has now been proved.

COR. *If P_1 and P_2 are any two points of the set of points common to R_1, R_2, ..., and S any simple polygon dividing P_1 from P_2, then there is at least one point of the set on the perimeter of S.*

This follows from Theorem 4, Cor., by the use of Cantor's Theorem of Deduction for a plane set.

122. A set of points which satisfies the condition that the set got by closing it has the property mentioned in Theorem 11, Cor., is said to be "*connected**." Thus the preceding Theorem and Cor. may be more shortly expressed by saying as follows :—

If R_1, R_2, ... be a countably infinite series of closed regions, each lying within the preceding, and if the span in two directions does not decrease indefinitely as n increases, there are points common to all the regions, and they form a perfect connected set.

It is clear that the set of common points may, as in Theorem 7, Cor. 4, generate a single region, or it may be that they form a set of regions, or that some of the points form a set of regions and other points are not included in these regions. It may be, however, that the set is dense nowhere; in this case it will be shewn that they generate a curved arc. In general the set will consist of a combination of regions and curved arcs, with limiting points, forming altogether a set with the properties enumerated.

123. Regions are divided into *finite regions* and *infinite regions*; the former are such as have a finite span, the others are such that the span in at least one direction is infinite.

The fundamental region may be the whole of the plane (space), or an infinite or finite region of the plane (space). Since, however, it is always possible to set up a continuous (1, 1)-correspondence of the whole plane (space) with a finite circle†, it will in general be assumed that the fundamental region is finite ; when however the discussion is dependent on the fact that the fundamental region is finite, this will be expressly stated ; in general it will be only necessary to give a proof in a finite fundamental region, the result in the general case following by the correspondence. Usually the fundamental region will be considered not to be closed, just as the whole plane is in general considered to be open. We can however consider the plane as closed if we postulate that there is one point at ∞ and we include this in the plane. We may more generally postulate a closed set of points, *e.g.* a circle at ∞ and include this circle in the closed plane. In like manner any fundamental set may for special purposes be regarded as closed.

124. Since the part of the plane covered by two triangular domains, which overlap without coinciding, as well as the area common to two such triangles, can be tiled over by a finite number of such triangles, it follows that

(1) *The part of the plane covered by two domains (sum of two domains) is a domain or domains.*

(2) *The common part of two domains is a domain or domains.*

(3) *The part of a domain left over after removing a domain contained in the first* (difference of two domains) is a domain or domains.*

The first of the above statements, but not the second and third, is true if for two domains we substitute a countably infinite number.

The first and second statements remain true if for *domain* we substitute *region,* the third statement is clearly not always true for regions, since two regions may only differ in their edge-points.

Thus we can work with the small domain as plane (space) element in much the same manner as we did in the straight line with the small interval; this is not the case if we cling to some particular form of region, *e.g.* the circle (sphere), since neither the sum nor the common part of two circles is itself a circle.

A slight complication is introduced by the fact that we may have more than one domain occurring as the sum or common part or difference of two domains.

125. THEOREM 12. *The boundary of a region is a closed set of points, dense nowhere.*

For by Theorem 3, there is at least one boundary point. If there are only a finite number, this is a closed set; if there are an infinite number, there is at least one limiting point, by Weierstrass's Theorem. Let P be a point internal to the region; a circle can be drawn with P as centre, lying entirely within a generating triangle, and therefore containing no point of the boundary. Similarly if Q be an ordinary external point of the region, we can draw a circle with Q as centre, lying entirely outside the region, and therefore containing no point of the boundary. Neither an internal nor an ordinary external point of the region can therefore be a limiting point of the boundary, the

---

* That is a domain whose points form a proper component of the first domain.

only points left, which can be limiting points of the boundary, are the points of the boundary itself, which is therefore a closed set.

Further, in any region containing a boundary point P, there is a triangle consisting only of internal points of the given region, and containing therefore no point of the boundary, thus the boundary is dense nowhere.

THEOREM 13.   *The rim is a perfect set dense nowhere.*

Take a region which has ordinary external points. Let L be a rim point, since we then know by Theorem 3 that at least one exists. Draw any circle with L as centre. In this circle there is an external point Q, and, since there is an internal point, there is a triangle consisting entirely of internal points of the given region. Join Q to any point of this triangle off the line QL, this line lies entirely within the circle, and it contains at least one rim point L', which is not L itself. Since the circle was any circle whatever with L as centre, it follows that L is a limiting point of the rim. Thus every point of the rim is a limiting point of the rim, that is the rim is dense in itself.

Fig. 34.

To shew that the rim is a closed set, take any point P which is a limiting point of the rim; then any circle with P as centre has a point L of the rim within it, and therefore contains a circle with L as centre; in this smaller circle there are both internal and ordinary external points of the given region, these lie therefore in the circle with P as centre. Therefore P is a rim point, so that the rim is closed, and being dense in itself, is perfect.

Again, since the points of the rim belong to the boundary it follows from the preceding theorem that the rim is dense nowhere.

126. A set of regions, regarded as a set of points, is not necessarily a closed set, even when all the regions are closed. There may be limiting points of the set of points which bear much the same relation to a set of regions that the external limiting points of a set of intervals on the straight line bear to that set of intervals.

The set consisting of all the internal points of a set of regions is, by the definition of a region, equivalent to the internal points of a set of triangles: if there is only one region the equivalent triangles may be taken as countable (Theorem 1): if however we try to include the boundary points of the region as points on the boundary of the triangles, we have seen that this is not always possible. When we come to sets of regions the same difficulty occurs; before investigating the general theorem the following definitions must be given.

127. The points of the plane are classified with respect to a set of regions in the following way :—

(1) *an internal point of the set of regions* is a point which is internal to at least one region of the set;

(2) *an external point of the set of regions* is a point which is an ordinary external point of every region of the set;

(3) *a boundary point of the set of regions* is a point which is neither internal nor external to it.

The external points are subdivided into (2*a*) *ordinary external points*, and (2*b*) *external limiting points*.

An *ordinary external point* of a set of regions is such that it is the centre of a circle containing only ordinary external points of every region of the set. An *external limiting point* is an ordinary external point of every region of the set, but such that it is the centre of no such circle.

The boundary points of a set of regions are further subdivided into (3*a*) *ordinary boundary points*, (3*b*) *superboundary points*, and (3*c*) *semi-external points*.

An *ordinary boundary point* of a set of regions is a boundary point which is the centre of a certain circle such that a *finite* number of the regions can be assigned, having the point as boundary point, and containing every point of every one of the regions inside the circle. If this is not the case, but a circle can be found with the point as centre such that an *infinite* number of the regions having the point as boundary point can be found, containing every point of every region inside the circle, the point is called *a superboundary point* (cp. Ex. 4, p. 199).

A boundary point which is neither an ordinary nor a superboundary point is called *a semi-external point*.

If there are only a finite number of regions it is clear that there are no superboundary or semi-external points.

Ex. 4. Take a square $ABCD$ as in the figure, and divide the side AB and the diagonal AC by continued bisection so that the points A and C respectively are the only limiting points. Now join C to all the points constructed on AB, so as to form right-angled triangles with CB as common side, and join B to all the points constructed on AC, so as to form triangles with the common side AB. If we add to these triangles the triangle ACD, we get a set of triangles having as internal points all the internal points of the square, except the points of the diagonal AC; the external points are all the points external to the square; the ordinary boundary points are

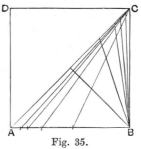

Fig. 35.

those on the periphery and on the diagonal AC, with the exception of the point C, which is a superboundary point.

If we omit all the triangles having B as vertex, every point of AC, except the superboundary point C, is a semi-external point.

128. THEOREM 14. *Given any set of regions, a countable set from among them can be chosen having the same internal points as the given set.*

Replace each region by a set of generating triangles; these, taken for all the regions, constitute a set of triangles, which, by Theorem 1, may be replaced by a countable number from among them, having the same internal points. Corresponding to each triangle let us take one region of which it was a generating triangle, we get a countable set of the given regions, having the same internal points as the given set of regions.

COR. 1. CANTOR'S THEOREM OF NON-OVERLAPPING REGIONS. *Any set of non-overlapping regions is countable*.*

COR. 2. *Given any set of regions, a (countable) set of non-overlapping regions is uniquely determined, having the same internal points as the given set.*

By the above theorem we may take the given regions to be countable, say $d_1, d_2, \ldots$.

Picking out in turn the first of the regions which overlaps with d_1 and then the first which overlaps with these, and so on, we get a certain determinate region, D_1 say, whose internal points are all internal points of the given set of regions. Similarly starting with d_2 we get a determinate region, which will by construction be D_1, or a new region D_2 not overlapping with D_1, according as d_2 does or does not belong to those forming D_1. Proceeding thus,

* See footnote, p. 38.

after at most a countably infinite number of regions D have been formed, there is no region d which does not form a part of a region D. Any internal point of the d's will then be internal to the D's and *vice versâ*, so that, as the D's do not overlap we have done what was required.

THEOREM 15. *Given any set of regions, the countable set of them having the same internal points can be so chosen that they have the same ordinary boundary points and the same super-boundary points as the given set of regions.*

Let P be an ordinary boundary point of the regions. Then we can describe a circle with P as centre, such that every point of the given regions inside this circle is internal to, or is a boundary point of, a certain finite number of the given regions having P as boundary point. If there is any other boundary point P' inside this circle, since P' is not internal to any region, it is a boundary point of one or more of the regions determined by P, and is an ordinary boundary point, since these regions contain every point of the given regions in a sufficiently small circle round P'. Take all such circles, round all such points P. By the preceding theorem we can

Fig. 36.

replace these by a countable number of such circles, having the same internal points; these will therefore contain as internal points all the ordinary boundary points of the given regions, and the countable set of the regions of which the centres of these circles are boundary points, as has been shewn, have all these points as boundary points, and contain all the points of the given regions in the neighbourhood of each such point. Adding this countable set of the regions to the countable set having the same internal points as the given regions, we have a countable set of the given regions having not only the same internal points but also the same ordinary boundary points as the given regions.

By a precisely similar argument we can get a countable set of the regions having the same superboundary points as the given regions, and adding these to the regions already determined we get such a countable set of the given regions as was contemplated in the enunciation.

Ex. 1 (p. 183) shews that it is not always possible to choose a countably infinite set of the regions having not only the same internal, ordinary boundary and superboundary points, but also the same semi-external points. We had there an equilateral triangle which revolved round its centre of gravity. The internal

points of the set of triangles are the internal points of the circum-
scribed circle; the external points of the set are the points external
to the circle; the boundary points are the points of the circumference, and, since each such point P is the vertex of one, and only
one, of the triangles, while in any circle with P as centre there are
an infinite number of points of the circumference, each of which
belongs to no triangle passing through P, it is clear that every
point of the circumference is a semi-external point of the set of
triangles. When, however, we take only a countable number of
the triangles, as already pointed out in Example 1, it is impossible
to include as boundary points more than a countable number of
the points on the circumference; thus a set of semi-external points
of potency c will become external limiting points.

THEOREM 16. *Any set of regions, not merely finite in number,
determines at least one point L, such that in any region, with L as
internal point, there are points of an infinite number of the given
regions.*

Assume first that all the regions lie in a finite square. Divide
it into four equal squares and number them 1, 2, 3, 4, beginning
at the top left-hand corner and going round clockwise (Fig. 31).

Then not all the four squares can contain only points of a
finite number of the given regions. Take the first of the four
which has not this characteristic and proceed with it as with the
original square. This process can be repeated indefinitely; in this
way we get a series of squares, each lying within the preceding,
and their sides decrease indefinitely.

By Theorem 7 there is one point L internal to all of them, this
point has evidently the characteristic in question.

In the general case the theorem now follows by correspondence
assuming the fundamental region to be closed.

N.B. If the regions do not overlap, L is clearly not internal
to any of the regions, nor an ordinary boundary point: it may be
an external limiting point or a semi-external point, or a superboundary point. In none of these three cases is it necessary that
in the neighbourhood of L the span of the regions should decrease
indefinitely; this constitutes an important difference between one-
dimensional space and plane or higher space; methods which were
fruitful when dealing with the straight line are, in consequence,
not applicable in higher space. The following theorem, for
instance, is not a direct consequence of Theorems 14 and 16.

THEOREM 17. THE EXTENDED HEINE-BOREL THEOREM FOR THE PLANE OR HIGHER SPACE. *Given any closed set, lying in a finite square, and, corresponding to each of its points, a region, having that point as internal point, there are a finite number of these regions which suffice to cover every point of the given closed set.*

Suppose the theorem were not true. Divide the square, in which the given set G lies, into four equal squares 1, 2, 3, 4, as in the proof of Theorem 16. Then in and on the sides of at least one of these four squares there must be points of G, and these points form a closed set. Suppose this to be the case with the square 1, and take those of the given regions which suffice to cover all the points of this latter closed set, omitting those of the given regions which contain no point of the given set in 1. If the theorem were true for these regions and this latter closed set, and it were true similarly taking each of the squares 2, 3, and 4, the theorem would be true for the given set, contrary to hypothesis: therefore we may assume that in one of the squares, say 1, it is not true, we then proceed with the points in that square and on its sides, and the corresponding regions, as with the given set and regions. This process can be repeated *ad infinitum*, since by the hypothesis, it cannot come to an end after a finite number of stages. We thus get a series of squares, each lying in the preceding, and having a side of half the length of the preceding, while two of these sides at most may lie on sides of a preceding square. Each of these squares contains points of G, and does not lie entirely in one region, since, by the hypothesis, the points of G in it and on its sides cannot be covered by any finite number of the regions. These squares determine one and only one point L, which is either internal to them all, or, after a certain stage, lies on a side of each. Since each square contains points of the given set G, and this set is closed, L must be a point of G, and is therefore internal to one of the regions. We can therefore draw a circle of radius r, say, with L as centre, lying entirely in one region; any square of side $\frac{1}{2}r$ passing through L, or containing L inside it, will then lie entirely in this circle, and therefore entirely inside one region; therefore one of the squares which served to determine L will, contrary to the hypothesis, lie entirely in one region. Thus the hypothesis was unallowable, and the theorem must be true.

Q. E. D.

Cor. 1. *This theorem is true, with proper conventions as to points at infinity and regions containing them, for any closed set.*

This follows from the above by correspondence.

Cor. 2. The Heine-Borel Theorem for the plane or higher space. *Given a set of regions such that every point of a closed region is internal to at least one of the regions, there are a finite number of the regions having the same property.*

Conversely it is easy to see that *if a set G has the property that, in whatever manner regions be described enclosing the points of G, a finite number of these regions suffice for the purpose, then G is a closed set*.*

For if P be any point not belonging to G, we can describe a triangle with each point of G as internal point and with P as external point. By hypothesis a finite number n of these suffice to enclose every point of G. Since P is external to each of these n triangles, it cannot be a limiting point of points internal to them, and is therefore not a limiting point of G. Thus no point not belonging to G is a limiting point of G, so that G must be closed.

It is clearly sufficient to prove G closed, if the theorem holds for some particular form of region, *i.e.* triangles or circles with the points of G as centres.

129. Theorem 18. *The points of any closed set which are not internal to a set of regions form a closed set.*

For if P be any point of the closed set internal to the set of regions, there is a circle with P as centre containing no point of the set of points in question, so that P is not a limiting point of that set, which is therefore closed.

Theorem 19. *Any closed set determines a set of non-overlapping domains, the "black regions of the closed set," consisting of all the points not belonging† to the set.*

For, if P be any point not belonging to the closed set, there is a circle, and therefore a triangle, containing P as internal point, and having no point of the set in it or on its periphery. Taking all such triangles for every point not belonging to the set, these generate one or more non-overlapping domains, of which the points of the closed set are the boundary points. By Theorem 14, Cor. 2, these regions are unique.

---

* Veblen, *Bull. of the Amer. Math. Soc.* x. p. 436 (1904).

† The edge of the fundamental region, if it has one, being excluded. Cp. footnote, p. 19.

130. DEF. A set of points such that, describing a region in any manner round each point and each limiting point of the set as internal point, these regions always generate a single region, is said to be *a connected set*, provided it contains more than one point. Hence if a set is connected the set got by closing it is connected and *vice versâ*.

This definition is equivalent to that given by Cantor[*], which runs as follows:—A set T is said to be *connected*, when, assigning any positive quantity e, there is, corresponding to each pair of points of the set t and t', a finite number of points of the set, $t_1, t_2, \ldots, t_n$, such that each of the distances $tt_1, t_1t_2, \ldots, t_nt'$, is less than e.

To prove the identity of the two definitions, let T be any set and T' the set got by closing T. Then describing regions round every point of T' as internal points, it follows by the generalised Heine-Borel Theorem that a finite number m of these regions suffice to cover every point of T'.

Let R be the region generated or one of the regions generated (Theorem 14, Cor. 2). Then since all the points of T' are internal to the regions generated, there are no points of T' on the boundary of R. Therefore the points of T' inside R form a closed set, and, if there are any points of T' outside R, these also form a closed set. If we can choose e less than the minimum distance between these closed sets, it is clear that T is not connected in Cantor's sense. But we always can do this, unless there is no point of T outside R, that is unless R is the only region generated. Thus if T is connected in Cantor's sense it is so by the definition first given.

On the other hand choosing the m regions described to be circles of radius $\frac{e}{3}$ with the points of T' as centres, it is clear that if T is connected in the former sense it is so in Cantor's sense. Thus the two senses are identical.

THEOREM 20. *If T is a connected set, it cannot be divided into two closed components without common points. Conversely a set which cannot be divided into two closed components without common points is, if closed, a connected set*[†].

First let T be a connected set, T_1 one of its closed proper components, and T_2 the complementary set of T_1 with respect to T.

---

[*] Cantor, *Math. Ann.* xxi. p. 575.

[†] This property has been used by Jordan and others to define a connected set in the particular case when it is closed. Jordan, *Cours d'Analyse*, Vol. i. p. 25.[11]

The minimum distance (§ 106) between T_1 and T_2 is only zero if T_2 is unclosed, since T_1 and T_2 have no common point. But, by Cantor's definition this minimum distance must be zero, and therefore T_2 is unclosed, which proves the first part of the theorem.

Next let T be a closed set which is not connected, and let us describe regions round the points of T in such a manner that they generate more than one region, and let R be one of the regions generated. Then, as before in proving the identity of the two definitions of a connected set, the points of T inside and outside R are both closed sets, they have no common point and together make up T. Thus if T is not connected it can be divided into two such components, which proves the second part of the theorem.

THEOREM 21. *A necessary and sufficient condition that T should be a connected set is that if P and Q be any pair of points of T, and Π any simple polygon separating P from Q, there is at least one point or limiting point of T on the polygon Π.*

That this is necessary follows at once from the preceding theorem, since otherwise T' (the set got by closing T) would be divided into two closed components by Π.

On the other hand if T is not connected, T' is not connected, and therefore, by the preceding theorem can be divided into two closed components T_1 and T_2.

If e be less than the minimum distance between T_1 and T_2, and we describe a finite number of triangles of span $\dfrac{e}{3}$ round the points of T_1 (which by the generalised Heine-Borel Theorem is possible), the boundary of the region or of the regions, generated by these triangles will be one or more simple polygons (see § 132) containing every point of T_1 inside it, or them, and every point of T_2 outside it, or them. Thus on such a polygon there is no point of T' and therefore none of T, and yet, since there are points of T' both inside and outside this polygon, there are certainly points of T both inside and outside the polygon. That is, if T is not connected the condition is not fulfilled. Thus the condition is sufficient to insure T being connected.

By Theorem 4, Cor., *the points of a region form a connected set, and in fact a perfect connected set.*

The points of two regions may also form a connected set, for instance the area bounded by a lemniscate consists of two regions, but they form together a connected set: the same is true of any number of regions having common boundary points, but not overlapping.

131. DEF. *A plane set of points, dense nowhere in the plane, such that, given any norm e, and describing round each point of the set a region of span less than e, these regions generate a single region R_e, whose span does not decrease indefinitely with e, is called a curved arc, or shortly, a curve.*

It follows from Theorem 7, Cor. 3, that a stretch or linear segment is a particular case of a curved arc; the perimeter of a triangle, or of a square, or circle, and indeed all the curves with which we are familiar in plane Geometry, ellipses, branches of a hyperbola, parabolas, and others are curved arcs.

It was already pointed out (§ 122) that *the inner limiting set of a series of closed regions, each lying inside the preceding, is a perfect connected set.* A special case of this is therefore the following :—

THEOREM 22. *If R_1, R_2, ... be regions lying each entirely inside the preceding, and if the span of these regions in two directions does not decrease indefinitely as n increases, the common points of these regions form a curved arc other than a stretch, provided only they are dense nowhere and, if R_1, R_2, be closed regions, the points of the curve form a perfect set.*

132. Simple polygonal regions. Let S be any closed connected set, and $e_1, e_2, ...$ a sequence of continually decreasing positive quantities, having zero as limit. Describe a triangle of span less than e_i with each point of S as internal point (*e.g.* as centre of gravity). By the generalised Heine-Borel Theorem a finite number n_i of these triangles suffice to cover every point of S, and, since S is connected, these generate a single region R_i.

The regions R_1, R_2, ... have then every point of S common, and, since S is a closed set, they have no other common point; that is to say S is the inner limiting set of this series of regions.

It follows that as a preliminary to the study of connected sets it is necessary to be quite clear as to the forms of regions generated by a finite number of triangles. It is not however necessary to discuss the most general form of such a region ; *we may make the restriction that no vertex A of one of the triangles lies on a side of another, and not more than two triangles have a common rimpoint B.* For if the triangles chosen do not obey this restriction, we can always remove any such inconveniences by suitably enlarging the triangles, without making the span of any triangle equal to or greater than e_i, if it was previously less than e_i; the inner limiting set will in this way be unaffected.[12] A region generated by a finite number of triangles obeying these restrictions will be called *a simple polygonal region.*

Thus a closed connected set is the inner limiting set of a series of simple polygonal regions.

Let R be a simple polygonal region, and P any rim point of R. Then, since P is a limiting point of internal points of a finite number of triangles, P must be a limiting point of points of one only of the triangles; therefore P is a rim point of one of the triangles, since P is not internal to any of them. It follows that (the triangles having none of the inconveniences specified) P lies on the rim of one only of the triangles, or on the rim of two, and, in the latter case, P is not a vertex of either triangle. Since P is an ordinary external point of all the remaining triangles, it follows that, in the former case, there are two stretches with P as end point, belonging to the rim of the single triangle on which P lies, every point of each of these stretches being a rim point of R.

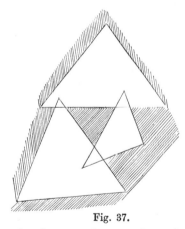

Fig. 37.

The same is true in the second case, since the two sides meeting at P do not lie in a straight line, and therefore, of the two stretches with P as end point determined on any one of them, as containing no point of any of the remaining triangles, one is internal to the other of the two triangles through P; thus there are again two stretches with P as end point forming part of the rim of R.

The two stretches determined by any rim point P are in a straight line if, and only if, P lies on only one triangle and is not one of its vertices; thus the rim of R is formed of a finite number of stretches, not more than two of which pass through any point, no point of intersection being internal to any of the stretches, while there is no point through which there passes only one of the stretches. Thus *the rim of R consists of a finite number of simple polygons, without intersections.*

If the rim of R consists of only one simple polygon, R is said to be *simply connected,* while, if there be r bounding polygons, R is called an *r-ply connected simple polygonal region.*

A single triangle, or a pair of overlapping triangles, forms a simply connected region. Three triangles may form a doubly connected region (Fig. 37).

Since the number of generating triangles is finite, and any point inside, or on the periphery of, a triangle can be joined to any other by a stretch lying inside the triangle, it follows that *any point on the rim of a simple polygonal region can be joined to any other point inside, or on the boundary of, the region by a simple polygonal path, every point of which, excluding the two end-points, is internal to the region.* Such a path is said to be *internal* to the region.

Hence, choosing out any one of the bounding polygons, the whole domain and all the remaining polygons either lie inside or outside it. The polygonal region, if finite, must, by Theorem 2, Cor., lie inside one of the polygons Λ; hence Λ, as well as the whole domain of the region, lies outside each of the other polygons. Λ is called the *outer rim,* and the other polygons the *inner rims* of the region.

This being so, we may shew that *any path Π, internal to the region, and joining two rim points P and P', divides the region, provided it is simply connected, into two parts, such that any two points in one part can be joined by an internal path, while any internal path joining a point in one part to a point in the other intersects the dividing path Π.*

For P and P' divide the simple polygon bounding the region into two simple polygonal paths, either of which with Π makes up a simple polygon. Since the points of Π are internal to the region, and every internal point of a region lies inside a triangle consisting entirely of such points, there are internal points of the region both inside and outside either of these polygons. Taking a pair of such points any simple polygon joining them must intersect the polygon. If therefore it does not intersect Π, it must contain a point of the rim other than one of its end points, which proves that Π divides the region into at least two parts. On the other hand any point inside one polygon can be joined to any point inside the same polygon by a simple polygonal path which does not intersect the polygon, and therefore, containing no rim-point of the region, consists entirely of internal points of the region. Thus Π divides the region into precisely two parts, as was stated.

133. The outer rim. Take any finite region lying in a square S. If it have any boundary points other than rim points, these are points in the neighbourhood of which there are no ordinary external points of the region, and, by regarding these points as internal points, we obtain a new region having no boundary points other than rim points. We may assume therefore, in examining the rim, that there are no other boundary points.

Surround each point of the rim by a triangle of span less than e having its centre of gravity at that point, and, as in the preceding section, choose a finite number of these, generating one or more simple polygonal regions.

Then outside the triangles round the points of the rim there are always ordinary external points of the region. Also, since we can draw a triangle consisting entirely of internal points of the region, so that no point of the rim is inside it or on its boundary, if we choose e sufficiently small, there are outside the triangles internal points of the given region.

It is then clear that, if n be the finite number of triangles of span less than e which suffice to cover every point of the rim, n is greater than 2, for outside the triangles there is an internal and an ordinary external point of the given region, and these can be joined by a simple polygonal path, not cutting any two given triangles; by Theorem 3, Cor. there would be a point of the rim on this path, and therefore there must be at least three triangles.

If C be the outer rim (§ 132) of one of the simple polygonal regions generated by the n triangles, either every point of C is internal to the given region R, or all the points of C are ordinary external points of R. Now there are internal points of the given region inside each of the n triangles; thus, unless all the n triangles lie inside C, we can take one of these internal points P inside C and one Q outside C. We can then, by Theorem 2, join P to Q by a simple polygonal path, every point of this path being an internal point of the region R. This path must have a point common with C, thus there must be a point of C internal to the region R; it follows that every point of C is internal to the region R. *Thus unless all the n triangles are inside C, C consists entirely of points internal to the region R.*

The points of the rim isolated from the rest by such a simple polygon C we shall call *an inner rim*. There is then certainly inside C an ordinary external point of the given region, and surrounding this a triangle consisting only of such points. Thus the points external to the given region but internal to C form a set of domains, the points of which are the ordinary external points of the given region inside C; conversely it is easily seen that the ordinary external points of this set of domains inside C are the internal points of the given region inside C. It follows that any rim point of one of these domains is a rim point of the given region, while any rim point of the given region inside C which is not a rim point of one of these domains, is an external limiting point of the set of domains. Thus *an inner rim consists of the rims of a finite number of non-overlapping domains, or the rims of an infinite number of such domains, together with the external limiting points of these domains, the internal points of these domains being ordinary external points of the given region.*

If now we " wipe out" this inner rim, that is, if we take every point inside C to be an internal point, we get a new or modified region having every internal point of the given region as internal point, and, apart from the inner rim wiped out, the same rim. In a certain sense the given region may be said to be got from the modified region by cutting out a certain set of regions; it must however be borne in mind that this is only literally true when this set of regions consists of a finite number only of regions, otherwise the external limiting points may belong to the inner rim.

If when we had wiped out all the inner rims determined by the n triangles, there were no rim points left, the modified region would have no ordinary external points; but those external points of the given region which were external to the square S are still external points of the modified region; therefore this is impossible, and there are rim points left after wiping out all the inner rims. Those of the n triangles which surround the rim points left are such that they form a single region, $K(e)$ say, since its outer rim must, by what has been shewn, contain all the original n triangles. The part of the rim left inside this region $K(e)$ we shall call $Rim(e)$.

If we draw a strip bounded by two parallel lines, enclosing every point of the region $K(e)$, there are certainly ordinary external points of the modified region outside this strip on either side; therefore all the internal points of the modified region lie inside the strip, since all the rim points of the modified region lie inside the strip. Since all the internal points of the given region are internal to the modified region, it follows that the whole of the given region is internal to the strip. Thus the span of $K(e)$ in any direction is greater than or equal to that of the given region. On the other hand if we draw such a strip enclosing every point of the given region, it is clear that, increasing the breadth of the strip on both sides by e, we enclose all the triangles of span less than e round the points of the rim, and therefore enclose the whole region $K(e)$. *Thus the span of $K(e)$ in any direction lies between s and $s + 2e$, where s is the span of the given region in the same direction.*

Now let us perform the same process, taking $\frac{1}{2}e$ instead of e and taking care that the triangles of span less than $\frac{1}{2}e$ round the rim points lie inside those of radius less than e. Any part of the rim that was wiped out before will then still be wiped out. The part of the rim left over after all the new inner rims have been wiped out, being denoted by $Rim(\frac{1}{2}e)$, it follows that $Rim(\frac{1}{2}e)$ is contained in $Rim(e)$, and it is also a perfect set, nowhere dense, and the rim of a certain modified region, while the triangles of span less than $\frac{1}{2}e$ round all its points form a single region $K(\frac{1}{2}e)$, lying inside $K(e)$, the span of $K(\frac{1}{2}e)$ in any direction lying between s and $s + \frac{1}{2}e$. Thus taking in turn $e, \frac{1}{2}e, \frac{1}{4}e, \ldots$ we get a countably infinite set of rims, $Rim(e)$, $Rim(\frac{1}{2}e)$, $Rim(\frac{1}{4}e), \ldots$ each lying inside the preceding, and a corresponding series of regions $K(e)$, $K(\frac{1}{2}e)$, $K(\frac{1}{4}e)$, $\ldots$ each lying inside the preceding, the span of these regions in no direction decreasing without limit. Therefore, by Theorem 22, there is a perfect set of points Λ internal to all these regions. If L be one of these points, it lies inside a triangle of span less than e containing points of $Rim(e)$, and inside a triangle of span less than $\frac{1}{2}e$ containing points of $Rim(\frac{1}{2}e)$, that is points of $Rim(e)$, and so on; thus L is a point of $Rim(e)$, since this is a closed set. It follows that Λ is dense nowhere, since $Rim(e)$ is so, and therefore by Theorem 22 is a curved arc.

Now Λ is the rim of a certain modified region which is said to be a *simply connected region,* since it has no boundary points other than rim points and no inner rims, therefore if any internal point of the modified region be joined to any of its ordinary external points by means of a simple polygonal path, there will be a point of Λ on this path between the two chosen points: in other words, every internal point of the modified region is separated from every ordinary external point by Λ. For this reason *the curve Λ is called the outer rim of the given region,* since it separates every internal point of the given region from all those ordinary external points which are not enclosed in inner rims, in particular from the boundary of the fundamental region, if it is finite, or from the infinitely distant part of the whole plane.

134. Returning to the simple polygon C consisting entirely of points internal to the given region R and surrounding an inner rim of R, let I be one of the domains formed by the ordinary external points of R inside C. *Then I is a simply connected domain.* For if I had an inner rim, or other rim point not belonging to its outer rim, we could, as before, find a simple polygon C' consisting entirely of points of I, containing inside it that inner rim, or rim point; C' would then contain inside it an internal point P of R, since every rim point of I was shewn to be a rim point of R. But this is impossible, since P can be joined to any point of C by a simple polygonal path consisting entirely of internal points of R, whereas there would be of necessity a point of C' on that path, and no point of C' is internal to R. Thus I has no such inner rim or rim point, and is therefore a simply connected domain.

Summing up the results of §§ 133 and 134, we have the following statement:—

Every finite region determines a perfect set of points dense nowhere called the rim, such that every point of the rim has both internal and ordinary external points of the region in its immediate neighbourhood.

The rim may consist entirely of one curved arc; in this case, if the region has no boundary points other than rim points, the region is said to be simply connected.

The most general finite region, having no boundary points other than rim points, is constructed by removing from one simply con-

nected region any finite or countably infinite set of simply connected domains.   The rim of the region so constructed consists of the rim of the first region, called the outer rim of the region, and the rims of the domains removed, called the inner rims of the region, together with all the external limiting points of the regions removed.

It is to be remarked that, contrary to the popular idea, the rim of a simply connected region is not necessarily a closed curve.

Ex. 3 (p. 184) is a case of a simply connected region, whose rim is not a closed curve.   If we omit the whole of the dotted line, except the origin, making every one of these points an external point, we get another simply connected region, whose edge satisfies the condition for a closed curve, but is not a closed set of points.[46] The rim consists of this closed curve, with " the tail."   If we take any point of the tail, it can be joined to any other point of the rim in only one way, if the second point also belong to the tail : if the second point belong to the edge, there are two curved arcs joining the two points, but they have more than two points common.

The general form of an infinite region, and the characteristics of its rim, are at once obtained by correspondence from the same for a finite region, regarding the plane as closed at infinity.   The outer rim may consist of the whole locus at infinity, or it may consist of a curve, one or more points of which lie at infinity. The inner rims of an infinite region are of precisely the same nature as those of a finite region.

THEOREM 23.   *A simply connected region is the inner limiting set of a series of simply connected simple polygonal regions.*

Describe as in § 132 triangles of span $< e_i$ round all the points of the given simply connected region R, so as to form a simple polygonal region R_i.   If this is not simply connected, let P be a point inside one of its inner rims, and Q a point outside its outer rim.   Then, since P and Q are both ordinary external points of R, they can be joined by a simple polygonal path external to R.   The points of this path form a closed set, let e_j be less than the minimum distance between it and the rim of R. Then Q and therefore every point of the path PQ is outside the outer rim of R_j.   Hence, if the simple polygonal region R_i was not simply connected, and we make it simply connected by wiping out all its inner rims, as in § 133, we shall introduce no extraneous points into the inner limiting set.   Thus the theorem follows.[47]

It is to be remarked that we cannot assert of the general

simply connected region that any point P of its rim can be joined to any internal point Q by a simple polygonal path internal to the region. Osgood's simply connected region (Ex. 2) and the Chow (Ex. 3) are both examples demonstrating this point; a point of the Chow's tail other than the origin, can, for instance, not be joined to an internal point of the Chow by a simple polygon which contains no rim points or ordinary external points of the region*.

THEOREM 24. *Any simple polygonal path* Π, *joining two rim points* P *and* P' *of a simply connected region, and passing through at least one internal point* M *of that region, divides the region into at least two parts.*

For let R_1, R_2, ... be a series of simply connected simple polygonal regions having the given region R as inner limiting set (Theorem 23). Let us produce the sides of Π which end in P and P' respectively, and let P_i be the point of the rim of R_i which lies nearest to P. Then P_1, P_2, ... form a sequence of points in order leading up to P as limiting point (Fig. 38). Similarly we construct a sequence P_1', P_2', ... with P' as limiting point.

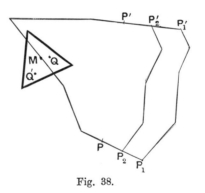

Fig. 38.

Let us describe with M as centre of gravity a triangle consisting entirely of internal points of R, and of span so small that it is divided into two parts by Π. Let Q and Q' be any two points one in each of the parts of this triangle.

Then, denoting the simple polygonal path $P_i P_i'$ by Π_i, Π_i divides R_i into two parts (§ 132) one of which contains Q and the

* Schoenflies has shewn that when a rim point of a simply connected region can always be joined to an internal point by an internal simple polygonal path, and to an external point by such an external path, the rim constitutes a Jordan curve, *Math. Ann.* LVIII. p. 230.

other Q'. Let that one which contains Q be denoted by S_i, and the other by S_i'. Then if S and S' denote the inner limiting sets of S_1, S_2, ... and S_1', S_2', ... respectively, S and S' together with Π will make up the whole of R.

Now a simple polygonal path joining Q to Q' (where these are the points already so designated, or any other pair, one in S and the other in S') and not intersecting Π, must either intersect the rim of R_i, or pass between P and P_i, or between P' and P_i', since Q and Q' are points of R_i separated by Π_i.

Suppose the path intersects PP_i, then we can assign an integer m, such that P_m lies between P and the point of intersection. The path would then have to intersect the rim of R_m, or to pass between P' and P_m'. In the latter case however we could in like manner determine n, so that the path does not pass between P' and P_n', and must therefore intersect the rim of R_n. Thus in any case, if the path does not intersect Π, it contains an ordinary external point of the given region R, so that any internal path QQ' must intersect Π. Thus R is divided into at least two parts by Π.

It does not follow that R is divided into only two parts, for since Π may itself contain rim points of R, there may be parts of the rim of R dividing up S, or S', into parts, the internal points of which cannot be joined by an internal path which does not intersect Π.

135. It is a direct consequence of the investigation of the preceding section, that *a closed plane set which contains no curves, is complementary to a single open region, whose rim is the rim of the fundamental region.*

This is so far remarkable, that it shows that the mode of construction of closed or perfect sets by means of black domains, (which is the extension of that habitual on the straight line, by means of open black intervals, which, in the former case may, and in the latter may not, abut), when applied to the plane, leads only to closed or perfect sets containing curves. Those which do not contain curves have a special interest in theory of functions of two variables, as was pointed out by Baire, although those sets of this kind which occur in Baire's work are not of the most general character*.

* See footnote, p. 230.

136. Having gained some insight into the meaning of the term *region*, we are now in a position to resume the discussion of the meaning of the term *dimensions* which was raised in Ch. VIII. We saw that the question of the dimensions of a region resolves itself into that of the possibility or impossibility of a continuous (1, 1)-representation of the region on some known form, *e.g.* on the straight line. It is necessary here to insert some preliminary theorems about such representations.

THEOREM 25. *In any continuous representation of a plane set on a linear set, if a closed set g on the straight line represent a plane set G, G is also closed.*[48]

For however we describe circles C_Q round each point Q of G as centre, a finite number of the corresponding intervals d_Q (§ 98) will suffice to enclose every point of g; therefore the corresponding circles, in finite number, will suffice to cover every point of G, so that G is closed (§ 128).[48]

COR. 1. *The points of any sequence on the straight line correspond to the points of a sequence of points of G, and the limiting point of the former sequence has for correspondent the limiting point of the second sequence.*

Conversely given a sequence of points of G, every limiting point of the corresponding points on the straight line is a correspondent of the limiting point of the sequence of points of G.

COR. 2. *If no point of G corresponds to every point of some segment (in particular if no point of G has c correspondents), then any point t of the straight line having T as correspondent is a limiting point of the correspondents of a sequence having T as limiting point.*

THEOREM 26. *In any continuous representation of a plane set on a linear set, if a connected set g on the straight line represent a plane set G, G is also connected, provided no point of the plane corresponds to a connected set of points on the straight line.*

[In particular the above holds if no point has an infinite set of correspondents on the straight line.]

Let g' be the set got by closing g, and G' that got by closing G, and let us represent G' on g', by extending the given representation so that every limiting point of a sequence of points of g corresponds to the limiting point of the corresponding sequence of points of G (Theorem 25, Cor. 1).

Now by the restriction imposed on the representation, G, and therefore G', contains more than one point. Also if two closed sets have together the same points as G', the two corresponding linear sets being closed (Theorem 25), and having the same points as the connected set g', have at least one common point (Theorem 20). Thus, by Theorem 20, G', being closed, is connected, so that G is also by definition a connected set.   Q. E. D.

THEOREM 27.   *A continuous* (1, 1)-*correspondence between a region of the plane and a segment of the straight line is impossible**.

For suppose it to be possible. Let A and B be any two internal points of the region and a and b the corresponding points.

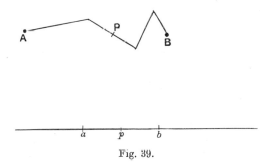

Fig. 39.

Then A can be joined to B by a finite number of stretches forming a polygonal line and lying entirely in the given region: the points of this polygonal line form a closed set which is connected. By Theorem 25 the representative set on the straight line is closed; also, since the correspondence is (1, 1), the argument used to prove Theorem 26 shews that it is also connected, and is therefore none other than the segment (a, b). Let P be any point of this polygonal line and p the corresponding point of (a, b), then since no point near P and not on the polygonal line can correspond to a point nearer to p than the nearer of a or b, we can assign a sequence with P as limiting point, whose representative points have not p as limiting point, which by Theorem 25 is contrary to the hypothesis that the correspondence is (1, 1): therefore the hypothesis that the correspondence is both (1, 1) and continuous is untenable.   Q. E. D.

---

* Jürgens, *Jahresbericht d. d. Mathvgg.* VII. p. 50 (1899) where quotations and criticism of earlier work will be found.   The following is not quoted by Jürgens:—
L. Milesi, *Riv. di Mat.* II. (1892) pp. 103—106.

137. Uniform Continuity. The interval d_p corresponding
to the circle C_P in § 98 may, if we please, be determined uniquely
as the largest interval (§ 16) with p as centre, such that all the
points of g *internal* to d_p have their correspondents inside C_P.
If d_p be taken smaller than this largest interval, the end-points
a and b of d_p are such that, if they belong to g, their correspondents A and B also lie inside C_P; the end-points of the largest
interval itself certainly belong to g and have no correspondents
inside C_P.

The length of the largest interval, when the radius of C_P is a
fixed quantity e, is finite for all points p, but the lower limit of all
the lengths may be zero. If this is not the case the representation
is said to be *uniformly continuous*. In other words the *representation is uniformly continuous if, and only if, when e is arbitrarily
assigned, we can determine d, so that, whatever point p be chosen, if
q is within a distance d of p, the corresponding point Q is within
a distance e of P.*

THEOREM 28. *If the set g is closed, the correspondence is always
uniformly continuous.*

For, by the generalised Heine-Borel Theorem (p. 41), a finite
number of the intervals d_p corresponding to circles of radius $\frac{1}{2}e$ with
the points of G as centres, suffice to enclose every point of g. Since
the end-points of a finite number of intervals form a closed set,
and g is a closed set having no point common with the former,
there is a positive minimum distance (§ 105) $2d$ between these
two closed sets. If P be any point of G, and p the corresponding
point of g, p will be internal to at least one of the given intervals,
and an interval of length d with p as centre will contain none of
the end-points, and will therefore lie entirely inside any of the
intervals containing p. It follows that the points of G corresponding to the points of g in this interval of length d, will lie
inside any one of the circles of radius $\frac{1}{2}e$ which correspond to the
interval, or intervals, containing p, and will therefore contain P.
Thus every one of these corresponding points is within a distance
e of P. This proves that the correspondence is uniformly continuous.

CHAPTER X.

CURVES.

138. A curve has already been defined in Ch. IX. The definition is repeated here.

DEF. *A plane set of points, dense nowhere in the plane, such that, given any small norm e, and describing round each point of the set a small region of span less than e, these small regions generate a single region R_e, whose span does not decrease indefinitely, is called a curved arc, or shortly a curve.*

The following then follow from the investigations on regions :—

A curve is never a point and never a region, and, only when the span of the region R_e in one direction diminishes without limit as e does so, is it a stretch (segment of a straight line).

The points of a curve form a connected set.

A closed connected set dense nowhere in the plane is a curve, and is said to be a complete curve.

The points of a curve may or may not form a closed set: the non-included limiting points may be finite, or countably infinite, or more than countable.

DEF. An arc, every one of whose points is a point of a certain curve, is called *an arc of that curve.*

The following property of a curve is an immediate consequence of the definition :—

Given any two points P and Q of a curve, there is at least one arc of the curve PQ not containing P nor Q, but having both these points as limiting points.

Such an arc is said to join P to Q.

THEOREM 1. *If we add to the points of a curved arc any or all non-included limiting points, the set so obtained will itself be a curve.*

Since the set got by closing a set which is nowhere dense is itself nowhere dense, this closed set, and therefore any component of it, is nowhere dense.

Describe round any non-included limiting point L a region of span less than e. Then, since L is a limiting point of the points of the arc, there are certainly points of the arc inside this small region, which, therefore, certainly overlaps with R_e. Thus adding any such small regions to R_e, we still get a single region whose span is not less than that of R_e. Thus the set has the two characteristic properties of a curve.

139. Branches, end-points and closed curves.

Given a point P of a curve, and describing round it as centre a circle of radius e, there are certainly points of the curve inside this circle, and, if Q be any one of these, there is at least one arc PQ, having P and Q for non-included limiting points (this arc does not necessarily lie entirely inside the circle, no matter how small e may be; cp. Ex. 3, Chap. IX, p. 184). It may be that, when e is chosen sufficiently small, there is a point Q in the circle such that there is only one arc PQ; if so, PQ is called a *branch of the curve from the point P*.

Ex. 1. If P be a point of a circular arc, not a complete circle, then from any other point Q of the arc there is only one arc to P, thus the circular arc is a branch from P.

It may be however that, however small e may be, there is always more than one arc PQ; if this be so, but if there is any arc of the given curve having a branch PQ, then PQ is still called *a branch of the given curve from the point P*.

Thus the expression "branch" must be taken to refer to the form of the curve in the neighbourhood of some particular point, and not to the form of some finite portion of the curve in its entirety.

Ex. 2. If P be a point of a complete circle, there are always two arcs PQ, wherever Q may be on the circle. But if we omit from the circle a small arc having P as non-included limiting point, and not containing Q, the remaining part of the circle is a circular arc, and therefore by Ex. 1, this is a branch of the circle from P. Similarly the omitted arc is another branch of the circle from Q to P. Thus there are two branches of the circle from every point P of the complete circle, these branches having no common point (except P, if we choose to include it).

Suppose that PQ is a branch of a curve, Q lying inside a circle of radius e with P as centre. Let Q' be any other point of the curve inside the same circle: then if, by choosing e sufficiently small, the arc PQ' is always either part of the arc PQ, or PQ is a part of PQ', then there is said to be only one branch of the curve from P, and P is called *an end-point of the curve.*

It is clear from these definitions that *a curve remains a curve if we omit one or more of its end-points.*

Ex. 3. The curve which forms the edge of the region in Ex. 3, Ch. IX, affords an example in which the end-points are of potency c. The complete curve with the tail has an end-point at every point of the tail. Any set of points of the tail may be omitted without the set losing the character of being a curve.

If P is a point of a curve but not an end-point of the curve there will always be at least two arcs of the curve, neither of which is a part of the other, both arcs having P as limiting point. P is said to be *an ordinary* point of the curve*, if there are two and only two such arcs PQ and PQ', when e is chosen sufficiently small.

If, however small e may be taken, there are more than two such arcs, P is called *a fork-point of the curve*, which is said to *fork* at P.

Ex. 4. A lemniscate forks at the double point; there are four branches of the lemniscate from the double point.

Ex. 5. Fig. 40 shews a fork-point O which is itself a limiting point of fork-points. However small a circle be described round O, if Q be a point of the curve inside this circle, there is a more than countably infinite set of arcs OQ, none of which is a part of another.

Fig. 40.

DEF. *A simple closed curve* is a curve such that given any two of its points P and Q the curve can be divided up into two curved arcs, which together with P and Q make up the whole curve, and are such that P and Q are limiting points of each arc, while no point of one arc is a limiting point of points of the other arc.

A simple closed curve has no end-points.

A closed curve is not necessarily a closed set of points†, nor is the complete curve got by adding to a closed curve all its non-

* In the theory of algebraic curves the word *ordinary* point is used in a somewhat narrower sense: a cusp for instance is in the above general sense an ordinary point, but not in the sense in which the term ordinary is used in the theory of algebraic curves.

† This is an objection to the use of the term "closed curve," but the expression is too familiar to be lightly displaced. In Italian the term "curva rientrante" has

included limiting points necessarily a closed curve. An example of this is the edge of the region given in Ex. 3, Ch. IX; if we include the origin, this is a closed curve, but the non-included limiting points form the "tail" which consists entirely of end-points.[49]

140. Jordan curves. A plane set of points which can be brought into continuous (1, 1)-correspondence with the points of a closed segment (a, z) of a straight line is called a Jordan curve*.

The end-points a and z of the segment may be exceptions to the uniqueness of the correspondence in so far that they may correspond either to different points or to the same point.

Analytically this definition may be expressed as follows.

Let $$x = f(t), \qquad y = \phi(t),$$

where f and ϕ are continuous single-valued functions of the independent variable t, for all values of t from t_0 to T_1, both inclusive, which do not assume the same pair of values for any two different values of t within the given limits: the locus of the point (x, y) is called a Jordan curve.

The pair of extreme values $f(t_0)$, $\phi(t_0)$ and $f(T_1)$, $\phi(T_1)$ may be identical.

By Theorem 25, Ch. IX, *the points of a Jordan curve form a closed set.* Consequently, since, by Theorem 27, Ch. IX, they do not form a region, they form a set which is dense nowhere. Further, by Theorem 26, Ch. IX, *the points of a Jordan curve form a connected set.* Thus it is clear that each point of a Jordan curve is a limiting point of the curve, *whence the points of a Jordan curve form a perfect connected set dense nowhere.*

It follows that *the points of a Jordan curve form a curved arc.*

---

been used instead of "curva chiusa"; in English, however, *reentrant* suggests a different idea. In German there is no confusion between "abgeschlossene Menge" and "geschlossene Kurve." Schoenflies uses "einfache geschlossene Kurve" in a special sense, which he shews to be equivalent to "simple closed Jordan curve" (*Math. Ann.* LVIII. p. 217).

* Jordan's original definition of a "continuous line" (*Cours d'A.* I. p. 90) is rather more general, since it does not postulate that the correspondence should be (1, 1). As already pointed out, it would not then follow that the points of the locus are dense nowhere. The presence of a finite or countably infinite set of multiple points of the correspondence, only necessitates our considering the locus as consisting of a finite or countably infinite number of Jordan curves: the presence of more multiple points however introduces so many complicated possibilities, that it is best, as is now customary, to reserve the term Jordan curve for one without multiple points. Jordan himself does not work with any other curves than those without multiple points.

THEOREM 2. *A Jordan curve does not fork; and it has two end-points if a and z have different correspondents, but is a closed curve if a and z have the same correspondent.*

Describe a circle with centre at any point P of the curve, and let (b, c) be the interval of length d with the corresponding point p of the segment (a, z) as centre, such that the circle contains all the points corresponding to (b, c). Then the points corresponding to (b, p) form a curved arc, and so do those corresponding to (p, c), and these two arcs have no common point except P. Thus from P there are at least two arcs PB, PC of the curve, so that P is not an end-point.

Again, if there were any sequence of points $Q_1, Q_2, \dots$ of the Jordan curve inside the circle, no point Q_i belonging to one of the arcs PB, PC, the corresponding points $q_1, q_2, \dots$ of (a, z) would not lie inside (b, c) and therefore would not have p as limiting point. Thus the limiting point of the sequence Q_1, Q_2, will be some point other than P, so that P is certainly not a fork-point.

The same reasoning applied to the points a and z shews that if they have different correspondents they are end-points, or, if the same correspondent, ordinary points of the Jordan curve.[50]

THEOREM 3. *A circle is a closed Jordan curve.*

This is most easily seen by projecting the points of the circle from any point S of it on to the tangent at the diametrically opposite point O. Each point P of the circle then corresponds to a single point p of the straight line, and *vice versa*, the point O corresponding to itself and the point S to the point at infinity on the straight line.

Since the whole straight line can be put into continuous correspondence with a segment OZ, the point ∞ corresponding to both O and Z and the correspondence being otherwise $(1, 1)$, this gives us a $(1, 1)$-correspondence between the circle and a segment OZ, the point S corresponding to both O and Z.

It is easy to see that this correspondence is continuous; analytically the correspondence between the points (r, θ) of the circle and x of the segment $(-1, 1)$ may be expressed by the formulae

$$\left. \begin{aligned} r &= 2a \cos \theta \\ \frac{x}{1+x} &= 2a \tan \theta \end{aligned} \right\} \left(-\frac{\pi}{2} \leqslant \theta \leqslant 0 \right), \qquad \left. \begin{aligned} r &= 2a \cos \theta \\ \frac{x}{1-x} &= 2a \tan \theta \end{aligned} \right\} \left(0 \leqslant \theta \leqslant \frac{\pi}{2} \right).$$

It is clear that any two Jordan curves are in (1, 1)-continuous correspondence.   Hence the last theorem enables us to phrase the definition of a closed Jordan curve rather more simply.

A closed Jordan curve is a plane set of points which can be brought into continuous (1, 1)-correspondence with the points of a circle*.

THEOREM 4.   *Given any positive quantity e, we can determine a positive quantity d, such that, if with any point p of the segment (a, z) as centre, we describe an interval d_p of length d, we can describe a simply connected region, bounded by a countable number of circular arcs, and of span less than e, containing as internal points every point of the Jordan curve corresponding to the internal points of d_p, and no other points of the Jordan curve.*

For, by § 137, starting with $\frac{1}{4}e$, we can determine d so that, if p be any point of (a, z) and d_p the interval of length d with p as centre, all the points corresponding to the points of d_p, lie inside a circle, with its centre at the corresponding point P and radius $\frac{1}{4}e$.

Let b and c be the end-points of d_p, and describe the small circle of radius $\frac{1}{4}e$ round the point P; the points B and C, corresponding to b and c, lie certainly inside this circle.

Now the points of (a, z) not internal to d_p correspond to points of the Jordan curve forming a closed set which includes b and c;

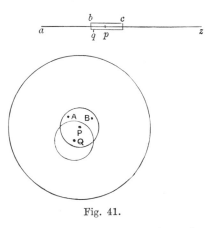

Fig. 41.

hence, if Q be the correspondent of any point q internal to d_p, the distance of Q from the points whose correspondents are not

* Hurwitz, Zürich Address, *Verhandlungen des ersten internationalen Mathematiker-Kongresses in Zürich vom 9. bis 11. August 1897*, Leipzig 1898, p. 102.

internal to d_p, has a definite minimum $2r_Q$, which is not greater than BQ or CQ, and is therefore less than $\tfrac{1}{2}e$.

Describing round each point Q as centre a circle of radius r_Q, a countably infinite set of these suffice to cover every point Q, and these generate a single region, since the points Q generate an arc of the Jordan curve. Further this region lies entirely within the circle with centre P and radius $\tfrac{1}{2}e$, since $r_Q < \tfrac{1}{4}e$. Now the points of the closed Jordan curve not internal to the region so constructed, form a single arc, and if e is sufficiently small it certainly lies partially outside the circle with centre P and radius $\tfrac{1}{2}e$. Thus if this region is not itself simply connected, there are certainly no points of the Jordan curve enclosed within its inner rims and we can, without enclosing any new points of the Jordan curve, make it simply connected by obliterating all inner rims. Let us then call it R_P. Then R_P is a simply connected region, of span less than e, and its rim consists of a countable set of circular arcs; R_P contains as internal points all the points of the Jordan curve corresponding to internal points of d_p, and no other points of the curve.

Cor. *The region R_P can be so constructed that it has only two points B and C of the curve on its rim, and so that, enclosing B and C in any regions whatever, the part of the rim of R_P not internal to these regions, consists of a finite number only of circular arcs.*

The region R_P, as already constructed, has clearly the first of these properties, since the only points of the Jordan curve internal to the circle of radius r_Q are points internal to R_P, so that the only points of the curve on the boundary of R_P are limiting points of points of the curve internal to R_P, and are therefore B and C only, which clearly are boundary points.

The second part of the corollary is a direct consequence of the fact that the points of the curve in R_P, not internal to the regions round B and C, form a closed set, so that only a finite number of the circles are needed to cover them.

THEOREM 5. (JORDAN'S THEOREM*.) *The plane is divided into two parts by a closed Jordan curve.*

* Jordan, *Cours d'Analyse*, Vol. i. pp. 91 to 99. The proof in the text is printed here for the first time. See also Veblen, *Trans. Amer. Math. Soc.* Vol. v. p. 365 (1905). Other proofs of this theorem, with restrictions as to the constitution of the Jordan curve, have been given by Schoenflies, *Gött. Nachr.* 1896, p. 79, Bliss, *Bull. Amer. Math. Soc.* Ser. 2, Vol. x. p. 398 (1904), and Ames, *Amer. Journ.*

Assuming any small positive quantity e, determine d, and choose out a finite number of intervals of length d covering up the whole of (a, z), so that no interval overlaps with more than two of the others, and no two intervals abut (Fig. 42). These intervals may be numbered in order from left to right $d_1, d_2, \dots d_n$.

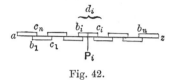

Fig. 42.

Let b_i and c_i be the left and right hand end-points of d_i, and p_i be a point of d_i not lying in any of the other segments, and let B_i, C_i, and P_i be the corresponding points of the Jordan curve.

Describe round each point P_i a region R_i, having the properties enunciated in Theorem 4 and the Corollary. Since the interval d_i has the segments (b_i, c_{i-1}) and (b_{i+1}, c_i) respectively common with d_{i-1} and d_{i+1}, it follows that the region R_i has in common with the regions R_{i-1} and R_{i+1} two parts (which are simply connected regions*) enclosing the arcs $B_i C_{i-1}$ and $B_{i+1} C_i$. If it overlaps with a third region, however, this can contain none of the points of the curve in R_i, and may therefore, since the common part is bounded by a finite number of circular arcs, be reduced till it has no common point with R_i, and no common boundary point, the newly introduced part of its boundary consisting still of a finite number of circular arcs. Similarly if the two regions which necessarily overlap with R_i, have other points common with R_i, we can reduce them till they have no such common points.[13]

Thus we may, without loss of generality, assume that the regions R_1, R_2, $\dots R_n$, each of span less than e, bounded by a countable number of circular arcs, and containing every point of the curve, are such that each R_i overlaps in a single part with each of the regions R_{i-1} and R_{i+1}, but with none of the others (Fig. 43). These regions form a single region R_e.

of Math. Vol. xxvii. (1905). The last of these, which is an arithmetic proof of great elegance and simplicity, is reproduced on pp. 130—141 of Osgood's *Lehrbuch der Funktionentheorie* (1906), the first half of the first volume of which has been received during the correction of the proof-sheets of the present chapter; a sketch of this proof had already been given in *Bull. Amer. Math. Soc.* Ser. 2, Vol. x. p. 301 (1904).

  * Since a simple polygon internal to such a part consists of points internal to both regions, and does not therefore enclose boundary points of either.

Now since the points $B_1 C_n B_2 C_1 B_3 C_2$, ... $B_n B_1$ are internal to R_e, they can be joined by a simple* polygon Π internal to R_e.

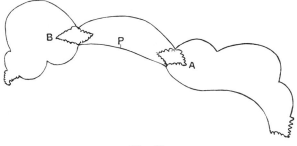

Fig. 43.

Also since R_i is simply connected, and B_i and C_i are points on its rim, R_i is divided into at least two parts by Π (Theorem 24, Ch. IX, p. 214). But, since R_{i-1} and R_{i+1} do not intersect, the rim of R_i is divided by R_{i-1} and R_{i+1} into precisely four parts, of which one is internal to R_{i-1} and intersects Π at B_i, another is internal to R_{i+1} and intersects Π at C_i. The remaining two parts form part of the boundary of R_e and are divided from one another by the parts in R_{i-1} and R_{i+1}, and therefore by Π. By Theorem 4, Cor. these consist of a finite number only of circular arcs: let that part which lies inside Π be l_i and the other m_i. Similarly the rim of R_{i+1} is divided into four parts, two of which, each consisting of a finite number of circular arcs, belong to the boundary of R_e, and one, and one only, of these, l_{i+1}, will abut with l_i, while the other, m_{i+1}, abuts with m_i. In this way we get inside Π a finite number of circular arcs, l_1, l_2, ... l_n, forming a simple circular polygon L_e, and outside Π another such polygon M_e from the m's: these two polygons form the complete boundary of R_e. M_e is therefore the outer rim (§ 133) of R_e, while L_e is an inner rim, and is the only inner rim of R_e. Thus R_e is a doubly connected region, got by cutting a simply connected region I_e with rim L_e out of a simply connected region with rim M_e.

Denoting the part of the plane outside this latter region by O_e, I_e is completely divided from O_e by the region R_e, and the whole plane consists of I_e, O_e and R_e.

Let e describe a sequence e_1, e_2, ... having zero as limit; the points common to R_{e_1}, R_{e_2}, ... form the Jordan curve, while if O denote the set of all the points belonging to O_{e_1}, O_{e_2}, ... and I of

* Footnote †, p. 179.[14]

all the points of $I_{e_1}, I_{e_2}, \ldots, O$ and I have no common points, and the whole plane consists of O, I and the Jordan curve.

Given any two points of O, we can assign a value of n such that the two points lie in O_{e_n}, and can therefore be joined by a path lying entirely in O_{e_n} and containing therefore no point of the Jordan curve: similarly for two points in I.

On the other hand if P is a point of O and Q of I, any polygon of a finite number of sides joining P to Q must cut the region R_{e_n}, for every value of n, in a closed set of points, and must therefore, by Cantor's Theorem of Deduction, contain at least one point of the Jordan curve.

Thus any polygon joining P to Q contains a point of the Jordan curve, so that the Jordan curve separates all the points of O from the points of I, while, as has been shewn, the points of O may be joined by polygons containing no points of the curve, and so may the points of I. That is to say the Jordan curve divides the plane into two parts, O and I. Q. E. D.

COR. *O and I are each simply connected domains with the Jordan curve as common rim.*

For every point of O, being a point of some O_{e_n}, is internal to a triangle consisting entirely of points of O, so that O consists of one or more domains. Also given any two triangles belonging to O, we can assign m so that e_m is less than the minimum distance between the Jordan curve and the pair of triangles, these latter will then belong to the transitive set of triangles generating O_{e_n}. Thus the triangles generating O form a transitive set, so that O is a single region.

That the Jordan curve forms the rim of O is easily seen, since any point of I, being a point of some I_{e_n}, is an ordinary external point of O_{e_n} and therefore of O, while every point of the Jordan curve is clearly a limiting point of points both of I and O. Thus the Jordan curve being the rim of O, O is by definition (§ 133) a simply connected region. Similarly I is a simply connected region with the Jordan curve as rim.

THEOREM 6. *If P be a point of a closed Jordan curve and Q a point not belonging to the curve, P and Q can be joined by a simple polygonal line of a finite number of sides, no point of which lies on the Jordan curve, or by one of a countably infinite number of sides having only P as limiting point.*

For let p be the point of the segment (a, z) corresponding to P, and describe a small interval d_p, of length less than an assigned norm d, having p as middle point, and the corresponding region R_P of span less than e, as in Theorem 5 of the present chapter.

Let O and I denote the two parts into which the plane is divided by the Jordan curve, and of these let O be that part to which Q belongs. Then since P is a limiting point of the points of O, there is a point P_1 of O inside R_P. Q can be then joined to P_1 by a simple polygonal line of a finite number of sides.

Now let e take all the values of a sequence

$$e_1, e_2, \ldots$$

of continually decreasing positive quantities having zero as limit, then d also describes such a sequence, and therefore, by Theorem 8 of Ch. IX, P is the only point internal to all the regions R_P for each value of e, and is therefore the sole and only limiting point of the points $P_1, P_2, \ldots$ of O chosen inside these regions; thus the polygon of a finite number of sides joining Q to P_1, together with the simple polygonal lines $P_1P_2, P_2P_3, \ldots$, (each drawn so as not to have any point except its first end-point common with any of the preceding), forms such a polygon as was contemplated in the enunciation of the theorem. Q. E. D.

The boundary of Osgood's region (Ex. 1, Ch. IX) is an example of a curve which divides the plane into two parts but is such that there are points of the curve which cannot be joined to points not on the curve by polygons such as were contemplated in the enunciation of the preceding theorem; such a curve is of course not a Jordan curve.

Schoenflies[*] has shewn that a complete curve (closed connected set) is a closed Jordan curve if it divides the plane into two parts, and is such that any point of it can be joined to any point in either part by a finite number of stretches consisting only of points of that part.

141. Sets of arcs and closed sets of points on a Jordan curve. It is now clear that the theory of sets of intervals on a straight line, gives us a corresponding theory of sets of arcs on a Jordan curve. It is not necessary to recapitulate the details.

In the same way we get the theory of closed sets of points on a Jordan curve, such a set being always complementary to a set of open arcs.

---

[*] *Loc. cit.*, pp. 221–2, footnote †.

The following is an example of a closed set of points on a Jordan curve, and its complementary set of arcs.

This closed set presents the interesting feature that, though it is dense nowhere in every region and on every Jordan or other curve, its projection on either of the coordinate axes consists of every point of a segment *.

Ex.   Take a unit square, and dot the ends of a diagonal : all dotted points are to belong to the pattern (set) and are called *primary points*.

Next divide the square into nine equal squares and dot the four corners of the middle square.

Our pattern is to lie entirely in those of the small squares which contain the chosen diagonal ; in the others we may erase the dividing lines. The

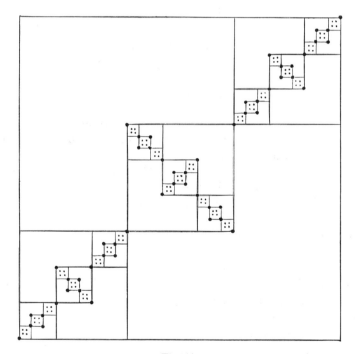

Fig. 44.

<hr />

* *Math. Ann.* LXI. pp. 281—286. This was supposed by Schoenflies to be impossible (*Bericht*, p. 87), and is interesting in relation to Baire's work, see § 135. Baire shewed that if $f(x, y)$ is a function which is continuous with respect to x and also with respect to y, and K is the set of points at which the oscillation of $f(x, y)$ is greater than or equal to k, then K is a set containing no stretches or curves. It follows that such a function cannot be totally discontinuous on any stretch or curve. The converse is not true, as is shewn by the above example.

two extreme squares are to contain the same pattern as the large square on a smaller scale, and the middle square is to contain the same pattern turned through a right angle round the centre of the square. These indications suffice to construct all the primary points. Fig. 44 shews the lie of the primary points constructed up to the point where the small squares have a side $\frac{1}{27}$. The subsequent primary points all lie in the small squares drawn in the figure, each of these small squares contains the main pattern on a small scale, either not turned through any angle or turned through a right angle.

The secondary points consist of all the limiting points of the primary points, and are therefore internal points of the small squares at every stage.

The pattern consists of all the primary and secondary points.

If l be any line parallel to a side of the square, it is either one of the bounding lines of the first central square, or it meets one of the first three small squares and one only. In the former case it contains a primary point, but no secondary point : since the pattern repeats itself in the small squares the same will be true if l is a bounding line of any small square. In the latter case, when l is not a bounding line of any small square, it cuts a single small square at every stage, and therefore contains the single secondary point which is internal to all these squares. Thus every such line l contains a point of the pattern.

In other words :—

The pattern projects into a side of the square: the projection is (1, 1) *except at the primary points which correspond in pairs to a countable set of points everywhere dense on the side of the square.*

A secondary point P has the property that, given any small quantity ϵ, a square can be described of side $< \epsilon$, having P as internal point, such that the points of the pattern in this square project into the same segment as the square itself.

By construction the pattern is *a closed plane set of points.* Any point P of the set is such that, given any small quantity ϵ, one of the small squares can be found of side $< \epsilon$ containing P as internal point or corner, and since this square contains a pattern similar to the whole pattern, it follows that P is a limiting point of the pattern. *Thus the pattern is a perfect set.*

It is easy to see also that it is *dense nowhere*, since in every small square there are squares containing no point of the set.

The pattern is not a connected set, for a straight line drawn perpendicular to the chosen diagonal and dividing it in the ratio 7 : 11 does not contain any point of the pattern although there are points of the pattern on each side. Since the pattern is reproduced in every small square it follows that the part of the pattern in any square, however small, is not connected. No arc of a curve, therefore, however small, consists entirely of points of the pattern.

If however we join the two primary points on any ordinate by a straight line, we get a Jordan curve, on which the pattern consists of the end-points and external points of a set of arcs. This is most easily proved as follows :—

Divide the segment (0, 1) of the x-axis into five equal parts, and make the parts $(\frac{1}{5}, \frac{2}{5})$ and $(\frac{3}{5}, \frac{4}{5})$ correspond projectively to the two largest segments of ordinates joining primary points, viz. the segments $(\frac{1}{3}, \frac{1}{3})$ to $(\frac{1}{3}, \frac{2}{3})$ and $(\frac{2}{3}, \frac{1}{3})$ to $(\frac{2}{3}, \frac{2}{3})$. The order of the points in these segments is to be maintained, viz. from left to right on the x-axis and upwards in the square (Fig. 45). The points

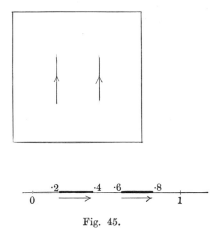

Fig. 45.

of the pattern in the three squares about the chosen diagonal in order are then to correspond to points in the three remaining intervals of the x-axis in order.

In each of the small squares and small segments this correspondence is carried further, care has only to be taken that the proper directions in the small squares are chosen, as indicated for instance by the arrow-heads in Fig. 46.

Continuing this process in each of the small squares *ad infinitum*, we have set up a (1, 1)-correspondence maintaining the order between the primary points and the end-points of the black intervals on the x-axis.

Extending this correspondence in the usual way to the secondary points and the external points of the black intervals, we have established a (1, 1)-correspondence between the points of the pattern and the points of a perfect set on the straight line, the parts of the ordinates between pairs of primary points corresponding to the black intervals of the perfect set, and the order being throughout maintained. This correspondence is clearly continuous. *Thus the pattern is a perfect set dense nowhere on a certain Jordan curve, and projects into the side of the square.*

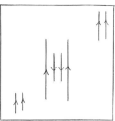

Fig. 46.

CHAPTER XI.

POTENCY OF PLANE SETS.

142. The Theory of Potency in higher space is in all essentials identical with that in linear space, since, as has been shewn in Ch. VIII, all the points of a plane, or of space of any finite (or indeed countably infinite) number of dimensions, are of potency c, so that any set of points in the plane or higher space has the same potency as a certain linear set. Thus the only potencies which can occur are those which occur on the straight line, and, as there, the only known potencies are, beside finite numbers, that of countable sets a, and that of the continuum c.

143. Countable sets. A countable set is, as before, one such that the elements of it can be brought into (1, 1)-correspondence with the natural numbers. The coordinates of a countable set of points are, by Theorem 3, Ch. IV, clearly countable; conversely, any set of points whose coordinates are countable, is itself countable; thus *the rational points in the plane or higher space are countable, and so are the algebraic points.*

When arranged in countable order a countable set will be said to form a *progression*, precisely as on the straight line.

Cantor's Theorem, that *a set of non-overlapping regions is always countable*, has been proved in Ch. IX, as well as the theorem that a set of overlapping regions may be replaced by a countable number of them having the same internal points as the whole set.

Theorem 9 of Ch. IV, which states that *any isolated set of points is countable*, is equally true in the plane or higher space, and may be proved in precisely the same way as on p. 42, using regions instead of intervals. The corollaries follow as a matter of course.

The extended Heine-Borel Theorem has already been proved.

As already mentioned in Ch. IX, the whole theory of derivation and deduction, adherences and coherences, is true in space of any finite number of dimensions, and requires no fresh proof for the plane. In particular *any set which has no component dense in itself is countable.*

The proof given on p. 43 of the theorem that *a countable set is never perfect* (Ch. IV, Theorem 11), applies equally for the plane or higher space, using regions instead of intervals, and the span instead of the length.

144. The potency c. It has been shewn that the potency of all points of the plane, or higher space, is c, and that the whole plane can be brought into $(1, 1)$-correspondence with the interior of a circle ; it follows, since any internal point of a region can be surrounded by a circular area lying entirely inside the region, that *the potency of all the points of any region is c.*

The theorem that *the potency of any perfect set is c*, may be most simply proved as follows by projection.

Project the set on to a straight line. If there be any isolated point P in the projection, and Q be any one of the points of the perfect set which project into P, since Q is a limiting point of the given set, there are points of the given set as near as we please to Q, and these cannot lie off the line PQ, because there are no points of the projection as near as we please to P. Thus Q is a limiting point of the section of the given set by PQ, and therefore this section is dense in itself. On the other hand, if the projection contains no isolated point it is dense in itself, thus in any case either the projection or the section is dense in itself. Now both the projection and the section are closed sets (§ 104), thus one or other is perfect, and has therefore the potency c. In either case the perfect set has a component of potency c, either the section itself or the component consisting of the nearest point on each ordinate to the corresponding projection, and is therefore of potency not less than c; on the other hand since the potency of the plane is c, the perfect set is of potency not greater than c, and is therefore of potency c^*.

---

* This important theorem, which had been enunciated by Cantor, was first proved for plane sets by Bendixson, *Bih. Svensk. vet. Handl.* 9, No. 6 (1884), using continued fractions. Another proof may be found in Schoenflies, p. 86. The *Quarterly Journal of Pure and Applied Math.* No. 143 (1905) contains a proof *ab initio* in which the potency of the plane, or of n-dimensional space is not pre-assumed to be c, a perfect set being here considered as a special case of an ordinary inner limiting set.

145. As in the straight line, limiting points may be classified
into those of *countable* and those of *more than countable degree*,
according to the potency of the component of the given set inside
a small region containing the point. The whole theory of the
Nucleus is precisely the same as for linear space, and may be
developed as in Ch. IV, § 25, using regions instead of intervals.
In like manner the discussion of §§ 26—30 of the same chapter
is applicable to the plane and higher space.

146. Ordinary inner limiting sets*.

DEF. Given a finite, or countably infinite series of sets of
regions, overlapping in any way, the set of all the points each of
which is internal to at least one region of each set is called *the
inner limiting set of the series of sets of regions, or an ordinary
inner limiting set*.

THEOREM 1. *Any ordinary inner limiting set may be defined
by means of normal regions, that is, by means of a series of sets of
regions, such that* (a) *the regions of each set are non-overlapping and
therefore countable, and* (b) *each region is internal to a region of the
preceding set, possibly however with common boundary points.*

The property (a) of the normal regions follows from Cor. 2
of Theorem 14, Ch. IX ; (b) is evident, since we only have
to cut off from the regions of the second set all parts external to
the regions of the first set, and so with each successive set ; it is
clear that, in so doing, a region of the second set which is reduced
will have boundary points common with a region of the first set,
and so for the other sets.

It now follows at once from Theorems 7 and 11 of Ch. IX, that
unless the span of the normal regions decreases without limit, the
inner limiting set will contain a perfect component, and will therefore have the potency *c*. We have therefore only to investigate the
case when the span of the normal regions decreases without
limit, in which case, each series of normal regions, one from each
successive set, lying each inside its predecessor, defines one and
only one point, which is a point of the inner limiting set, unless,
from and after some stage, all the regions have one common
boundary point.

LEMMA. *If an inner limiting set is such that the span of its
normal regions decreases without limit, and also that we can assign
a series of constantly increasing integers*

$$r_1, r_2, \ldots\ldots$$

* *Proc. L. M. S.* Ser. 2, Vol. III. Part 5, p. 371 (1905).

and corresponding to these a region of the r_1th normal set which contains two regions of the r_2th normal set entirely internal to it, each of which contains two regions of the r_3th normal set entirely internal to them, and so on, then the potency of the inner limiting set is c.

To prove this, let us denote the normal region of the r_1th set by d_{01}, and the two regions of the r_2th set which it contains by d_{001} and d_{011}, and continuing this system of notation, let us denote by d_{N01} and d_{N11} the two normal regions referred to in the enunciation, which lie inside the region already denoted by d_{N1}, where N denotes any combination of zeros and ones with n figures.

Since these regions have no common boundary points to complicate the issue, every series of them lying one inside the other defines a point of the inner limiting set, and conversely. To each such series of regions, however, by our system of notation, corresponds one non-terminating binary fraction, which, if d_{N1} be any region of the series, begins with the figures denoted by N; conversely to each non-terminating fraction there corresponds such a series of the regions. Thus there is a (1, 1)-correspondence between the points of our inner limiting set and the non-terminating binary fractions, so that the inner limiting set is of potency c.

THEOREM 2. *An ordinary inner limiting set which has a component dense in itself has the potency c; otherwise it is countable.*

By the theorem of the Nucleus, due to Cantor, any set which has no component dense in itself is countable, therefore we only have to prove the first part of the theorem.

Suppose the given ordinary inner limiting set to have a component U which is dense in itself. Let P be any point of U, then there is a region d_{01} of the first normal set containing P as internal point, and since P is a limiting point of U, there will be another point Q of U inside the region d_{01}.

Round these two points P and Q as centres we can then, since they are both internal points of d_{01}, describe circles external to one another, and entirely internal to d_{01}. Since, as has been remarked, we only have to prove the theorem when the span of the normal regions decreases without limit, we can assign a stage at which the span of all the normal regions is less than the radius of the smaller of these two circles; at this stage the normal regions containing P and Q will be distinct, and both internal to d_{01}. Denoting these two regions by d_{001} and d_{011}, there will be in each a component of U which is therefore dense in itself, and the argument can be repeated.   Thus the condition of the Lemma is satisfied, and the inner limiting set is therefore of potency c.

COR. *Since the component of an ordinary inner limiting set internal to any region is itself an ordinary inner limiting set, it follows that the degree of any point of an ordinary inner limiting set is either finite, a or c.*

147. The theorems given in Ch. IV now follow immediately for ordinary inner limiting sets in the plane or in n-dimensional space. The proofs there given only require verbal modification, " region " being used instead of " interval," and " span " for " length." These theorems may be enunciated in the following form. *If any set of points E be taken, and regions described round the points of E, so as to form a series of sets of regions, whose inner limiting set contains E as a component,*

(1) *The inner limiting set consists of E, together with certain points of the first derived set E' : the latter points may sometimes be absent:*

(2) *The regions may be so constructed that the inner limiting set contains every point of E':*

(3) *If the content of the regions is ever less than that of E', there is a more than countable set of points of E' not contained in the inner limiting set:*

(4) *The potency of the inner limiting set is the same as the potency of E', unless E contains no component dense in itself, while E' is more than countable.*

(5) *If E contains no component dense in itself, we can so arrange the regions that the inner limiting set consists of E alone; if however E' be more than countable, the regions can be so arranged that the inner limiting set is either countable or has the potency c.*

(6) *In general, we can so arrange the regions that those points of the inner limiting set which are not points of E are limiting points only of U, the greatest component of E which is dense in itself.*

(5) in conjunction with Theorem **2** gives *the necessary and sufficient condition that a countable set should be an inner limiting set, viz. the countable set must contain no component dense in itself.*

CHAPTER XII.

PLANE CONTENT AND AREA.

148. The Theory of Content, as developed in Chap. V, is again unaltered in all its main features when we come to deal with two or more dimensions; it receives, however, a vast extension from the fact that we have to distinguish between linear content, plane content, three-dimensional content, and so forth.

149. The theory of plane content in the plane. This theory is practically the same as that of linear content in the straight line, and generally as n-dimensional content in space of n dimensions. The definitions and properties given in Ch. V require little more than the substitution of the word *regions* for *intervals*, to be valid as they stand. As however there are one or two alterations of a more than verbal character which slightly complicate the issues, the discussion is here shortly given for the plane, and can then without difficulty be modified to suit higher space.

The principles which govern this theory are the same as in the straight line :—

(1) The content of a plane set of points, where it exists, is to be a non-negative quantity, and depends only on the relative position of the points of the set, not on its position as a whole in the fundamental region.

(2) The content of the sum of two sets having no common points is to be the sum of their contents.

As in the straight line we started with the content of intervals and sets of intervals as the basis of the whole discussion, so now we start with regions, and, as simplest form of regions, with triangles.

150. Content of triangles and regions. *The content of a triangular domain is taken to be its area,* measured in the usual way, by the product of its base and altitude.   Since then a triangle with an edge contains the same triangle without an edge and lies inside a triangular domain, whose content differs by as little as we please from that of the original triangle without its edge, it follows that, by principle (2), *the content of any triangle is the same whether, or no, any of its boundary points be included, and is the area of the triangle.*   The content of any finite number of triangles, without common points, is then the sum of their areas.

When we come to an infinite number of non-overlapping triangles, we know, by Cantor's Theorem, that they are countable, and it may easily be shewn, as on the straight line, that their areas form an absolutely convergent series if the fundamental region is finite.   *We define the content of a set of non-overlapping triangles to be the sum of their areas,* and this, whether, or no, some, or all, of the edge points of each triangle be included.

We have now to shew that neither of our principles is violated by this definition.   It is clear that the content is not negative, but it is not clear that it is unique, that is that it depends only on the set of points, not on the mode in which this set is divided up into triangles; if this is so, however, it is then clear that neither of our principles is violated.   This doubt can be set at rest precisely as on the straight line, but the reasoning is somewhat complicated by the fact that a triangle, unlike an interval, has a more than countable set of boundary points.

LEMMA 1.   *If a triangle contain a set of non-overlapping triangles, such that the content of the set of triangles is less than that of the first triangle, there is a more than countable set of internal points of the first triangle which are external to every triangle of the set.*

For let *e* be any small positive quantity less than the difference between the content of the set of triangles and the area of the first triangle.

Then we can enclose all the points on the perimeter of the first triangle in three triangles the sum of whose areas is less than $\frac{1}{2}e$, and we can increase the area of each of the triangles of the set inside the first triangle by so little that the sum of the parts added is less than $\frac{1}{4}e$; finally, we can, unless there is a more than countable set of points of the first triangle which are not internal to these triangles, enclose every one of them in a countable set of triangles of content less than $\frac{1}{4}e$.   When this has been done, since the points of the first triangle including its periphery form a closed set, a finite number of the triangles so constructed will suffice, so that every point of that closed triangle is internal to at least one of the triangles; but the sum of the areas of these triangles will be greater than that of the given set of triangles by less than e, and therefore less than the area of the triangle, which, by elementary geometry, is impossible.   Thus there will be a more than countable set of the internal points of the first triangle not internal to the triangles we had constructed and therefore external to the given set of triangles.   Q. E. D.

COR.   *However a triangle be divided up into triangles, so that no internal point of the first triangle is external to the small triangles, the content of the small triangles is equal to that of the first triangle.*

THEOREM 1.   *The content of a set of non-overlapping triangles is unique.*   For suppose two different sets of non-overlapping triangles to be given, such that every point which is not external to the one set is not external to the other, then, in order that the content of the set of points not external to either set may be unique, it is necessary and sufficient that the sum of the areas of the one set of triangles should be the same as that of the other.

The parts of one of the first set of triangles each of which belongs to one of the second set of triangles, are either themselves triangles, or polygons of at most six sides, which can therefore be divided up into at most three triangles; supposing this done, we have a third set of triangles, the points of which are again the same as those of either of the other sets.   By the preceding Lemma the sum of the areas of those of the third set of triangles which lie inside each of the first triangles is the area of the latter triangle, and therefore the sum of the areas of all the triangles of the third set is the same as that of all the triangles of the first set.   Similarly it is the same as that of all the triangles of the second set, therefore the sum of the areas of all the triangles of the second set is the same as that of all the triangles of the first set.   Q. E. D.

We have now proved that the content so far as it has now been defined is in accordance with our two principles.   *We have incidentally defined the content of a set of overlapping triangles, and the content of a domain, or of any region which is, as a set of*

points, equivalent to a set of triangles, and the content of a set of
such regions, viz. the content is in all these cases that of any
equivalent set of non-overlapping triangles.

THEOREM 2.   ADDITION THEOREM FOR CONTENT OF SETS OF
DOMAINS.   *Given two sets of domains of contents I_1 and I_2, and*
calling the content of their sum I and the content of their common
parts I',

$$I_1 + I_2 = I + I'.$$

The proof of this theorem is identical with that for the
corresponding Theorem 2 of Ch. V, using triangles instead of
intervals.

151. Content of a closed set. Since the complementary
points of a closed set form a set of domains, the black regions of
the set (p. 203), it follows that we have now, in accordance with
our second principle, defined *the content of a closed set as the
difference between the content of the fundamental region and that of
the black regions.*

It was already pointed out in Ch. VIII that the content of
the perfect set there defined as the boundary of a certain region
in Ex. 4 was zero, while that in Ex. 5 lies between $\frac{3}{10}$ and $\frac{2}{5}$.

Exactly as in Ch. V, using regions instead of intervals, it
may then be shewn that *the content I_P of a closed set of points is
the lower limit of the content of a set of regions containing the
points, and the upper limit of the content of closed components of the
closed set.*

The original definition of the plane content of a closed set, or
of n-dimensional content in space of n dimensions, was given by
Cantor (*Math. Ann.* XXIII. p. 473), and is only in form different from
that given above. Cantor describes a circle, or n-dimensional
sphere of radius e round every point of the closed set, and defines
the content of the set of points as the limit of the content of
these spheres when e is indefinitely decreased. As in Chap. V,
it may be shewn that this definition agrees with that given
in the text.

As on the straight line, the set consisting of all the points
common to two sets being called *their sum*, and the sum consisting of all their common points *their common part*, we have
the same connection between the two closed sets and their sum
and common part as we had in the case of regions.

THEOREM 3. *If G_1 and G_2 be two closed sets of points having no point common, the set consisting of G_1 and G_2 together is a closed set of content equal to the sum of their contents.*

THEOREM 4. *If a closed set G of content I contain a closed component of content J, it contains a closed component of content $I - J - \epsilon$ (where ϵ is as small as we please), having no point common with the former component.*

THEOREM 5. ADDITION THEOREM FOR THE CONTENT OF CLOSED SETS.

If G_1 and G_2 be two closed sets of points of contents I_1 and I_2, and G their sum, and G' their common part, G and G' are both closed sets, and, if I and I' be their contents,

$$I_1 + I_2 = I + I'.$$

152. Area of a region. It is to be noticed that a domain always has a content, and so has a closed region, but we have not defined the content of the most general region, and the same doubt remains as to whether every region has a content as in the case of any other open set of points.

The concept of the *area* of a region, though intimately connected with that of plane content, is not identical with it. If two regions which do not overlap, together form a single region, they must have a curved arc, as common boundary; this arc then becomes part of the interior of the single region which is their sum. It is one of the fundamental properties of the *area* of a region that the latter region is to have as area the sum of the areas of the two regions into which it is divided; the curved boundaries however are not to be allowed to complicate the issue.

Now it is clear that area cannot be identical with plane content, since, if the common boundary arc were of positive plane content, the sum of the areas of the two regions would be greater than the area of the region which is their sum.

Hence the following definition :—

If the content of a certain region without any edge is the same as the content of the same region together with all its boundary points, the common value of the content is called the area of the region, and the region is then said to be quadrable*.[51]

We have already had an example (Ex. 5 of Ch. VIII) of a region which is not quadrable, the following is another such region ;

---

* Jordan, *Cours d'Analyse,* I. p. 29 ; Lebesgue, *Intégral, Longueur, Aire,* § 12.

the boundary of this region gives us another example of a plane perfect set of points nowhere dense and of positive content.

Ex. 1.   Take a unit square, divide it into nine equal squares, and blacken those $(3+3-1)$ of them which lie about the diagonals of the unit square (Fig. 47).

Fig. 47.

Next divide each of the remaining $(3-1)^2$ squares into 3^4 equal squares, and blacken those (3^2+3^2-1) of them which lie about the diagonals, and so forth.

The blackened squares form at the end of the process a single region, whose rim is the periphery of the square ; the remaining boundary points form a set which is not only dense nowhere in the plane, but also on every straight line or curve, since none of the points lie on the diagonals of any of the squares used in the construction, and indeed every point on one of those diagonals can be surrounded by a small region lying entirely inside the black region, as is clear from the construction.

The black region together with all its boundary points constitutes the unit square.   The content of the black region without its boundary points is however less than the area of the unit square, for, denoting it by I_d, it is clearly given by the following expression :—

$$I_d = 1 - \left(1 - \frac{1}{3}\right)^2 + \left(1 - \frac{1}{3}\right)^2 \left\{1 - \left(1 - \frac{1}{3^2}\right)^2\right\} + \dots$$

$$= 1 - \underset{n=\infty}{\text{Lt}} \left\{\left(1 - \frac{1}{3}\right)\left(1 - \frac{1}{3^2}\right)\left(1 - \frac{1}{3^3}\right) \dots \left(1 - \frac{1}{3^n}\right)\right\}^2.$$

Thus $\frac{3}{5} < I_d < \frac{7}{10}$.

153. The region constructed in the preceding example is not simply connected. The following is an example of two simply connected regions, neither of which is quadrable. The common boundary of these regions is a Jordan curve, and therefore a closed set of points, and has positive plane content. Thus, we see that not only the general perfect set nowhere dense may have positive plane content, but even a curve, and still more specially a Jordan curve, may have positive plane content[*].

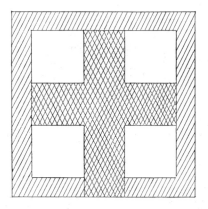

Fig. 48.

[*] The property given in Theorem 4, Ch. XIII, gave occasion to the first construction of a curve of this type by Osgood, *Trans. of the Amer. Math. Soc.* IV. p. 107 (1903), and Lebesgue, "Sur le problème des Aires," *Bull. de la Soc. Math. de France* (1903). The method, which is essentially the same in both papers, is a slight modification of that given by Hilbert, *loc. cit. supra*, p. 168; conversely, Ex. 2 (p. 245) may be adapted to give a new representation of the plane on the straight line. The points of Osgood's curve are constructed in isolated stretches or segments of straight lines; looking on these as beads, they are to be supposed first arranged in order, more and more beads, smaller and smaller in size, being inserted between each pair, and finally the curve threaded through them, without disturbing the order. This process, though quite satisfactory for demonstrating the existence of such a curve, is not altogether favourable for enabling us to picture its form. It would be easier to picture the curve if the beads were threaded one by one consecutively. The example in the text was constructed with this end in view.

Ex. 2*. Take a unit square : divide it up into regions as in Fig. 48, viz.:

(1) a cross of area x_1 (double shaded);

(2) a band round the rim, from one side of the base of the cross to the
other, of area y_1 (single shaded) ;

(3) four small squares, each of area $\frac{1}{4}s_1$ (*e.g.* $s_1 = 1 - \frac{1}{3}$).

Here $x_1 + y_1 + s_1 = 1.$

In the figures the parts which are double shaded will be referred to as if
they were coloured green, and single shaded parts as if they were blue. Thus
(1) will be supposed to be coloured green, (2) blue, and (3) to be left white.
The regions (1) and (2) are simply connected.

In each of the small white squares the construction is to be repeated,
using numbers x_2, y_2 and s_2 $\left(e.g., s_2 = 1 - \dfrac{1}{3^2}\right)$ for the ratios of the contents of
the regions in each small square to the area of that square, and arranging that
each small green cross has its base on the large green cross, so that together
the green crosses form a single simply connected region : the blue bands also
form together a single simply connected region.

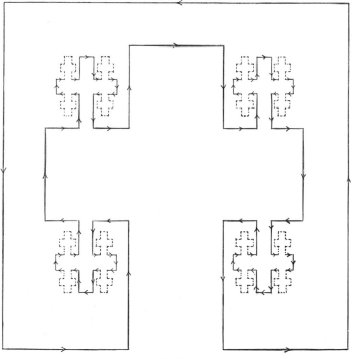

Fig. 49.

This construction is to be carried on *ad infinitum*. We then get green and
blue regions which ultimately leave no white regions over in the unit square

* *Quart. J. of Math.* No. 145, pp. 87—91 (1905).

at all. The unit square, without its rim, is thus divided up into three parts :
(1) the blue region without its rim, (2) the green region without its rim,
(3) the common part of the rims of the green and blue regions.

The rim of the blue region consists of this curve (3) together with part of
the periphery of the square ; these together form a closed curve, drawn in
Fig. 49, where the dotted part is, of course, only an approximation, since the
curve has an infinite number of tiny crosses, invisible in their details to the
naked eye, grouped about the dotted portions inside the small white squares
of Fig. 50.

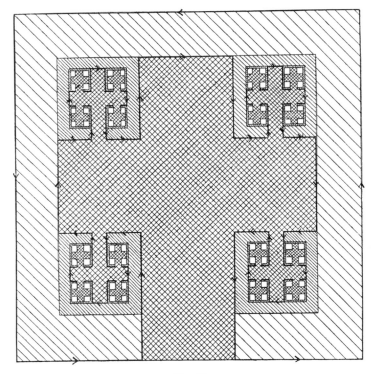

Fig. 50.

The curve (3) is easily shewn to be a Jordan curve. We have, in
fact, only to divide the segment (0, 1) of the x-axis into nine parts, and
blacken the alternate parts, beginning with the extreme parts. The five
black segments we now make correspond point for point continuously from
left to right with the five parts of the common rim of the largest green cross
and the blue region. The four white segments are to correspond to the four
white squares in Fig. 48. The correspondence is now continued in like manner,
by repeated division of the white segments into nine parts. When this has
been done *ad infinitum*, the straight parts of the common rim of the green
and blue regions correspond to the set of black intervals in the segment (0, 1).
By the usual extension, any other point of the curve, being the single point

internal to a series of white squares, each lying inside the preceding, corresponds to the single point internal to the corresponding segments of the x-axis. Thus the remaining points of the curve (3) are in (1, 1)-correspondence maintaining the order with the points of the perfect set complementary to the black intervals in question. Thus the points of the common rim, being in continuous (1, 1)-correspondence with the points of the segment (0, 1), form a Jordan curve. Since the remaining part of the rim of the blue region consists of a finite number of straight lines, this rim is a closed Jordan curve*.

The closed Jordan curve, being a closed set of points, has a content I, the green and blue regions without rims also have contents I_g and I_b respectively: the sum of these three sets being the unit square without its rim,

$$I + I_b + I_g = 1.$$

Now
$$I_g = x_1 + x_2 s_1 + x_3 s_2 + \ldots,$$
$$I_b = y_1 + y_2 s_1 + y_3 s_2 + \ldots,$$

where
$$x_i + y_i = 1 - s_i \; ;$$

therefore
$$I = 1 - [1 - s_1 + s_1(1 - s_2) + s_1 s_2(1 - s_3) + \ldots]$$
$$= \underset{n=\infty}{\mathrm{Lt}} \, s_1 s_2 s_3 \ldots s_n.$$

Thus
$$0 \leqslant I < 1,$$

and by suitably choosing the proportions, I may have any value between these limits ; for instance, when $s_i = 1 - \frac{1}{3}$,

$$I = \left(1 - \frac{1}{3}\right)\left(1 - \frac{1}{3^2}\right)\left(1 - \frac{1}{3^3}\right) \ldots > \frac{5}{9}.$$

This choice does not depend on the values of any finite number of the s's, but on the mode of their ultimate formation, in so far as the content is to be other than zero: the values of the s's at the beginning of the series will only modify the actual value of the content. The form of the curve will not be altered in its essential features by the choice, but the smaller the content the more of the curve will be distinguishable to the naked eye, and the less part of the plane will be apparently completely covered by the windings of the curve.

154. On the straight line we had nothing corresponding to the difficulty of conceiving the most general form of region, in the same way we had nothing corresponding to the difference between the area and the content. The only form of region on the straight line was the segment or interval, and its length was the same as its content. It is due to this distinction between sets of points on a straight line and sets in the plane or higher space that the properties of the content given in § 41 of Ch. V are not immediately generalisable, or rather the obvious generalisations are of no particular use. If we enunciate similar theorems for

---

* Similarly the rim of the green region is a closed Jordan curve.

sets of domains we gain no fresh information, since the complementary set of points, including as it does the rim of each region, certainly has the potency c. On the other hand a closed region, being a closed set of points, has a definite content, and, in accordance with the principles already laid down, *we shall define the content of a set of non-overlapping closed regions to be the sum of their contents.*

We can now prove the following theorem, which corresponds to Theorem 1, Ch. V.

THEOREM 6.   *If the content of a set of non-overlapping closed regions be less than the content l of the fundamental region*, there is a set of points of potency c external to all the regions.*

Let the fundamental region be denoted by F, and the regions of the set by $R_1, R_2, \ldots$. Then the points which are external to $R_1, R_2, \ldots$ constitute the inner limiting set of the regions F, $F - R_1, F - R_1 - R_2, \ldots$ and have therefore the potency c if they are not countable.

Suppose, if possible, that these external points are countable, then we can enclose the first in a triangle of area $\frac{1}{2}e$, the second in one of area $\frac{1}{4}e$, and so on, where e is as small as we please. These triangles together with $R_1, R_2, \ldots$ form a set of regions whose content differs from that of $R_1, R_2, \ldots$ by less than e, and is therefore, if e has been chosen sufficiently small, less than that of the fundamental region, which is impossible : therefore the hypothesis that the external points were countable is untenable, and they must therefore have the potency c.

As in Chap. V the converse of this theorem is not true, since it is possible to construct a set of closed regions, dense everywhere in the fundamental region, and of content as small as we please, for instance as follows.[52]

Ex. 3.[52] Take the unit square and divide it into m^2 equal squares and blacken the central square. Divide each of the small squares except the central one into m^4 equal squares and blacken the central squares in each, and so on. The sum of the areas I_d of the small squares is

$$1 - \left(1 - \frac{1}{m^2}\right) + \left(1 - \frac{1}{m^2}\right)\left\{1 - \left(1 - \frac{1}{m^4}\right)\right\} + \left(1 - \frac{1}{m^2}\right)\left(1 - \frac{1}{m^4}\right)\left\{1 - \left(1 - \frac{1}{m^6}\right)\right\} + \ldots$$

$$= 1 - \operatorname*{Lt}_{n=\infty} \left(1 - \frac{1}{m^2}\right)\left(1 - \frac{1}{m^4}\right) \ldots \left(1 - \frac{1}{m^{2n}}\right).$$

* Without rim, or closed, or described by means of a countable set of non-overlapping triangles.

Thus
$$\frac{1}{m^2} < I_\delta < \frac{1}{m^2 - 1},$$
although the small squares are dense everywhere in the unit square.

There is then a more than countable set of points external to the small squares and internal to the unit square, but they do not fill up any region, however small.

155. The theorems of § 46 are of general application and the proofs there given only require trifling verbal alterations to make them valid for space of any number of dimensions. The enunciations are here given for reference.

THEOREM 7.　*The content of a countable closed set is zero.*

THEOREM 8.　*The content of a closed set is the same as that of any one of its derived or deduced sets.*

COR.　*A closed set which is more than countable has the same content as its nucleus.*

THEOREM 9.　*The contents of two closed sets are equal if the points of both which are not common to both sets are countable.*

156. The investigations and theorems of § 48 are again of general application and the proofs only require verbal alteration; we have now, however, to distinguish more carefully between regions in general, closed regions and domains, and it must be noted that the analogous theorems in question, the enunciation of which is given below, must be taken to refer to closed regions.

THEOREM 10.　*Given a countably infinite series $D_1, D_2, \ldots$ of sets of closed regions, each of which contains only a finite number of regions, such that each region of D_{n+1} is contained in a region of D_n (with possibly one or more common boundary points), there is at least one point common to a region from each set; and the common points form a closed set.*

THEOREM 11.　*If to the hypothesis of the preceding theorem we add that the content of each D_n is greater than some positive quantity $\geqslant g$, the common points form a closed set of points D' of content $\geqslant g$, so that they have the potency c.*

THEOREM 12.　*Given a countably infinite series $D_1, D_2, \ldots$ of sets of closed regions, such that (1) each region of D_{n+1} is contained in a region of D_n for every value of n, and (2) the content I_n of each set D_n is greater than some positive quantity g, then (a) there is a* *set of points such that each is internal to a region of D_n, for every value of n, and* (b) *it contains closed components of content $> g - e$, where e is as small as we please; so that the potency of these points is c.*

In the proof of this theorem, we must replace the *intervals* $D_{n,r}$ by *triangles* (or other regions having areas) $D_{n,r}$, so as to be sure that we can cut off their rims without appreciably altering their contents. The same care must be taken in making the necessary verbal alterations in the proof of Theorem 14, Ch. V, so as to apply to the following theorem.

THEOREM 13. *Given an infinite number of sets of closed regions, in a finite fundamental region of content L, such that the content of each set of regions is greater than some positive quantity g, then a set of points of potency c exists, which is internal to an infinite series of these sets of regions, and contains closed components of content $> g - e$, where e is as small as we please.*

157. As in Ch. V the corresponding theorems for closed sets now follow, and so do the whole of the following articles (§§ 52—67) to the end of the chapter, which give the definition and properties of the content of measurable sets and of the inner and outer measures of the content.

158. Calculation of the plane content of closed sets. Just as the actual calculation of the area of a region has become a problem in simple integration, so the problem of determining the content of a plane set of points can be reduced to one of simple integration, when the content of its sections in one direction, and that of one section in the perpendicular direction, are known. This determination is obtained by the intermediate process of expressing the plane content as the upper integral of the content of its sections in any one direction (Theorem 16). The reduction of the problem of upper integration to that of simple integration is properly one of the applications of the theory of sets of points to other mathematical subjects; but, since it is not at present embodied in any text-book, it is reproduced in the next article.

In the same way the n-dimensional content of a closed set of points in n dimensions can be calculated as an n-ple upper integral of the (linear) content of its sections (§ 163), or—since the n-ple upper integral can be replaced by an n-fold upper integral—by the repetition n times of the process of finding a single upper integral (§ 159).

Each stage of the process may then, if we please, be effected by an ordinary integration. In this way we avoid the introduction of the content of any but linear sets.

The expression for the upper n-ple integral in a region Ω as an ordinary single integral is $K\Omega + \int_K^{K'} I\,dk$, where I is the content of those points of the region for which the maximum of the function is greater than or equal to k, and the limits are suitably chosen. This expression gives us the content of any closed set in space of n dimensions in the form $\int_0^\infty I\,dk$, where I is the content of points of an S_{n-1} for which the section of the set by straight lines perpendicular to the S_{n-1} has content $\geqslant k$.

159. A few preliminary definitions and explanations are necessary. We shall, for convenience, suppose that the regions in which the functions we are dealing with are defined, are finite and simply connected; also that the functions themselves are everywhere finite, and have finite upper and lower limits.

Consider any function. Take any point of the region for which it is defined; describe round this point an n-dimensional sphere having the point as centre; the upper limit of the values of the function in this sphere will itself tend towards a definite lower limit as the radius of the sphere is diminished: we call this the *maximum* of the function at the point. The *minimum* at the point is similarly defined. The excess of the maximum at P over the minimum at P is called the *oscillation* of the function at P.

Taking any function whatever of n variables, defined for a region of space of n dimensions, divide the region up into any finite number of quadrable partial regions, and multiply the content of each such part by the upper limit of the values of the function in that part, and sum for all the parts; the limit of this sum, when the content of each part is indefinitely diminished, and the number of parts accordingly increased, is called the *upper n-ple integral* of the function. The *lower n-ple integral* is similarly defined.

Now suppose, for simplicity of wording, that $n = 3$; and, to further simplify the ideas, let the region considered be a rectangular parallelepiped, having edges a, b, and c along the three axes. Find the upper integral of the function with respect to z, regarding x and y as constant, the limits being 0 and c. Find the upper integral with respect to y of the function of x and y

so obtained, regarding x as constant, the limits being 0 and b. Finally find the upper integral with respect to x of the function of x so obtained between the limits 0 and a. This final upper integral is called the *three-fold upper integral* of the original function, taken over, or with respect to, the parallelepiped. It is clear how we may generalise this conception, and give the corresponding definition for the *n-fold upper integral* of a function of n variables with respect to any closed n-dimensional region.

The theory of ordinary multiple integration begins by shewing that, when an ordinary n-ple integral exists, it is always equal to the n-fold ordinary integral; so that, moreover, the order of integration is immaterial. The corresponding theorem in our case does not hold. It is, however, at once obvious that the upper n-ple integral is greater than or equal to the n-fold upper integral. It has been proved that the equivalence does exist in the case of upper semi-continuous functions*; a corresponding theorem holds, of course, for lower semi-continuous functions.

160. The consideration of semi-continuous functions is due to Baire†; they are defined as follows:—

DEF. *A function is said to be an upper semi-continuous function if its value at every point is the upper limit of the values assumed by the function in the neighbourhood of the point when this neighbourhood is indefinitely diminished.*

A corresponding definition holds for a lower semi-continuous function : we have only to replace the word "upper" by "lower" in the above definition. Baire has shewn not only that these functions are point-wise discontinuous, but that they possess the following characteristic property :—

The points at which an upper semi-continuous function has a value greater than or equal to k form, for each value of k, a closed set. Similarly the points at which a lower semi-continuous function has a value less than or equal to k form a closed set.

It is scarcely necessary to add that an upper semi-continuous function actually assumes its maximum in any interval or region,

---

* "On Upper and Lower Integration," *Proc. L. M. S.* Ser. 2, Vol. II. pp. 52—66, 1904, Theorem 6.

† Baire, *Ann. di Mat.* (3), Vol. III. (1899).

and a lower semi-continuous function its minimum. We note also that a function not otherwise semi-continuous may be so at a particular point.

161. Take any function whatever. At every point of the region for which it is defined the function possesses a maximum, in the sense explained above. This system of values determines, therefore, a new function, the *upper limiting function of the given function*, or simply the *associated upper limiting function*. Taking the minimum instead of the maximum, we have in the same way the definition of the *associated lower limiting function*. Finally, taking the excess of the maximum over the minimum—that is, the oscillation—we have a third function, the *associated oscillation function*.

It is proved by Baire that the first and third of these functions are upper semi-continuous functions, while the second is a lower semi-continuous function. Baire also proves that a function and its associated upper limiting function have, in any open region, the same maximum. The fact that this is not in general true of a closed region is the explanation of the slightly complicated character of the proof of the theorem of the next article.

THEOREM 14. *The upper n-ple integral of a discontinuous function of any number of variables is unaltered if we replace the discontinuous function by its associated upper limiting function. The lower n-ple integral is in like manner unaltered if we replace the discontinuous function by its associated lower limiting function.*

The detailed proof is given for two dimensions: it is, however, of a perfectly general character, and requires at most a few trifling verbal alterations to make it valid for space of any number of dimensions.

Assume any small positive quantity e, and let us determine a corresponding e', such that, if the region of integration be divided into a finite number of small regions d, each of span less than e', the following two properties hold: (1) $\Sigma \bar{F} d$ is greater than $\int F dw$ by less than e, (2) $\Sigma \bar{f} d$ is greater than $\overline{\int} f dw$ by less than e: here f is the function, F the associated upper limiting function, $\bar{F}$ and $\bar{f}$ the upper limits of F and f respectively in the region d, dw is the element of area, and $\overline{\int}$ denotes upper integral. This is evidently possible.

Suppose $ABCD$ to be one of the regions d (Fig. 51). Then, $\bar{f}$ being the upper limit of f in $ABCD$, $\bar{f}$ is $\geqslant$ the upper limit of f the open region $ABCD$, that is, $\geqslant$ the upper limit of F in the domain of that region, and therefore $\geqslant \bar{F}'$, where $\bar{F}'$ is the upper limit of F in a closed region $A'B'C'D'$, lying inside $ABCD$, but nearly coinciding with it.

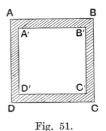

Fig. 51.

Suppose the boundary of $A'B'C'D'$ drawn so that the part of $ABCD$ outside $A'B'C'D'$ (shaded in the figure) has content less than k, where, if M be the upper limit of F in the whole region of integration and m the number of small regions d, $mMk < e$. Then, however we subdivide the shaded region and multiply each part d' by the upper limit of F inside it and sum, the sum will be less than e/m. Thus $\Sigma \bar{F}d$ over the now subdivided regions (i.e., the original m regions divided into shaded and unshaded parts in the manner indicated) lies between

$$\Sigma \bar{F}'(A'B'C'D') \quad \text{and} \quad e + \Sigma \bar{F}'(A'B'C'D').$$

But the spans of the new regions are still less than e'; therefore $\Sigma \bar{F}d$ over the new regions lies between $\bar{\int} Fdw$ and $e + \bar{\int} Fdw$, and may be denoted by $e_1 + \bar{\int} Fdw$, where e_1 lies between 0 and e.

Thus $\Sigma \bar{F}'(A'B'C'D') \leqslant e_1 + \bar{\int} Fdw \leqslant e + \Sigma \bar{F}'(A'B'C'D')$......(1).

Now, since, as was pointed out, $\bar{f} \geqslant \bar{F}'$,

$$\Sigma \bar{f}(ABCD) \geqslant \Sigma \bar{F}'(ABCD) > \Sigma \bar{F}'(A'B'C'D')......(2).$$

Comparing (1) and (2),

$$e_1 + \bar{\int} Fdw < e + \Sigma \bar{f}(ABCD);$$

that is, since that latter summation is greater than $\bar{\int} fdw$ by less than e, and e can be made as small as we please,

$$\bar{\int} Fdw \leqslant \bar{\int} fdw.$$

But, since F is never less than f at any point x, $\bar{\int} Fdw \geqslant \bar{\int} fdw$; therefore

$$\bar{\int} Fdw = \bar{\int} fdw. \qquad \text{Q. E. D.}$$

162. It is clear from the preceding article that we may in discussing upper and lower integrals confine our attention to semi-continuous functions. As, moreover, an upper semi-continuous function becomes a lower semi-continuous function if its sign be changed, and *vice versa*, we may confine our attention to upper semi-continuous functions. All the results will hold *mutatis mutandis* for lower semi-continuous functions.

THEOREM 15. *If I be the content of those points for which an upper semi-continuous function has a value greater than or equal to k, then I is an integrable (monotone) function of k, and the upper integral of the function (upper n-ple integral, if there be n independent variables), in any region of content S, is equal to $SK + \int_K^{K'} I\,dk$,*

where K' is any finite quantity greater than or equal to the greatest value assumed by the function in the region considered, and K is any finite quantity less than or equal to the lower limit of the values assumed by the function.

For, as k increases from K up to K', I never increases; it is therefore a monotone function. Since K and K' are finite, the points at which the function I makes a jump greater than or equal to e must, for each value of e, be finite in number, thus forming a closed set of zero content; hence I is an integrable function of k.

For simplicity the proof of the second part of the theorem is given only for one-dimensional space; the necessary verbal alterations for space of n dimensions can easily be made.

Since we know beforehand that an upper integral of the given function, say f, exists, we can determine a small positive quantity e such that, if the whole segment in which we are operating (region of integration) be divided into small segments, finite in number and each less than e, and if we multiply the length of each of these by the upper limit of the values of f in it, and sum for all the small regions, the result of the summation differs from the upper integral in question by less than some assigned small quantity e'. For brevity let us write e_n for $e/2^n$; so that

$$e = e_1 + e_2 + \dots.$$

Let us divide the segment (K', K) of the K-axis into $n+1$ equal parts, where n is any chosen integer, and denote the points of division by $K_1, K_2, \dots, K_n$. Then the points x of the segment (A, B) at which $f(x) \geqslant K_r$ form a closed set, say G_r of content I_r, contained in the closed set G_{r+1} of all the points of (A, B) at which $f(x) \geqslant K_{r+1}$, which is itself contained in the closed set G_{r+2} of all the points at which $f(x) \geqslant K_{r+2}$, and so on.

We can therefore enclose all the points of G_1 in a finite number of intervals, in general overlapping, each less than e', and the content of them lying between I_1 and $I_1 + e_1$.

The content of the remaining segment, or segments, of (A, B) lies between $L - I_1 - e_1$ and $L - I_1$; and the points of any one of the sets G_r which lie in them or on their boundaries form a closed set of content lying between $I_r - I_1 - e_1$ and $I_r - I_1$.

In this segment, or these segments, we can now, in like manner, enclose all the points of G_2 which lie in them, or on their boundaries, in a finite number of intervals, each less than e', so that their content lies between $I_2 - I_1 - e_1$ and $I_2 - I_1 - e_2$.

The segment or segments of (A, B) now left over have content lying between $L - I_1 - e_1 - I_2 + I_1 - e_2$ and $L - I_1 - I_2 + I_1$, that is, between $L - I_2 - e_1 - e_2$ and $L - I_2$; and inside these segments the points of G_r form a closed set of content lying between

$$I_r - I_2 - e_1 - e_2 \quad \text{and} \quad I_r - I_2.$$

Proceeding thus, we must, after at most n stages, have shut up all the points of (A, B) in a finite number of intervals, of course overlapping, each of length less than e'.

If we take only such parts of these intervals as do not overlap, and multiply the length of each part by the corresponding upper limit of f, we see that we get something less than

$$(I_1 + e_1) K' + (I_2 - I_1 + e_2) K_1 + (I_3 - I_2 + e_3) K_2 + \dots$$
$$\dots + (I_n - I_{n-1} + e_n) K_{n-1} + (S - I_n) K_n\,;$$

that is, less than $SK + (I_1 + I_2 + \dots + I_n + S) k' + eK'$, where k' is the nth part of the interval (K, K').

Since this is greater than the upper integral in question, but differs from it by less than e', the result at once follows, when we make n increase without limit.

163. THEOREM 16. *If $X(x)$ denote the content of the section of a plane closed set G by the ordinate through the point x, then $X(x)$ is a semi-continuous function of x, and the upper integral $\bar{\int} X(x)dx$ of $X(x)$ with respect to x is the content of the closed set.*

More generally, taking space of n dimensions, let $X(x)$ denote the content of the hyperplane section of a closed set by the S_{n-1} through the point x of the x-axis perpendicular to that axis; then the same is true.

Take then any set of lines, parallel to the axis of y, having a single limiting line, say p. Choose them in such a way that, as they approach p, the contents of the corresponding ordinate sections (which are, of course, also closed) of the given set G

have a definite limit, say I. Then we have to prove that the section of the set G by p has content greater than or equal to I.

We can evidently commence the set of lines at such a line that all the corresponding ordinate contents lie between $I - e$ and $I + e$, where e is as small as we please. Project all these ordinate sections on to the line p. Then, by Theorem 20, Chap. V, p. 96, there is a set, say C, of (inner) content greater than or equal to $I - e$, contained in an infinite number of these projections. That is, taking any point of C, and drawing its ordinate, this ordinate meets an infinite number of the lines in points of G; the limiting point of these points lies, by our choice of the lines, on p, and is therefore the point of C which we took. Since G, and therefore the ordinate section of G by p, is closed, this point belongs to the ordinate section by p. Thus this ordinate section contains closed components of content as near as we please to I; so that its own content is not less than I. Q. E. D.

Now consider the second part of the theorem. For simplicity let us take the region of integration to be a square of side L.

Let e be any chosen small positive quantity. Then, by the definition of content, I denoting the content of G, we can determine a small positive quantity e'', such that, if the square be divided up into small rectangles of span less than e'', the content of those small rectangles which contain points of G lies between I and $I + e$.

Also, by the definition of an upper integral $\int X(x)\,dx$, we can certainly find a small positive quantity e' less than e'', such that, if the segment (A, B) of the x-axis in which we are operating be divided in any way into a finite number of parts, then, provided the length of each part be less than e', the upper summation $\Sigma \overline{X}(x)dx$ is greater than $\int X(x)dx$ by less than e.

Now $X(x)$ lies between 0 and L: let us then choose any integer n, such that $ne > L$, and consider the closed [53] sets $G_1, G_2, \ldots$, where G_r consists of all the points x at which $X(x) \geqslant \dfrac{n-r}{n} L$.

Thus G_n is itself the section of G by the x-axis, and each G_r contains the preceding G_{r-1}, while the remaining points, viz., those of $G_r - G_{r-1}$, are such that at each

$$\frac{n-r}{n} L \leqslant X(x) < \frac{n-r+1}{n} L.$$

First, let us determine e_1', so that, if the points of G_1 be enclosed in intervals each of length less than e_1', the content of these intervals lies between I_1 and $I_1 + e$. Then let us enclose the ordinate section of G by the ordinate through x, in a finite number of small vertical intervals, each of length less than e'.

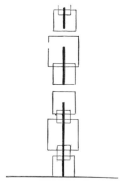

In the remaining parts of the ordinate marked black in Fig. 52 there are no points of G inside or on the boundaries. Let P be any point of such a black part; then, since G is closed, P is not a limiting point of G. We can therefore describe a small square, with P as centre, whose sides are parallel to the coordinate axes, and of length less than

Fig. 52.

e_1', so that there is no point of G inside this square or on its boundary. This being done for all points P in all the black parts, we can, by the Heine-Borel theorem, determine a finite number of the squares such that each point of the black parts is *internal* to one at least of the squares. There being now only a finite number of squares to consider, there is, of course, a definite smallest of them. Let the length of its side be $e_1(x)$, and let us continue its vertical sides up and down to bound off a strip parallel to the y-axis, of breadth $e_1(x) < e_1'$, having the ordinate through x as central line.

If we draw parallels to the x-axis through the ends of the vertical intervals originally drawn round the points of the ordinate section x, this strip is divided up into rectangles, of which those which contain the black parts of the ordinate x, shaded in

Fig. 53.

Fig. 53, contain no points of G inside or on the boundary, while the linear dimensions of the others are less than e', and their content lies between $X(x) \times e_1(x)$ and $Le_1 \times (x)$.

Now, n was chosen so large that $X(x)$ differs from L by less than e, and therefore from the upper limit of the values of $X(x)$ in $e_1(x)$ by less than e. Thus the content of the plain rectangles lies between $[\overline{X}(x) - e]\,e_1(x)$ and $[\overline{X}(x) + e]\,e_1(x)$.

Now let us construct such strips and rectangles for every point of G_1. By the extended Heine-Borel theorem (p. 202), there are then a finite number of these strips, each of breadth less than e_1', which enclose every point of the closed set G_1. Each strip is divided, as before, into plain and shaded rectangles.

Proceeding in like manner to enclose those points of G_2 in the remaining portions of the square (including those which lie on the boundaries), and then the points of G_3 not already enclosed, and so on, we eventually get the whole square divided up into plain and shaded rectangles; in the shaded rectangles there are no points of G, and the content of the plain rectangles lies between

$$\Sigma[\overline{X}(x) - e]d \quad \text{and} \quad \Sigma[\overline{X}(x) + e]d,$$

where d is the breadth of any vertical strip and $\overline{X}(x)$ the upper limit of the contents of the ordinate sections in d. Also, since the span of each rectangle is, by construction, less than e', and therefore less than e'', their content lies between I and $I + e$, and may be denoted by

$$I + \theta e \quad (0 < \theta \leqslant 1)$$

Again, $\Sigma \overline{X}(x)d$ differs from $\int X(x)dx$ by less than e, and may be denoted by $\int X(x)dx + \theta'e$. Thus

$$\int X(x)\,dx + \theta'e - Le \leqslant I + \theta e \leqslant \int X(x)\,dx + \theta''e + Le.$$

Since e can be chosen as small as we please, it follows that

$$I = \int X(x)\,dx. \qquad\qquad \text{Q. E. D.}$$

COR. *If the sections of a plane closed set by straight lines be translated parallel to themselves according to any law by which the plane set remains closed, the content of the set will not be altered.*

164. From Theorems 15 and 16 *we at once obtain the following expression, giving the content of a closed plane set as a single definite integral, viz.,* $\int_0^\infty I\,dk$, where I stands for the content of those points of the x-axis at which the ordinate sections of the given plane set are greater than or equal to k.

It is clear that this result may be extended in various ways to sets in space of higher dimensions. It should be noticed that these formulae are exactly analogous to the ordinary formulae in the integral calculus for areas, volumes, &c., and include them as particular cases.

From this formula a number of particular consequences are at once deduced. For example, *the necessary and sufficient condition that a plane closed set should have zero content is that the points x whose ordinate sections of the plane set have content greater than or equal to k should form a set of zero content.*

165. The function X of x is not necessarily what is usually called an integrable function. If it is an integrable function the upper integral $\bar{\int} X\,(x)\,dx$ is the ordinary or Riemann integral $\int X\,(x)\,dx$. An example of this was given in Ex. 4, Ch. VIII, where the section of the plane set by the axis of x, being a perfect set of content zero, $X(x)$ is an integrable function. In that example $X(x)$ was zero everywhere and therefore the content of the plane set was itself zero ; if, instead of making the ordinate sections of content zero, we make them perfect sets nowhere dense of constant content k (which is effected by suitably modifying the ordinate width of the arms of the crosses), the content of the plane set would still be

$$\int_0^1 X(x)\,dx = k.$$

In Example 5, $X\,(x)$ is not an integrable function, since the set of points x at which $X(x)$ is continuous (viz. the points where $X(x) = 0$), Ch. VIII, has not unit content. $X(x)$ is however what is now called a summable* function, since the points at which $X(x) \geqslant k$ form a measurable set, and has therefore a *generalised or Lebesgue integral*†. This generalised integral is in the case of an upper semi-continuous function identical with the upper integral‡ ; thus we have the following form of Theorem 16.

THEOREM 16 *a. If $X(x)$ denote the content of the section of a plane closed set by the ordinate through the point x, the content of the closed set is the generalised integral of $X(x)$ with respect to x.*

---

* "General Theory of Integration," *Phil. Trans.* Series A, Vol. 204 (1905), p. 243.

† Let the fundamental segment, or set, be divided into measurable sets in any conceivable way, and let the upper limit of the values of the function in each partial set be multiplied by the content of that set, and let the sum of these products be formed ; this is called an *upper summation* of the function ; similarly using the lower limits of the function in the partial sets we get a *lower summation* of the function. If the lower limit of the upper summations is equal to the upper limit of the lower summations, the function is said to be *summable*, and the common value is its generalised integral. No functions other than summable functions are known.

‡ "General Theory of Integration," p. 241.

Hence the content of the plane set, Ex. 5 of Ch. VIII, is

$$\int_0^\infty I\,dk = I \int_0^I dk = I^2,$$

where I is the content of the typical ternary set on the straight line; this agrees with the value of the content found in Ch. VIII by direct calculation.

THEOREM 17.   *If X^o and X^i be the outer and inner measures of the content of the ordinate section of a measurable set, such that the set got by closing it is of finite content, by the ordinate through the point x, X^o and X^i are both summable functions, and the generalised integral of either is the content of the measurable set.*

Let I be the content of the set, and e any assigned small positive quantity.   Let us take a closed component of the given set of content greater than $I - e$.   Denoting by X' the content of the ordinate section of this set, we have, by Theorem 16,

$$I - e < \overline{\int} X'\,dx.$$

Since X' is an upper semi-continuous function, the upper integral is the Lebesgue integral, hence

$I - e <$ the upper limit of the lower summations of X',

or, since X' is not greater than X^i,

$I - e <$ the upper limit of the lower summations of X^i.

Since e may be as small as we please,

$I \leqslant$ the upper limit of the lower summations of X^i,

$\leqslant$ the upper limit of the lower summations of X^o.

Next let the content of the set got by closing the given set be denoted by S, and that of the complementary set by J, so that

$$I + J = S.$$

Denoting by Y' and Z' the quantities for the complementary set and the whole closed set corresponding to X' for the given set, we have, as before,

$J \leqslant$ the upper limit of the lower summations of Y'.

Now $(Z' - Y')$ is the content of the difference of two closed sets, that is of an inner limiting set*, containing the set X, therefore

$$Z' - Y' \geqslant X^o \geqslant X^i.$$

---

* Theorem 39, p. 73.

Hence

$J \leqslant$ the upper limit of the lower summations of $(Z' - X^o)$,

$\leqslant$ the upper limit of (a lower summation of Z' minus an upper summation of X^o).

But the upper limit of the lower summations of Z' is the generalised integral of Z', that is, S; therefore

$J \leqslant S -$ the lower limit of the upper summations of X^o,

that is,

$I \geqslant$ the lower limit of the upper summations of X^o,

a fortiori,

$I \geqslant$ the lower limit of the upper summations of X^i.

Now every upper summation of a function is greater than, or at least equal to, any lower summation, so that a quantity cannot be less than the upper limit of the lower summations without being less than the lower limit of the upper summations; neither can it be greater than the lower limit of the upper summations without being greater than the upper limit of the lower summations. Thus I must be actually equal to the upper limit of the lower summations as well as to the lower limit of the upper summations in the case of either X^o or X^i; that is to say, I is the generalised integral of either X^o or X^i, and both these functions are summable. Q. E. D.

COR. 1. *Each ordinate section of a measurable set being moved on its ordinate in such a manner that the (linear) content of the section is unaltered, and that the whole set remains measurable, the content of the whole set is unaltered.*

COR. 2. *At each point of a set of points of content A draw an ordinate, and on it take any set of points of (linear) content B, the content of the whole set so formed is AB.*

Here, as elsewhere, the fundamental set need not be a linear set, but may have a content of any number of dimensions.

166. Calculation of the content of any measurable set.
The preceding section gives us $\int_0^\infty I dk$ as the content of any measurable set (provided the set got by closing it has finite content), here I is the content of the set of points of the fundamental set

at which the inner (or the outer) measure of the content is $\geq k$. This, together with the preceding section, gives the solution of the problem of the reduction of the calculation of n-dimensional content to that of $(n-1)$-dimensional content, and so ultimately to that of linear content.

Bearing in mind the definition of a generalised integral, we have the following rule for finding the content of an n-dimensional set :—*Take any hyperplane section of the set, project the set on to this hyperplane, and take any measurable set containing this projection as the fundamental set S. Divide S up in any way into a finite or countably infinite set of measurable components, and multiply the content of each component by the upper (lower) limit of the values of the (linear) inner or outer content of the corresponding ordinate sections of the given set; summing all such products, the lower (upper) limit of all such summations is the content of the given set.*

CHAPTER XIII.

LENGTH AND LINEAR CONTENT.

167. Length of a Jordan curve. Let o, p_1, p_2, ... p_n, z be points of a segment (o, z) in order from left to right, such that the distance between each adjacent pair of points is less than some assigned norm d. Then the corresponding points

$$O, P_1, P_2, \ldots P_n, Z$$

on the Jordan curve are such that the distance between each adjacent pair of points is less than e, where e depends on d, and diminishes indefinitely with it. Let the perimeter of the simple polygonal path $OP_1 \ldots P_n Z$ be denoted by π_d. Then it may be shewn, as follows, that π_d has a definite (finite or infinite) limit, when d is indefinitely decreased; this is called *the length* of the Jordan curve.

THEOREM 1. *Every Jordan curve has a definite length, which may be either finite or infinite.*

For let d_1, d_2, ... be a sequence of decreasing positive quantities having zero as limit, and e_1, e_2, ... the corresponding sequence of radii for points of the Jordan curve, so that all the points of the segment (o, z) within an interval of length $\leqslant d_i$ correspond to points of the Jordan curve within a circle of radius $\leqslant e_i$. Then the e's also form a sequence of decreasing positive quantities having zero as limit.. Let A_1, A_2, ... be a series of polygonal paths of lengths a_1, a_2, ..., inscribed in the Jordan curve, the vertices of A_i being

$$P_1, P_2, \ldots P_{m_i},$$

so that the distance between consecutive corresponding points

$$p_1, p_2, \ldots p_{m_i}$$

is less than d_i, and therefore the segments $P_r P_{r+1}$ are each less than e_i.

Then, fixing i, we can determine j so that for this and all greater values of j,

$$2\,(m_i+1)\,e_j < \epsilon,$$

where ϵ is any assigned small positive quantity.

Let the vertices of A_j be denoted by $Q_1, Q_2, \dots Q_{n_i}$, and the corresponding points by $q_1, q_2, \dots q_{n_i}$: and let the first and last of the latter points internal to the interval $(p_r p_{r+1})$ be q_λ and q_μ. Then the path A_j consists of the m_i portions $Q_\lambda Q_\mu$, together with the missing sides $Q_{\lambda-1}Q_\lambda$, etc.: as to these latter there is either one, or a pair, corresponding to each point p_r, according as $q_{\lambda-1}$ does not, or does, coincide with p_r (Fig. 54).

Fig. 54.

Now $\qquad\qquad p_r q_\lambda < d_j,$

therefore $\qquad\qquad P_r Q_\lambda < e_j\,;$

similarly $\qquad\qquad P_{r+1} Q_\mu < e_j\,;$

and since $P_r Q_\lambda$ and $P_{r+1} Q_\mu$ and the part of A_j between Q_λ and Q_μ are together greater than the side $P_r P_{r+1}$ of A_i, it follows that

$$a_j + 2\,(m_i+1)\,e_j > a_i,$$

and therefore $\qquad\qquad a_j + \epsilon > a_i.$

Since ϵ may be made as small as we please, this shews that, given any integer i we can determine j so large that, for that and all subsequent integers, $a_j \geqslant a_i$.

Thus the quantities a_i approach a definite limit a, which is their *upper* limit, and may, of course, as such be finite or infinite. By exactly the same argument, using B_j and b_j instead of A_j and a_j it follows, that if we have a second series of such polygons $B_1, B_2, \dots$ and b is the upper limit of their lengths, we can determine j so that, for that and all subsequent integers,

$$b_j \geqslant a_i,$$

and therefore $\qquad\qquad b \geqslant a\,;$

but similarly $\qquad\qquad a \geqslant b,$

therefore $\qquad\qquad a = b,$

and the upper limits are the same.

Thus however the polygonal paths be constructed, their lengths have the same finite or infinite upper limit, or limits, so that the Jordan curve has a definite length.

THEOREM 2. *The necessary and sufficient condition that a Jordan curve*

$$x = f(t), \quad y = \phi(t)$$

should have finite length, is that the functions f and ϕ should have a bounded variation*.

For let a, a_1, ... a_{n-1}, b be the values of t corresponding to the vertices of an inscribed polygonal path. Then since a side of a path is not less than its projection on one of the axes, and not greater than the sum of its projections, the length of the path lies between the greater of

$$| f(a_1) - f(a) | + | f(a_2) - f(a_1) | + \dots + | f(b) - f(a_{n-1}) |,$$

$$| \phi(a_1) - \phi(a) | + | \phi(a_2) - \phi(a_1) | + \dots + | \phi(b) - \phi(a_{n-1}) |,$$

and the sum of these two quantities : the former of these is the variation of f and the latter the variation of ϕ with respect to the system of values a, a_1, ... a_{n-1}, b. If both f and ϕ are functions with bounded variation, we can assign some finite quantity which the sum of these two quantities cannot exceed, and therefore the upper bound of the length of the path being finite, the curve has finite length. Conversely if one of f and ϕ be not such a function, one of the two quantities may be made greater than any assignable quantity, and therefore the curve will have infinite length. This proves the theorem.

A Jordan curve of finite length is said to be *rectifiable*†; the term rectifiable may also be extended to a Jordan curve of infinite length, provided the only partial arcs of the curve which have not finite length are such that the corresponding points of the segment (o, z) always include one or both of the end points o and z, $(t = t_0$, or $t = T)$.

A circle, ellipse, hyperbola, straight line are, in this sense, all rectifiable Jordan curves.

---

* Jordan, *Cours d'Anal.*, §§ 105 *seq.* If $f(x)$ is a function of x, defined from a to b, and we divide the interval (a, b) into n parts at the points $a, a_1, a_2, \dots a_{n-1}$, b, the sum of the oscillations

$$| f(a_1) - f(a) | + | f(a_2) - f(a_1) | + \dots + | f(b) - f(a_{n-1}) |$$

is called the variation of $f(x)$ for the system of values $a, a_1, \dots a_{n-1}, b$. If, for every choice of these values, the variation is always less than some finite quantity λ, $f(x)$ is said to be a function of x with bounded variation.

† Jordan, *loc. cit.* Lebesgue, *Intégral, Longueur, Aire*, § 42.

The argument of the preceding proof can be used to prove the following theorem :—

THEOREM 3. *The length of the arc of a rectifiable Jordan curve of finite length from t_1 to t, where $t > t_1$, is a continuous function of t which never decreases. The same is true of any arc of a rectifiable Jordan curve of infinite length, provided the arc does not contain a point corresponding to o or z.*

THEOREM 4. *A rectifiable Jordan curve, whether of finite or infinite length, has always zero plane content.*

Suppose first that the curve itself has finite length L. Divide it up into n equal arcs; then the distance of any point of any such arc from any other point of the same arc is, by the definition of the length, less than $\dfrac{L}{n}$, therefore we can certainly cover every point of the curve by means of n circles of radius $\dfrac{L}{n}$. The area covered by these circles is not greater than $n \cdot \pi \cdot \dfrac{L^2}{n^2} = \pi \dfrac{L^2}{n}$. Since n may be made as large as we please, this area may be made as small as we please; thus in this case the curve has zero plane content[*].

Next let the curve itself have infinite length. In the corresponding segment (o, z) take a sequence of points $o_1, o_2, \ldots$ having o as limit, and another $z_1, z_2, \ldots$ having z as limit. Then the corresponding arcs $(O_1 Z_1)$, $(O_2 Z_2)$, have each finite length, and therefore, by the above, we can enclose the first in small circles covering an area less than $\dfrac{e}{2}$, the second in small circles covering an area less than $\dfrac{e}{2^2}$, and so on; and therefore the whole curve in small circles covering an area less than e, so that the whole curve has plane content zero. Q. E. D.

THEOREM 5. *In calculating the length of a Jordan curve, we may take instead of a simple polygonal path of a finite number of sides, one of an infinite number of sides, or a finite or countably infinite set of such paths, provided the limiting points of the vertices of the path or paths are countable.*

For let the limiting points arranged in countable order be $P_1, P_2, \ldots$.

* Jordan, *Cours d'Anal.*, I. p. 107.

Then, as on the straight line, each countably infinite set of abutting chords of the Jordan curve forming a single path determines a pair of points of the set $P_1, P_2, \ldots$ so that we have two consecutive limiting points, say P_i, P_j. The path P_iP_j has a certain definite length, and we can determine a finite number of the chords whose sum differs from this length by less than $\dfrac{\epsilon}{2^{i+1}}$: let Q_i be the end point of these chords which lies nearest to P_i, then the sum of all the chords from Q_i to P_i is less than $\dfrac{\epsilon}{2^{i+1}}$: similarly we determine a point Q_i' on the other side of P. Now replacing all the chords from Q_i to Q_i' by the two chords Q_iP and PQ_i' for all values of i, we get a finite path, whose length differs from the sum of the lengths of the polygons in question by less than ϵ. This shews that the upper limit of the length of such paths or sets of paths is the same as the upper limit of the lengths of paths of a finite number of sides.   Q. E. D.

168. Calculation of the length of a Jordan curve.

Suppose that f and ϕ are both differentiable. Then the length of the chord joining the points of the curve whose parameters are t_1 and t_2, lies between $(t_2 - t_1)$ multiplied by the upper and lower limits of $\sqrt{f'^2 + \phi'^2}$ in the interval (t_1, t_2).

Thus the length of the path lies between the upper and lower summations of $\sqrt{f'^2 + \phi'^2}$, and therefore, in the limit, the length of the curve lies between the upper and lower integrals of $\sqrt{f'^2 + \phi'^2}$.

Thus, *if $\sqrt{f'^2 + \phi'^2}$ is an integrable function* (which will be the case if f' and ϕ' are each integrable),

$$L = \int_{t_0}^{T} \sqrt{f'^2 + \phi'^2}\, dt \quad \ldots\ldots\ldots\ldots\ldots(1).$$

Next suppose that f' and ϕ' are not integrable functions but $\sqrt{f'^2 + \phi'^2}$ is a perfect differential, say

$$\sqrt{f'^2(t) + \phi'^2(t)} = \frac{d}{dt}\,\sigma(t),$$

we can then prove that*

$$L = \sigma(T) - \sigma(t_0) \ldots\ldots\ldots\ldots\ldots(2)$$

as follows.

---

* Lebesgue, *Leçons sur l'Intégration* (1904), p. 61.

Taking any point P on the curve of parameter t_1, we can, since f' and ϕ' both exist, determine a small quantity d_1, such that for any point Q of the curve whose parameter t_2 lies between $t_1 \pm d_1$, the length of the chord PQ differs from

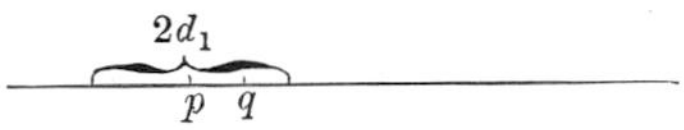

Fig. 55.

$$\sqrt{f'^2(t_1) + \phi'^2(t_1)}\,(t_2 - t_1)$$

by less than $\epsilon\,(t_2 - t_1)$.

Further we can determine a small quantity d_2 such that for any point Q of the curve whose parameter t_2 lies between $t_1 \pm d_2$, $\sqrt{f'^2(t_1) + \phi'^2(t_1)}$ differs from

$$\frac{\sigma(t_2) - \sigma(t_1)}{t_2 - t_1}$$

by less than ϵ.

Thus for any point Q of the curve whose parameter lies between $t_1 \pm \delta$, where δ is the lesser of d, d_2 and d_1 (where d is any fixed small quantity), the chord PQ differs from $\sigma(t_2) - \sigma(t_1)$ by less than $2\epsilon\,(t_2 - t_1)$.

Let us start from t_0 and draw such a chord, this determines t_1; then from t_1 draw such a chord, this determines t_2, and so on. Either after a finite number of such chords have been drawn, we have reached T_1, or else a countably infinite number of such chords determine a limiting point t_ω, which is either T, or some point to the left of T. In the latter case, starting afresh from t_ω, we repeat the process, and obtain $t_{\omega+1}$, $t_{\omega+2}$, ... and finally, if we do not arrive at T, a limiting point $t_{\omega.2}$ and so on.

Since a set of intervals is always countable, we cannot exhaust all the Cantor ordinal numbers of the second class before we arrive at a point of arrest, which cannot be other than T, since from any other point we could proceed further. The set of intervals so constructed is dense everywhere in (t_0, T), and has no external points, so that their content is the same as that of (t_0, T). Further, by Theorem 5, the corresponding series of chords of the Jordan curve may be used to calculate the length of the curve.

Now the sum of the lengths of these chords differs from $\sigma(T) - \sigma(t_0)$ by less than $2\epsilon\,(T - t_0)$, and therefore decreasing d and ϵ indefinitely the result (2) follows.

169. Linear content on a rectifiable Jordan curve.
The *linear content of a set of arcs on a rectifiable Jordan curve with respect to that curve* can now be defined as the sum of their

lengths; the linear content of a closed set of points on such a curve as the difference between the length of the smallest segment in which the set lies, and the content of the complementary arcs in that segment, precisely as on the straight line. The theory of linear content on a rectifiable Jordan curve can then be extended to open sets, defining the inner and outer measures of the linear content as respectively the upper limit of the linear contents of closed components of the set, and the lower limit of the linear content of a set of arcs containing the set. When the inner and outer measures of the linear content are equal, the set will be said to be linearly measurable, and the common limiting value will be called the content of the set. Everything said in the chapter on content of points on a straight line will now have its correlative in the theory of linear content on a rectifiable Jordan curve with respect to that curve; it is unnecessary to go further into details on this point.

170. General notions on the subject of linear content. It is to be noticed that, by Theorem 4, all sets of points on a rectifiable Jordan curve are measurable sets and have the plane content zero. On the other hand, given a plane set G of plane content zero, it is not always easy to decide whether or not G lies on a Jordan curve, or indeed whether it lies on any curve at all, still less whether it lies on one, and only one, Jordan curve, and whether that curve is rectifiable. We want a definition of linear content in the plane from the point of view of the plane itself, and to develop this theory independently of considerations as to curves on which the set may lie.

It will be seen that this theory is only in its earliest infancy, so that there is very little which can at present be said about it; it is clear however from the results to be given that it is by no means a very simple or obvious theory*.

Two definitions of linear content have been given. The first presents itself naturally, and we may assert with confidence that it must have occurred independently to all workers in the subject. It will be shewn however that it leads to surprising and even paradoxical results:—

(1) *Even the linear content of such a simple set as a countably infinite set of concentric circular arcs, with no other limiting point*

* *Proc. L. M. S.* Ser. 2, Vol. III. pp. 461—477 (1905).

except the centre, is not necessarily the sum of the lengths of the arcs, but may be greater than this.

(2)   *The linear content of a closed countable set of points in the plane is not necessarily zero.   It may even be infinite.*

The second definition of linear content does not affect these anomalies.

The question naturally arises : Is it possible to give a different definition of linear content, one, in fact, more consistent with what our knowledge of sets on the straight line would lead us to expect? Careful consideration suggests that such a definition is impossible. Content, whatever its dimensions, is a property of a set with respect to the fundamental region R.   If the set lies in a fundamental region R' of lower dimensions m, there is every reason to suppose that the m-dimensional content of the set with respect to R will depend not only on the m-dimensional content with respect to R', but also on the lie of R'.   In particular, the linear content of a plane closed set of points lying on a curve whose arcs can be measured may or may not be the same as the linear content of the set with respect to the curve, according to the form of the curve.

171.   The first definition of the linear content of a closed plane set of points is as follows :—

DEF.   Let small regions be described round the points of a closed set G, as in the process of finding the plane content, and let them be chosen to be circles of diameter d with the points of the set as centres, and let $F(d)$ be their content.   Let $f(d) = F(d)/d$; then, if $f(d)$ has a definite limit I when d is indefinitely decreased, I is called *the linear content of G.*

It is not clear from this definition of linear content that the linear content I in the case of a closed set always exists.   In other words, it is conceivable that $f(d)$ may oscillate between certain positive limits of variation.   In the case of the plane content in the plane or the linear content on the straight line this difficulty did not exist, since $F(d)$ is clearly a monotone function of d which is continuous.   It is, however, important to devise a definition by which the content of a closed set is always definite, since this would enable us to extend the theory of linear content to open sets, precisely as was done in the case of the ordinary content.   If it could be shewn that the limit of $f(d)$ is not always definite for a closed set, it might be convenient to take the lower and upper limits of variation of the limit as measures of the content.

There is another point which cannot be passed over. In the case of the ordinary content (whether linear content on the straight line, plane content in the plane, or n-dimensional content in space of n dimensions) it can be shewn that, if domains of span less than d be described in any manner so as to enclose all the points of the closed set, and the content of these regions be $\Phi(d)$, then $\Phi(d)$ approaches a definite limit as d is indefinitely decreased, independent of the form of the regions. Thus the choice of circles with the points of the set as centres and of equal diameters d is merely apparently arbitrary, and the ordinary content as originally defined by Cantor by means of circles (or spheres) can be found by using any convenient domains.

In the case of the linear content, however, if $\Phi(d)/d$ have a definite limit as d is indefinitely decreased, this limit may be quite different from the linear content I. It is easy to see that the limit may be zero, even when the linear content I is definite and positive (for instance, when the set lies on a straight line, we only have to take the span of the regions in the direction perpendicular to the straight line to decrease without limit compared to d), but the limit may also be greater than the linear content I.

If the set lie on a straight line, it is easy to shew that the regions may be so constructed that we get for $\Phi(d)$ the limit $2dI$, but they cannot be constructed so as to get a greater limit. Supposing for simplicity the set to be a closed segment AB, we only have to divide AB into two sets G_1 and G_2, both dense everywhere in AB, and describe circles round the points of G_1 with their centres on one of the parallels to AB at a distance $\frac{1}{2}d - e$, where e is as small as we please compared to d, and circles round the points of G_2 with their centres on the other parallel. Such sets of circles have a content differing by as little as we please from $2dI$.

On the other hand, *whatever set G we have in the plane, of linear content I, the content of a set of small regions of span d enclosing the points of the set cannot exceed $2d(I + d')$, where d' vanishes with d.* For each such region lies within a circle of *radius d* with its centre at any point internal to it. Therefore the content of the small regions is not greater than that of a set of circles of diameter $2d$ with their centres at the points of the set. By the definition of the linear content, the content of these circles is $2d\,(I + d')$, where d' vanishes with d; which proves the statement made above.

It is not, however, always possible so to choose the regions as to reach near the limit $2dI$. Ex. 2 is a case in which, whatever form of region we take, we cannot get a content much, if at all, greater than dI. (See pp. $276 - 281$.)

172. These considerations suggest the second definition of linear content.

DEF. Let small regions of span d be described in any manner round the points of the given closed set. Let their content be $\Phi\,(d)$, and let $\Phi\,(d)/2d$ be denoted by $\phi\,(d)$. As d is indefinitely decreased, let J be the upper limit of the variation of $\underset{d=0}{\mathrm{Lt}}\ \phi\,(d)$; J is called *the linear content of G*.

By this definition *the linear content J of a closed set is always definite, and the content of a set of regions of span d round the points of the set is less than $2d\,(J + d')$, where d' vanishes with d, but can, by proper choice of regions, be made as near as we please to $2dJ$.*

If the linear content I exists, $\tfrac{1}{2}I \leqslant J$.

173. THEOREM 6. *The linear content, whether I or J, of a finite arc of a rectifiable Jordan curve is not greater than its length.*

Let C be a finite arc of a rectifiable Jordan curve, and let the length of C be L.

Let A and B be two points of C whose distance apart is less than e, and a, b, the corresponding points of the segment (o, z) of the straight line; further let p be any point of (a, b) and P the corresponding point of C. Then since the distance of a from p is less than the distance of a from b, the distance AP is not greater than e.

Let Q be any point of the chord AB, then

$$QP < AQ + AP < AB + AP < 2e.$$

Thus, if we describe a region round P of span less than d, it will lie between the parallels to AB at a distance $d + 2e$ on each side of AB. Hence, if we describe such a region round every point of the arc (AB being a side of the polygonal path of length π_e the limit of whose length when e is indefinitely decreased is L), the content of these regions will be not greater than

$$2\,(d + 2e)\,\{\pi_e + 2d\}$$

where d and e are independent. Let us choose e so small that $\dfrac{e}{d}$ vanishes with d; then $\phi(d)$ will not be greater than a quantity which differs from π_e by a quantity which vanishes with d, and therefore the upper limit J of $\phi(d)$ when d is indefinitely decreased is not greater than L.

On the other hand we may divide the arc into two sets G_1, G_2 of points, both dense everywhere in the curve, and describing circles of diameter d, so that those containing G_1 lie almost entirely on one side of the curve, while those containing G_2 lie almost entirely on the other side, it may be inferred, as before, that the area covered by these regions differs from $2dL$ by a quantity which may be made as small as we please, and that therefore the upper limit of $\phi(d)$ when d is indefinitely decreased is not less than L. Hence in all ordinary cases the linear content J is also the same as the length.

Whether this is true in the case of a Jordan curve making oscillations of a complicated character remains uncertain.

A similar argument applies to I, since a circle of diameter d, with P as centre, lies between the parallels to AB at a distance $\tfrac{1}{2}d + 2e$.

Cor. *In the case of an arc of any ordinary curve (straight line, circle, conic section, etc.) the linear contents I and J coincide and are equal to the length.*

For I this is an immediate consequence of the above proof, since, by the roulette property of the circle etc., the area covered by the circles will actually differ from dL by a quantity which vanishes with d.

174. Linear content of a set of arcs on a rectifiable Jordan curve. It is however certain that the linear content of a set of arcs on a rectifiable Jordan curve of infinite length may differ from the linear content of those arcs with respect to the curve. This is proved by the following example in which the linear content with respect to the curve is *less* than the linear content.

Ex. 1.   Take a countable infinite series of concentric circles of radii

$$1, \quad \frac{1}{2}, \frac{1}{3}, \quad \dots, \quad \frac{1}{n}, \quad \dots,$$

for all values of n, having the origin O as centre.

Draw any radius of the first circle through O, meeting the nth circle in

P_n for all values of n. On the nth circle, at equal distances all round the circumference, place $4(n+1)$ arcs of equal length

$$x_n = \frac{e}{2^{n+2}(n+1)} :$$

the sum of all these arcs for all values of n is then e.

It is easy to see that these arcs lie on a rectifiable Jordan curve of spiral form; we can indeed construct such a curve by means of the concentric circles and the radius OP_1, rounding off the fork points P_1, P_2, ... by means of small semicircles. The linear content of the set of arcs with respect to such a rectifiable Jordan curve will then be e.

It may however be shewn that the linear content of this set of arcs (whether J, or I if definite), is certainly greater than this, if e is chosen less than $\frac{\pi}{2}$; the linear content with respect to the Jordan curve is thus certainly different from the linear content.

To prove this consider that the arcs between consecutive arcs of the given set on the kth circle are of length y_k, where

$$y_k = \frac{1}{4(k+1)}\left\{\frac{2\pi}{k} - \frac{e}{2^k}\right\} = \frac{\pi}{2}\,\frac{1}{k(k+1)} - \frac{e}{2^{k+2}(k+1)} \quad \ldots\ldots\ldots(1);$$

the distance between the kth and $(k+1)$th circles is $\frac{1}{k(k+1)}$; therefore, when k is sufficiently large, y_k is greater than the distance between the kth and $(k+1)$th circles.

Thus, if d is sufficiently small, and the integer l is determined for which

$$\frac{1}{l(l+1)} \leqslant d < \frac{1}{l(l-1)} \quad \ldots\ldots\ldots\ldots\ldots\ldots\ldots\ldots\ldots\ldots(2),$$

the bean-shaped regions generated by small circles of diameter d, with the points of the given set as centres, will not overlap with one another outside the lth circle. The area of such a bean-shaped region is

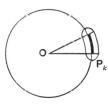

Fig. 56.

$$\tfrac{1}{2}kx_k\left[\left(\frac{1}{k}+\frac{d}{2}\right)^2 - \left(\frac{1}{k}-\frac{d}{2}\right)^2\right] + \tfrac{1}{4}\pi . d^2 = x_k . d + \tfrac{1}{4}\pi d^2.$$

Thus, inserting the value of x_k, the area covered by those bean-shaped regions which cover arcs on the kth circle $(0 < k \leqslant l)$ is

$$\frac{de}{2^k} + (k+1)\,\pi d^2,$$

and the area covered by all the bean-shaped regions outside the $(l+1)$th circle is

$$de\left\{\frac{1}{2}+\frac{1}{4}+\ldots+\frac{1}{2^l}\right\} + \pi . d^2\left\{2+3+\ldots+(l+1)\right\} = de\left(1-\frac{1}{2^l}\right) + \pi d^2\tfrac{1}{2}l(l+3).$$

Now, by (2), when d is sufficiently small, dl^2 is as near unity as we please; thus the content of the small circles described divided by $2d$ can be made as near as we please to $\frac{e}{2}+\frac{\pi}{4}$, by taking d sufficiently small. Hence it is clear that, if the set of arcs has a definite linear content I, $\tfrac{1}{2}I$ cannot be less than

$\frac{e}{2} + \frac{\pi}{4}$. The linear content I, therefore, if definite, is greater than the linear content with respect to the Jordan curve, and, if $e < \frac{\pi}{2}$ the linear content J is also greater than the linear content with respect to the Jordan curve.

This example shews that the value of the linear content of a set of Jordan arcs depends not only on their *length* but on their *lie*.

175. Linear content of a countable closed set of points. In the theory of content (plane, n-dimensional content) we had the theorem that a countable closed set of points has zero content, a theorem which was followed by many important consequences, and which holds for the linear content of a closed countable set on a rectifiable Jordan curve as defined with respect to that curve. It is another anomaly of the theory of linear content that this theorem is no longer true. The following is an example of a plane countable closed set of points, having only one limiting point, and whose linear content I is approximately $5\frac{1}{2}$ while J is about $2\frac{3}{4}$.

Ex. 2. *A Plane Countable Closed Set of Positive Linear Content.* Describe round the origin O as centre a countably infinite series of concentric circles of radii 1, $\frac{1}{2}$, $\frac{1}{3}$, ..., and place k' points at equal distances round the

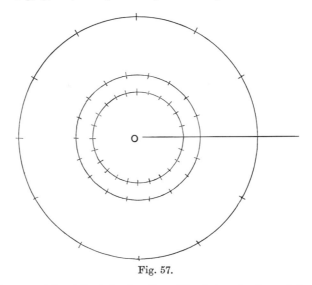

Fig. 57.

circumference of the kth circle, for all positive integral values of k, where k' is defined as an integer by the inequality

$$2\pi (k+1) < k' < 2\pi (k+1) + 1 \quad \dots\dots\dots\dots\dots\dots(1),$$

and let the points be so placed that on a fixed straight line OX through O there is one point of the set on each of the circles (Fig. 57).

These points together with the origin form a countable closed set whose linear content is about $5\frac{1}{2}$. Now, since (Fig. 58)

$$2\frac{1}{k}\sin\frac{\pi}{k'} = \left(\frac{k'}{\pi}\sin\frac{\pi}{k'}\right)\left(\frac{2\pi(k+1)}{k'}\right)\frac{1}{k(k+1)} < \frac{1}{k(k+1)} \quad\ldots\ldots\ldots(2),$$

the distance between consecutive points of the set on the kth circle (represented by the first of these expressions) is less than the distance from the kth to the $(k+1)$th circle (represented by the last of these expressions).

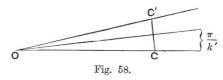

Fig. 58.

If therefore we describe, round the points of the set as centres, circles of diameter d, these will, for the earlier values of k, be quite separate, but will begin to overlap, in such a way as to form rings covering over the circumferences of the successive circles, as soon as k reaches the value l, defined by the following inequality :—

$$2\frac{1}{l}\sin\frac{\pi}{l'} < d \leqslant 2\frac{1}{k}\sin\frac{\pi}{k'} \quad (k<l) \quad\ldots\ldots\ldots\ldots(3).$$

Let OX be the fixed radius, C the point in which it meets the kth circle $(k \geqslant l)$, C' the point in which it meets the $(k+1)$th circle, and C_1 the point on the kth circle consecutive to C when we move anti-clockwise on the circle.

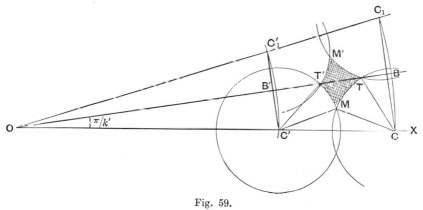

Fig. 59.

Fig. 59 shews the part of the plane between the kth and $(k+1)$th circles cut out by the radii OX and OC_1.

The point in this part in which the circles of radius d and centres C and C_1 meet is T, and OT meets the circle of centre C' in T', and the chord CC_1 and the parallel chord $C'C_1'$ in B and B' respectively.

The circles of radius d and centres C and C' meet, if, and only if, d is greater than CC', that is, if

$$d > \frac{1}{k(k+1)}.$$

Also, since CC' is the minimum distance between a point of the set on the

kth circle and one on the $(k+1)$th circle, no one of the circles whose centres are on the kth circle can overlap with one of the circles whose centres are on the $(k+1)$th circle unless those with centres C and C' do so. Thus the rings corresponding to values of k lying between l and $(m-1)$, both inclusive, will not overlap, where the integer m is defined by the following inequality :—

$$\frac{1}{m(m+1)} < d \leqslant \frac{1}{m(m-1)} \qquad \dots\dots\dots\dots\dots(4).$$

The circles whose centres are on the kth circle $(k \geqslant l)$ entirely cover over a simple circular ring bounded by two circles of centre O and passing through the points of intersection of consecutive small circles whose centres are on the kth circle ; one of these points is T in Fig. 59. Denoting the length of the common chord of the small circles of centres C and C' in Fig. 59 by t_k,

$$BT = \tfrac{1}{2} t_k,$$

and t_k is the width of the simple circular ring in question.

On the $(k+1)$th circle there are at least k' points of the set, and therefore the distance apart of consecutive points of the set on the $(k+1)$th circle is less than it was on the kth circle. Therefore

$$t_{k+1} > t_k.$$

It follows that the simple circular rings round the kth and $(k+1)$th circles will certainly overlap, provided k is not less than n, where the integer n is defined by the following inequality :—

$$\frac{1}{n(n+1)} < t_n \leqslant \frac{1}{n(n-1)} \qquad \dots\dots\dots\dots\dots(5).$$

Thus the whole area of the nth circle (at least) is tiled over by the small circles.

Now $$BT^2 + BC^2 = CT^2,$$

that is, $$\tfrac{1}{4} t_k{}^2 + \left(\frac{1}{k} \sin \frac{\pi}{k'}\right)^2 = \tfrac{1}{4} d^2,$$

or $$\left(\frac{t_k}{d}\right)^2 + \left(\frac{2}{kd} \sin \frac{\pi}{k'}\right)^2 = 1 \qquad \dots\dots\dots\dots\dots(6).$$

Now, as d approaches the value zero, l, m, n, l', m', n' all approach infinity. Moreover, since, by (1),

$$\frac{2\pi(k+1)}{k'} < 1 \leqslant \frac{2\pi(k+1)}{k'} + \frac{1}{k'},$$

for all values of k, we have, for l and l',

$$\operatorname*{Lt}_{d=0} \frac{2\pi(l+1)}{l'} = 1,$$

the same equation holding for m and m', and for n and n'.

Again, since, by (3), $\dfrac{2}{kd} \sin \dfrac{\pi}{k'}$ is greater than 1 when k is l, and less than 1 when k is $(l-1)$, we have

$$\operatorname*{Lt}_{d=0} \frac{2}{ld} \sin \frac{\pi}{l'} = 1,$$

that is, $$\operatorname*{Lt}_{d=0} \left(\frac{l'}{\pi} \sin \frac{\pi}{l'}\right) \left(\frac{2\pi(l+1)}{l'}\right) \left(\frac{l}{l+1}\right) \frac{1}{l^2 d} = 1,$$

whence, using (4),
$$\text{Lt}_{d=0} l^2 d = 1 = \text{Lt}_{d=0} m^2 d \quad \dots\dots\dots\dots\dots\dots(7).$$

Similarly, using (5),

$$\text{Lt}_{d=0} \frac{2}{nd} \sin \frac{\pi}{n'} = \text{Lt} \left(\frac{n'}{\pi} \sin \frac{\pi}{n'} \right) \left(\frac{2\pi (n+1)}{n'} \right) \left(\frac{n}{n+1} \right) \frac{1}{n^2 d} = \text{Lt} \frac{1}{n^2 d} = \text{Lt} \frac{t_n}{d},$$

hence, by (6), each of these limits has the value $1/\sqrt{2}$; so that

$$\text{Lt}_{d=0} n^2 d = \sqrt{2} \dots\dots\dots\dots\dots\dots\dots\dots(8).$$

The content of the small circles is best calculated in three parts:—

$$F(d) = L + M + N \dots\dots\dots\dots\dots\dots\dots(9),$$

where L is the content of all those small circles whose centres lie outside the lth circle, M of those of the remaining circles whose centres lie outside the mth circle, and N is the content of the rest of the circles. Thus

$$L = \sum_{k=1}^{k=l-1} k' \frac{\pi d^2}{4};$$

therefore, by (1),

$$\frac{\pi d^2}{4} 2\pi \sum_{k=1}^{k=l-1} (k+1) < L < \frac{\pi d^2}{4} 2\pi \sum_{k=1}^{k=l-1} (k+2),$$

that is,

$$\frac{\pi^2 d^2}{4} (l-1)(l+2) < L < \frac{\pi^2 d^2}{4} (l-1)(l+4),$$

whence, by (7),

$$\text{Lt}_{d=0} \frac{L}{d} = \text{Lt} \frac{\pi^2 l^2 d}{4} = \frac{\pi^2}{4} \dots\dots\dots\dots\dots\dots(10).$$

The ring formed by the small circles round the kth circle, when $k \geqslant l$, is less in area than the ring between the two circles with centre O touching the ring all round, that is, it is less than

$$\pi \left(\frac{1}{k} + \frac{d}{2} \right)^2 + \pi \left(\frac{1}{k} - \frac{d}{2} \right)^2 < 2\pi d \frac{1}{k}.$$

Thus*, using (7),

$$\text{Lt}_{d=0} \frac{M}{d} \leqslant \text{Lt}_{d=0} \pi \sum_{k=l}^{k=m-1} \frac{1}{k} = 0 \dots\dots\dots\dots\dots(11).$$

To calculate N, we will first calculate the content of the parts of the plane which are covered by the small circles between the circumferences of the kth and $(k+1)$th circles, where $m \leqslant k < n$.

By (1) the number of points of the set on the $(k+1)$th circle is less than $k' + 8$. Thus, if we replace the points of the set on the $(k'+1)$th circle by k'

---

* Here we use a particular case of the general theorem

$$\text{Lt}_{l=\infty} \sum_{k=l}^{k=m-1} \frac{1}{k} = \text{Lt}_{l=\infty} \log \frac{m-1}{l}.$$

To prove this we have

$$\sum_{k=l}^{k=m-1} \frac{1}{k} = \frac{\Delta x}{1+x},$$

where $\Delta = \frac{1}{l}$, and x has the values $0, \frac{1}{l}, \frac{2}{l}, \dots, \frac{m-l-1}{l}$; thus, if l, m are made infinite in the ratio $1 : t$, we get as the limit of the summation

$$\int_0^{t-1} \frac{dx}{1+x} = \log t.$$

points at equal distances, beginning on the fixed radius OX, we shall alter the area covered between the kth and the $(k+1)$th circles by at most $8\,(\tfrac{1}{4}\pi d^2)$ or $2d^2\pi$ *.

Thus, if we calculate the content between the kth and $(k+1)$th circles on the supposition that there are only k' points of the set on the $(k+1)$th circle, and then sum from $k=m$ to $k=n-1$, we shall introduce an error in N of at most $2\pi\,(n-m)\,d^2$, and therefore an error in the linear content of at most $\underset{d=0}{\mathrm{Lt}}\,2\pi\,(n-m)\,d=0$. Thus it is allowable to calculate N in this manner. We have then k' small uncovered areas between the kth and $(k+1)$th circles, all equal, and each bounded by four circular arcs, as in Fig. 59, where one of these areas, $TMT'M'$, is shaded.   Now

$\tfrac{1}{2}$ area $TMT'M'$

$=$ quadrilateral $BCC'B'$ $-$ triangles $BCT,\,B'C'T',\,CC'M$ $-$ sectors $TCM,\,T'C'M$

$$=\tfrac{1}{2}\left\{\frac{1}{k^2}-\frac{1}{(k+1)^2}\right\}\sin\frac{\pi}{k'}\cos\frac{\pi}{k'}-\tfrac{1}{8}d^2\left\{\sin\theta_k\cos\theta_k+\sin\theta_{k+1}\cos\theta_{k+1}\right.$$

$$\left.+2\sin\psi_k\cos\psi_k+\frac{\pi}{2}-\frac{\pi}{k'}-\theta_k-\psi_k+\frac{\pi}{2}+\frac{\pi}{k'}-\theta_{k+1}-\psi_k\right\},$$

where θ_k is BCT and ψ_k is MCC'; so that

$$\cos\theta_k=\frac{2\sin\dfrac{\pi}{k'}}{kd},\qquad\cos\psi_k=\frac{1}{k\,(k+1)\,d}.$$

Thus, neglecting small quantities which vanish in the limit,

$$\frac{N}{d}=\frac{\pi}{m^2d}-\sum_{k=m}^{k=n-1}\frac{2k'}{d}\left[\tfrac{1}{2}\left\{\frac{1}{k^2}-\frac{1}{(k+1)^2}\right\}\sin\frac{\pi}{k'}\cos\frac{\pi}{k'}\right.$$

$$\left.-\tfrac{1}{8}d^2\left\{\pi-\theta_k-\theta_{k+1}-2\psi_k+\sin\theta_k\cos\theta_k+\sin\theta_{k+1}\cos\theta_{k+1}+2\sin\psi_k\cos\psi_k\right\}\right].$$

Now, writing $\cos\phi=1/k^2d$, we see that θ_k, ψ_k, and ϕ_k all have the same limit. Thus, writing π for $k'\sin\pi/k'\cos\pi/k'$, $2\pi k$ for k', and ϕ_k for θ_k, θ_{k+1}, and ψ_k, we introduce no error in the limit,

---

* Let R_k denote the content of the circles whose centres lie on the kth circle, and R_k' the content of $(k'-q)$ circles of diameter d placed evenly round the circumference of the kth circle.   Then

$$R_k=k'\,\frac{\pi d^2}{4}-2k'\,\frac{d^2}{4}\,(\theta-\sin\theta\cos\theta),\qquad\qquad\cos\theta=\frac{2}{kd}\sin\frac{\pi}{k'},$$

$$R_k'=(k'-q)\,\frac{\pi d^2}{4}-2\,(k'-q)\,\frac{d^2}{4}\,(\theta'-\sin\theta'\cos\theta'),\qquad\cos\theta'=\frac{2}{kd}\sin\frac{\pi}{k'-q}>\cos\theta,$$

Therefore $\qquad\qquad\qquad\qquad\qquad\qquad\theta'<\theta,$

and $\qquad\qquad\qquad R_k'>(k'-q)\,\dfrac{\pi d^2}{4}-2k'\,\dfrac{d^2}{4}\,(\theta-\sin\theta\cos\theta),$

whence $\qquad\qquad\qquad\qquad R_k-R_k'<q\,\dfrac{\pi d^2}{4}.$

$$\underset{d=0}{\mathrm{Lt}}\,\frac{N}{d} = \underset{d=0}{\mathrm{Lt}}\left(\frac{\pi}{m^2 d} - \overset{k=n-1}{\underset{k=m}{\Sigma}}\left[\left\{\frac{1}{k^2} - \frac{1}{(k+1)^2}\right\}\frac{\pi}{d} - \frac{\pi^2}{2}\,dk\right.\right.$$
$$\left.\left. + 2\pi dk\,\{\phi_k - \sin\phi_k\cos\phi_k\}\right]\right)$$

$$= \underset{d=0}{\mathrm{Lt}}\left(\frac{\pi}{m^2 d} - \left[\frac{\pi}{m^2 d} - \frac{\pi}{n^2 d} - \frac{\pi^2 d}{4}\,(n^2 - m^2)\right.\right.$$
$$\left.\left. + \pi\overset{k=n-1}{\underset{k=m}{\Sigma}}2kd\,(\phi_k - \sin\phi_k\cos\phi_k)\right]\right)$$

$$= \underset{d=0}{\mathrm{Lt}}\left(\frac{\pi}{n^2 d} + \frac{\pi^2}{4}\,(n^2 - m^2)\,d - \pi\overset{k=n-1}{\underset{k=m}{\Sigma}}2kd\,(\phi_k - \sin\phi_k\cos\phi_k)\right).$$

Now $$\sec\phi_{k+1} - \sec\phi_k = 2kd + d^2 = \Delta\,(\sec\phi_k);$$

thus, neglecting under the sign of summation small quantities whose sum vanishes in the limit, we may write $\Delta\,(\sec\phi_k)$ for $2kd$, and sum with respect to ϕ from $+0$ to $\frac{1}{4}\pi - 0$: in the limit the summation becomes the integral. Now

$$\int(\phi - \sin\phi\cos\phi)\,d\sec\phi = (\phi - \sin\phi\cos\phi)\sec\phi - \int\sec\phi\,(1 - \cos 2\phi)\,d\phi$$
$$= (\phi - \sin\phi\cos\phi)\sec\phi - 2\int(1 - \cos^2\phi)\sec\phi\,d\phi.$$

Thus

$$\int_0^{\frac{1}{4}\pi}(\phi - \sin\phi\cos\phi)\,d\sec\phi = \left(\frac{\pi}{4} - \frac{1}{2}\right)\sqrt{2} - 2\log(\sqrt{2}+1) + 2\frac{1}{\sqrt{2}}$$
$$= \left(\frac{\pi}{4} + \frac{1}{2}\right)\sqrt{2} - 2\log(\sqrt{2}+1),$$

whence

$$\underset{d=0}{\mathrm{Lt}}\,\frac{N}{d} = \frac{\pi\sqrt{2}}{2} + \frac{\pi^2}{4}\,(\sqrt{2}-1) - \frac{\pi^2}{4}\sqrt{2} - \frac{\pi\sqrt{2}}{2} + 2\pi\log(\sqrt{2}+1)$$
$$= -\frac{\pi^2}{4} + 2\pi\log(\sqrt{2}+1).$$

Now, by (10), $$\underset{d=0}{\mathrm{Lt}}\,\frac{L}{d} = \frac{\pi^2}{4}.$$

Thus $I = 2\pi\log(\sqrt{2}+1) = 5\cdot538$ approximately.

It remains to prove that J is very nearly, if not exactly equal to $\frac{1}{2}I$.

To prove this, consider that the regions of span d, however constructed, cannot possibly cover over the whole circumference of the lth circle. Thus we get an upper limit for the content of the small regions if we suppose the whole area of a circle of radius $(1/l + d)$ to be tiled over, and the remaining small regions to be non-overlapping. Since **10** the area of a region of span d is at most $\frac{1}{4}\pi d^2$, we get as upper limit

$$\pi\left(\frac{1}{l}+d\right)^2 + \overset{k=l}{\underset{k=1}{\Sigma}}\frac{1}{4}\pi d^2 k' < \pi\left(\frac{1}{l}+d\right)^2 + \frac{1}{4}\pi d^2\overset{k=l}{\underset{k=1}{\Sigma}}\{2\pi\,(k+1)+1\}$$
$$< d\left\{\frac{\pi}{l^2 d} + \frac{1}{4}\pi^2 dl^2 + d\right\}$$
$$< d\,\{\pi + \frac{1}{4}\pi^2 + d'\}$$
$$< d\,\{3\cdot1416 + 2\cdot467413 + d'\}$$
$$< d\,\{5\cdot609013 + d'\}$$
$$< d\,\{I + 0\cdot071013 + d''\},$$

d' and d'' being small quantities which vanish with d.

Thus $$\tfrac{1}{2}I \leqslant J < \tfrac{1}{2}I + 0\cdot071013.$$

Ex. 3. *A Plane Closed Countable Set of Infinite Linear Content.* Take a set of concentric circles of radii $1, \frac{1}{2}, \frac{1}{3}, \ldots, 1/k, \ldots$ On each circle, of radius say $1/k$, place 2^k points at equal distances, beginning on a fixed radius, the same for all values of k. *The points so constructed, together with the centre of the circles, form a closed countable set of infinite linear content.*

To prove this, consider that, since, for all values of k greater than 5,

$$\frac{2}{k} \sin \frac{\pi}{2^k} < \frac{1}{k(k+1)},$$

the distance between consecutive points of the set on the kth circle (represented by the former of these expressions) is less than the least distance between points of the set on the kth and on the $(k+1)$th circles (represented by the latter of the two expressions). Thus, if d be sufficiently small, small circles of diameter d, with the points of the set as centres, will be distinct from one another for small values of k, and will first begin to overlap when $k = l$, where the integer l is defined by the following inequality :—

$$\frac{1}{l} \sin \frac{\pi}{2^l} < \frac{d}{2} \leqslant \frac{1}{l-1} \sin \frac{\pi}{2^{l-1}} \ldots\ldots\ldots\ldots\ldots\ldots(1).$$

From the lth to the nth circle, where n is defined by the inequality

$$\frac{1}{n(n+1)} < d \leqslant \frac{1}{n(n-1)} \ldots\ldots\ldots\ldots\ldots\ldots(2),$$

the small circles form rings round the circumference of each of the circles of radii $1/k$, but the rings are distinct from one another; as soon as we get to the nth circle, however, the rings begin to overlap. It is only necessary to consider the content N of these rings in order to prove the content to be infinite.

The area of the ring round the kth circle is greater than that of the ring bounded by the two concentric circles through the points of intersection of the small circles ; that is, it is greater than

$$\pi \left(\frac{1}{k} + \tfrac{1}{2} t_k\right)^2 - \pi \left(\frac{1}{k} - \tfrac{1}{2} t_k\right)^2 = \frac{2\pi}{k} t_k \ldots\ldots\ldots\ldots\ldots(3),$$

where

$$\tfrac{1}{4} t_k^2 = \tfrac{1}{4} d^2 - \left(\frac{1}{k} \sin \frac{\pi}{2^k}\right)^2 \ldots\ldots\ldots\ldots\ldots\ldots(4).$$

Let m be the integer defined by the following inequality :—

$$\frac{1}{m} \sin \frac{\pi}{2^m} < \tfrac{1}{4} d \leqslant \frac{1}{m-1} \sin \frac{\pi}{2^{m-1}} \ldots\ldots\ldots\ldots\ldots(5);$$

then, since

$$\frac{1}{m} \sin \frac{\pi}{2^m} < \tfrac{1}{4} d < \tfrac{1}{2} d,$$

m is not less than l. Also, since

$$\frac{1}{m(m+1)} = \left\{\frac{4}{m-1} \frac{\pi}{2^{m-1}}\right\} \frac{2^{m-3}}{\pi} \frac{m-1}{m(m+1)},$$

which, if m is sufficiently large, is greater than

$$\frac{4}{m-1} \frac{\pi}{2^{m-1}} > \frac{4}{m-1} \sin \frac{\pi}{2^{m-1}},$$

it follows that

$$\frac{1}{m(m+1)} > d ;$$

so that m is less than n.

By (4) and (5), for all values of k greater than m, t_k^2 is greater than $\frac{3}{4}d^2$; and therefore t_k is greater than $\frac{1}{2}d$. Thus, after $k = m - 1$, the area of each of the rings is, by (3), greater than π/kd. Thus

$$\operatorname*{Lt}_{d=0} \frac{N}{d} \geqslant \operatorname*{Lt}_{d=0} \frac{1}{d} \sum_1^{n-1} 2\pi t_k \frac{1}{k} \geqslant \operatorname*{Lt}_{d=0} \pi \sum_m^{n-1} \frac{1}{k} \geqslant \operatorname*{Lt}_{d=0} \pi \log \frac{n-1}{m}$$

$$\geqslant \tfrac{1}{2} \operatorname*{Lt}_{d=0} \pi \log \frac{n\,(n+1)}{(m-1)^2}.$$

Now, by (2) and (5),

$$\frac{n\,(n+1)}{(m-1)^2} > \frac{1}{(m-1)^2\,d} > \frac{1}{(m-1)\,4\sin\dfrac{\pi}{2^{m-1}}} > \frac{1}{(m-1)\,\dfrac{\pi}{2^{m-3}}} > \frac{2^{m-3}}{\pi\,(m-1)},$$

which can be made as large as we please by sufficiently decreasing d. Thus

$$\operatorname*{Lt}_{d=0} \frac{N}{d} = \infty\,;$$

à fortiori the linear content of the whole closed countable set of points is infinite.

APPENDIX.

(1) pp. 55—63.

The object of the present note is to shew that there is no foundation for the doubt expressed on p. 530 of the *Fortschritte der Mathematik* as to the mode of treatment adopted in "The Analysis of Sets of Points" (W. H. Young, *Quarterly Journal of Mathematics*, 1902) for the theory of derived and deduced sets, or adherences and coherences, without the use of Cantor's numbers. The memoir quoted has to all intents and purposes been reproduced in the present volume, and in the chapter on Cantor's numbers reference is made to this theory to elucidate the matter there treated of. A synopsis of the proof of Theorem 25, Ch. IV (Theorem 7 of the memoir quoted) is appended, in which the language used is such as to shew that the use of such terms as "series of derived and deduced sets" (p. 27) and "progress a stage further" (p. 28) does not in any way require a previous knowledge of the theory of well-ordered sets or of Cantor's ordinal numbers for its comprehension or justification, and may, on the other hand, if it is desired, be entirely avoided.

In Cantor's treatment of the subject (*Math. Ann.* XXIII. and *Acta Math.* VII.) free use is made of Cantor's numbers, not only of the first and second, but also of the third classes. It has been pointed out in the text that the use of Cantor's numbers of the third class is not universally accepted as legitimate. Under these circumstances, apart from the help which the theory under discussion affords in the subsequent development of the theory of Cantor's numbers, the fact that the existence and properties of the derived and deduced sets, and of adherences and coherences, stand on an independent basis, is one of too great importance to be left in doubt.

Synopsis of Proof of Theorem 25, Ch. IV, p. 56.

(1) We define *a limiting point, closed set, first derived set.*

THEOREM 1. *A closed set E contains its first derived E'.*

THEOREM 2. *If F is contained in G, F' is contained in G'.*

(The word "contained" is used to mean that a set is either identical with another or is a proper component of it.)

THEOREM 3. *A first derived set is a closed set.*

(2) We define *deduction* as the process of taking all the common points of a series of sets $E_1, E_2, \ldots E_n \ldots$ for all positive integers n.

THEOREM 4. (CANTOR'S THEOREM OF DEDUCTION.) *If E_1, E_2, ... are each of them closed sets, and E_n is always contained in E_{n-1}, then there is always at least one point in the deduced set, and the deduced set is a closed set.*

(3) We define *a limiting point of countable degree, and the nucleus.*

THEOREM 5. *If F^* is the nucleus of F, $F - F^*$ is a countable set.*

THEOREM 6. *If F is a closed set and F' its first derived, F^* is the nucleus of F'.*

THEOREM 7. *If F_1, F_2, ... all have the same nucleus, the deduced set has the same nucleus.*

(4) We define the set $K(E)$ of sets of points as follows :—

(*a*) $K(E)$ contains the first derived set E';

(*b*) $K(E)$ contains the first derived F' of every one of its sets F;

(*c*)   If $K(E)$ contains each of the sets F_1, F_2, ... F_n, ... it contains their deduced set F;

(*d*)   If F be any set of points, then F belongs to $K(E)$ if, and only if, either we can assign a set F_0 *belonging to* $K(E)$, such that F is its first derived F_0' or else we can assign a series of sets *belonging to* $K(E)$

$$F_1, F_2, ... F_n, ...$$

such that F is their deduced set.

THEOREM 8. *If F is a set of $K(E)$, F is a closed set.*

For if $F = F_0'$, F is a closed set by Theorem 3.

If F is the deduced set of F_1, F_2, ..., assume the theorem true for these sets. Then if F_1 is contained in all the others, F is identical with F_1 and therefore is a closed set. If this is not the case, let F_i be the first of the sets not containing F_1, and let f_2 be the set of all the common points of F_1 and F_i. Then f_2 contains F, and is easily shewn to be a closed set. By the same argument, either F is identical with f_2, and is therefore closed, or we determine the first of the sets not containing f_2, and take the common component of this set and f_2 as f_3. In this way either we shew that F is closed, or we get a series F_1, f_2, f_3, ... having F as deduced set, and such that each set is closed and contained in its predecessor ; it then follows by Cantor's Theorem of Deduction that F is closed. Thus if the Theorem is true at all it is true always. It is true for E' by Theorem 3, therefore it is true always. Q.E.D.

THEOREM 9. *If G is any set of $K(E)$, G is contained in the first derived set E'.*

Suppose first that the theorem is true for G_0, then, since G_0 is closed and therefore contains G_0', the theorem is true for G_0'.

Next suppose that the theorem is true for G_1, G_2, ... then it is true for their deduced set.

Thus if the theorem is true at all it is true always. But it is true for the second derived set, and therefore it is true always.

COR. *If G is any set of $K(E)$ except E', it is contained in E''.*

THEOREM 10. *If F and G are any two sets of $K(E)$, such that F is not contained in G, then G is contained in the first derived set F''.*

First let $G = G_0'$, and assume the theorem true for G_0 and F.

Now, if G_0 were not contained in F, F would be contained in G_0' which is G. But F is not contained in G, therefore G_0 is contained in F. Thus, by Theorem 2, G_0' is contained in F', that is G is contained in F'.

Next let G be deduced from G_1, G_2, ..., and assume that the theorem holds for these. Then since F is not contained in G, F is not contained in G_n for every value of n. Let m be the first integer for which F is not contained in G_m. Then since the theorem is true for G_m, G_m is contained in F'. But G is contained in G_m, therefore G is contained in F'.

Thus, if the theorem is true at all it is true always, but by Theorem 9, Cor. it is true if G is the second derived set and therefore it is true always.

THEOREM 11.   *All the sets of $K(E)$ have the same nucleus.*

Assuming the theorem for G_0, it follows for G_0', by Theorems 8 and 6.

Assuming the theorem true for G_1, G_2, ... it is true for their deduced set by Theorems 8 and 7.

Also it is true for the derived sets of E, and therefore it is always true.

THEOREM 12.   *The sets of $K(E)$ are countable.*

For let the points of $E' - E^*$ be arranged in countable order (Theorem 5) and let them be P_1, P_2, ...

Let F be any set of $K(E)$, then, if E^* be the nucleus of E', it is also the nucleus of F, and therefore contained in F. Similarly it is the nucleus of F'' (Theorem 11). Also by Theorem 9, both F and F'' are contained in E'; therefore $F - E^*$ is contained in $E' - E^*$ and $F - E^*$ contains $F'' - E^*$, and therefore $F - F''$, which is $(F - E^*) - (F'' - E^*)$, is contained in $E' - E^*$. Let P_i be the first of the points of $E' - E^*$ in $(F - F'')$, then F determines the index i, and we will now shew that no other set of $K(E)$ determines the same index.

If G be any set of $K(E)$ such that F is not contained in G, then, by Theorem 10, G is contained in F', and therefore does not contain P_i: thus the index determined by G is different from i.

If G is a set of $K(E)$ such that F is contained in G, then F is contained in G' by Theorem 10; therefore P_i, which is a point of F, is a point of G', and not of $G - G'$. Thus the index determined by G is different from i. Thus there is a definite index i determined by and determining F. This proves the theorem.

Since this note was composed we perceive that M. Lebesgue, who was undoubtedly unacquainted with the Analysis of Sets of Points, has of late given a discussion of the matter, emphasizing this very point of view. Lebesgue, like the present authors, regards the theory of derivation and deduction as the natural basis for the study of the transfinite numbers, not *vice versa*. He also uses freely the notions of "before" and "after" (*Leçons sur l'Intégration*, 1904, pp. 131 *seq.*). He proves from first principles that no countable set of symbols can suffice to characterise the sets (including the null-set, if it occurs, or repetitions of the nucleus) obtained by the processes of derivation and deduction, and uses this theorem, in conjunction with an argument which is tantamount to the use of the nucleus, to deduce the theorem that *a closed set is the sum of a countable set and a perfect set* (the latter or the former may of course in special cases be absent). He adds: "On remarquera que la démonstration ne suppose connus, ni la notion, ni même le môt de nombre transfini."

(2) p. 55. Vivanti (*Rend. d. Circ. Mat. di Palermo*, 1898) points out that *the decomposition of a closed set into a perfect set and a countable set is unique.* The proof depends on the following Lemma :—

The points of a perfect set S not belonging to one of its closed components U form a set V which is dense in itself and of potency c.

For $$S = U + V.$$

Any point therefore of V, being a limiting point of S, since S is perfect, but not of U, since U is closed, is a limiting point of V. Hence V is dense in itself.

Again, any point v of V, not belonging to the closed set U, must be internal to one of the black intervals of U (footnote, p. 19 *supra*). Taking any interval containing v and with both end-points internal to the same black interval as v, the points of S belonging to this closed interval form a component of V, and it is a closed set (Theorem 7, p. 84 *supra*). By the same argument as that used in proving that V is dense in itself, it follows that this component is dense in itself and therefore perfect: the potency of this component is therefore c (p. 50 *supra*), and therefore that of V is also c. Q. E. D.

Now, if possible, let there be two decompositions of a closed set P,
$$P = R + S = R' + S',$$
where R and R' are countable, and S and S' perfect.

Then, since S is of potency c and R' is countable, S cannot be a component of R', and therefore S and S' have in common a closed set (Theorem 7, p. 84 *supra*). The remaining points of S, if any, are therefore, by the Lemma, of potency c. But, being points of P but not of S', they are points of R', and therefore countable, which is a contradiction. Thus S and S' can only be identical and the decomposition is unique.

(3) p. 127. The distinction between "ordinal" and "actual" pointed out in footnote †, is not simply one of measurement. Given any geometrical form, it determines not only all the elements (points) in it, but also all possible fundamental regions passing through it. If the elements are simply ordered and we neglect the fundamental regions passing through the form, we get the point of view of Cantor in *Math. Ann.* XLVI. and XLIX. Without considering all possible fundamental regions passing through the form we cannot discover whether an ordinal limiting point is, or is not, an actual limiting point, whether the form, regarded as a set of points, is not merely *ordinally* closed but *actually* closed, and so forth. The *actual* properties of a form, regarded as a set of points, are those which are invariant for all transformations of the fundamental region which leave the form itself unaltered ; the *ordinal* properties are those which are invariant for certain transformations of the form into itself without reference to points other than those of the form itself. The distinction is therefore exactly parallel to that between Cremona transformation of the plane (space) and birational transformation of a curve.

It is one of the tacit assumptions made in considering the geometrical straight line, that, no matter what fundamental region passing through it be taken as standard of reference, its points constitute a perfect set.

(4)   p. 137.   The definition given by Cantor of a well-ordered set is more cumbrous and is as follows:—

A simply ordered set S is said to be well-ordered if it has the following properties :—

(a)   there is a first element;

(b)   if S' is any component of S, and if S possesses one or more elements which come after all those of S', then there is an element of S such that between it and the elements of S' there is no element of S.

The identity of this definition with that given in the text is however easily proved by Cantor (Math. Ann. XLIX.).

(5)   p. 141.   It follows from Theorem 6 that, if two well-ordered sets E and F are similar, there is one, and only one, (1, 1)-correspondence between them by which the orders are maintained (Beppo Levi, loc. cit. p. 150). Hence E and F are simply equivalent (see footnote $\ddagger$, p. 147, and below, note 11).   The question as to whether every set can be well-ordered is still unsettled, no proof that this is so, or that it is not so, which has yet been published, has been accepted as satisfactory.   In reading Cantor's original papers it must be borne in mind that Cantor was biassed by the opinion that every set could be well-ordered.

(6)   p. 147.   The potency of the set of all sets of points on the straight line (if such a potency is recognised) must, by the definition of exponentiation (§ 86, p. 152) be denoted by c^c.   This is also the potency of the set of all those functions of a single real variable which assume only the values 0 and 1, since such a function is determined by, and determines, the set of its zero points. It is not difficult to prove that the same is the potency of all functions of a single real variable, and therefore, by §§ 96 and 97, of any countable number of real variables (Borel, Leçons sur la Théorie des Fonctions, 1898, pp. 107—110). Cantor's proof that this potency is greater than c is as follows :—If possible let there be a (1, 1)-correspondence between the points x of a straight line and all sets of points.   Then we can characterise each set as S_x, where x is the corresponding point.   These sets are to include the null-set S_{x_0}, which contains no point, and the whole straight line S_{x_1}, which contains every point. Now S_x either contains the point x, or not.   Let us form a set S as follows :— If S_x contains x, then S does not contain x, but if S_x does not contain x, then S does contain x.   This law completely determines the set S, which will certainly contain the point x_0, and certainly not contain the point x_1, and will therefore be a proper component of the straight line other than the null-set.   By hypothesis there is a point y corresponding to S, so that S is the same as S_y.   But, by the law of formation of S, S and S_y are such that one does, and one does not, contain the point y, so that S cannot be the same as S_y.   Thus the hypothesis leads to an absurdity, whence we conclude that it is not possible to set up a (1, 1)-correspondence between all points of the straight line and all sets of points on the straight line.

Since, however, there are c sets of points each consisting of a single point only, it follows from the definition of § 85 that c^c is greater than c.   As

Cantor himself pointed out (*loc. cit.* p. 147 *supra*) this argument can be applied to shew that the set of all sets of elements of any fundamental region has a greater potency than that of the set of all those elements themselves; from this it follows that there is no maximum potency.

No objection has been found to this argument, except that, when applied to "the set of everything" it seems to lead to a contradiction which has not yet been conclusively disposed of. Some references have been given in the text (pp. 147, 156); since passing the proof-sheets for press some more literature on the subject has appeared (Schoenflies, *Jahresber. d. d. Mathvgg.* 1906, and several papers in *Proc. L. M. S.*, Ser. II. Vol. IV. Part 1). It is not proposed to enter here into this discussion, but only to speak of the doubts which have, in consequence of this difficulty, been thrown on the propriety of using the symbol c^c at all. Borel (*loc. cit.*) descants on the extreme vagueness of our conception of the fundamental region S consisting of all sets of points, and this has been adopted as a war-cry by a certain party. It is, however, to be pointed out that our knowledge of the fundamental region S has been considerably enlarged since Borel wrote in 1898. He himself pointed out that the set of all closed sets is of potency c (*loc. cit.* p. 50), a detailed proof of this and some analogous theorems being given by Bernstein in his Dissertation (1900) (cp. also Beppo Levi, *loc. cit.* p. 150 *supra*). A number of such theorems with regard to components of S are given on p. 164 *supra*. Lebesgue (*loc. cit.* note 6 *supra*) has shewn that all measurable sets have the same potency as S itself.

Parallel researches in the fundamental region S' of all functions of a real variable have shewn that the potency of all functions of a real variable for which an existence theorem is applicable is c (Baire, *loc. cit.* p. 70 *supra*, pp. 70 —79), and that, contrary to a statement made by Cantor, the potency of all integrable functions is c^c (Jourdain, *J. f. reine u. angew. Math.* 128, p. 179). Under these circumstances the reproach of vagueness must be felt to have been to a great extent removed from the sets S and S'. The objection that we have no means of realising the most general set of points or function is one which is equally applicable to the most general point or decimal fraction, and, since it is not felt to be sufficiently weighty to prevent us using the potency c, it cannot be admitted as condemning the potency c^c.

(7) pp. 147 and 156. **On simple and multiple equivalence, and the mathematical law of arbitrary choice.**

In Ch. II. the straight line, being considered as lying in a plane, the geometry of the plane was used to shew that we can, by arbitrarily choosing the points P, Q, on the straight line, C outside the line, and A on PC, construct a scale of points on the straight line corresponding to all the rational and irrational numbers. Since the points P, Q, C and A can each be chosen in c ways, the scale can be set up in this manner in $c^4 = c$ ways, no one of which is more likely to be chosen than another ; thus we say that *the scale is only determined with multiplicity c.*

On the other hand, if we do not use the geometry of the plane, but assume as an axiom that the straight line is a simply-ordered set of ordinal type θ (§ 78), and call its potency c, it is easily seen that we can choose out the countable component dense everywhere in it, which is necessary for the setting up of a scale, in $c^a = c$ ways, all equally likely, so that again the scale is only determined with multiplicity c.

It follows that all proofs which depend on such a scale have the multiplicity c, and that proofs of theorems for linear sets which do not assume such a scale, but assume the principle of arbitrary choice any countable number of times, have precisely the same multiplicity. This is, for instance, the case with the proofs of Theorem 2 on p. 18 and Theorem 1 on p. 175 ; a similar proof may be given of the same theorem in space of any countable number of dimensions.

In *Math. Ann.* XLVI. Cantor states that any transfinite set whatever has a component of potency a, that is $\aleph_0$. If M is the set, and m its potency, the proof of this theorem, depending essentially on a arbitrary choices, has the multiplicity m^a. This theorem justifies the use of the expression " more than countable." The same principle shews, with a multiplicity $m^{\aleph_1}$, that, if m be not less than $\aleph_1$, M has a component of potency $\aleph_1$, and generally that there is a first Aleph, say $\aleph$, which is not less than m, the proof involving $\aleph$ arbitrary choices and being therefore of multiplicity $m^{\aleph}$. This does not however prove that the set M can be well-ordered, nor does it prove the identity of m and $\aleph$.

The mathematical principle of arbitrary choice here used is not to be confused with the ordinary or human principle of arbitrary choice, or with the throwing of dice. These are processes which, like all human affairs, are essentially finite, indeed it would not be difficult to assign a finite number which the number of such choices could never exceed. *The mathematical principle of arbitrary choice however is axiomatic*, and, in order to be of universal application, *is weighted with a variable multiplicity*, depending on the conditions of the problem, and which ought strictly to be explicitly worked out. These facts may assist the student in studying the literature of the subject, references to which will be found in the index.

Du Bois Reymond, in a much discussed passage †, points out the difficulties connected with the conception of the most general decimal fraction, and suggests that the principle of arbitrary choice, as such, cannot be assumed for more than a finite number of choices. Here, however, he is referring to human choice, which is unique, and he speaks of the impossibility of determining all the digits by successive throws of the dice. The mathematical law of arbitrary choice would be quite useless for this purpose, since the digits would be determined each time with a multiplicity 10, whereas in order to actually set up a decimal fraction the digits must be determined uniquely.

Bernstein's diagrammatic representation of the simple order of countable sets (§ 73 *supra*) is such that to each countable set corresponds a set of c order-diagrams, or binary order-fractions.

If Π be the potency of the set of all countable ordered sets, we have therefore, since the use of binary fractions involves a multiplicity c,

$$\Pi c \leqslant c \quad \text{(Mult. } c) \quad \dots\dots\dots\dots\dots\dots\dots\dots\dots(1).$$

Cantor takes any binary fraction F and between the points 0 and 1 he places a set of type $*\omega + \omega$, or a single point, according as the first figure of F is 0 or 1. Generally in the segment $(\cdot 1^{n-1}, \cdot 1^n)$ he places a set of type $*\omega + \omega$, or a single point, according as the nth digit of F is 0 or 1. He thus obtains a certain simply ordered set S_r, and he easily shews (Bernstein's

† *Allg. Funktionentheorie*, p. 91.

Dissertation, § 6 of reprint in *Math. Ann.* LXI.), that two sets S_F and $S_{F'}$, obtained from different fractions F and F', are different. Now S_F can be chosen uniquely after setting up a scale, thus we have c countable simply ordered sets. Hence

$$\Pi \geqslant c \quad \text{(Mult. } c\text{)}\dots\dots\dots\dots\dots\dots\dots(2).$$

Therefore $\qquad\qquad\qquad \Pi c \geqslant c \quad \text{(Mult. } c\text{)},$

whence, by (1), $\qquad\qquad \Pi c = c \quad \text{(Mult. } c\text{)}\dots\dots\dots\dots\dots\dots\dots(3).$

In other words, *it is possible to divide the continuum up into Π sets, each of potency c.*

Since the potency of all sets of points is c^c, it follows that

$$\Pi \leqslant c \quad \text{(Mult. } c^c\text{)}\dots\dots\dots\dots\dots\dots\dots(4).$$

Hence, by (2) $\qquad\qquad \Pi = c \quad \text{(Mult. } c^c\text{)}\dots\dots\dots\dots\dots\dots\dots(5).$

Since $\aleph_1 c \leqslant \Pi c$ and $\geqslant c$, it follows from (3) and (5) that

$$\aleph_1 c = c \quad \text{(Mult. } c\text{)}\dots\dots\dots\dots\dots\dots(6),$$

$$\aleph_1 \leqslant c \quad \text{(Mult. } c^c\text{)}\dots\dots\dots\dots\dots\dots(7).$$

It will be noticed that *the commutative law for multiplication of potencies is also weighted with a variable multiplicity.*

(8) p. 165. **Peano's arithmetical representation of the square on the straight line.** The representation of the square on the straight line given in § 99 was originally given by Peano (*Math. Ann.* XXXVI. p. 157, 1890) in arithmetic form, the geometrical discussion was first given by Hilbert (*Math. Ann.* XXXVIII. p. 459, 1891).

Using the ternary notation, let

$$T = 0 \cdot t_1 t_2 t_3 \dots, \quad X = 0 \cdot x_1 x_3 x_5 \dots, \quad Y = 0 \cdot y_2 y_4 y_6 \dots,$$

and let us define the operation k performed on a, where a is 0, 1, or 2, by the identity $\qquad\qquad ka = 2 - a \quad\dots\dots\dots\dots\dots\dots\dots(1).$

Thus $\qquad\qquad\qquad k^2 a = k^4 a = k^{2n} a = a$ $\Big\}$

and $\qquad\qquad\qquad ka = k^3 a = k^{2n+1} a \Big\}\quad \dots\dots\dots\dots\dots\dots(2).$

The Peano correspondence between the points T of the segment $(0, 1)$ and the points (X, Y) of the unit square is then given arithmetically as follows:

$$x_1 = t_1, \qquad\qquad\qquad y_2 = k^{t_1} t_2,$$

$$x_3 = k^{t_2} t_3, \qquad\qquad\qquad y_4 = k^{t_1 + t_3} t_4,$$

$$x_{2n+1} = k^{t_2 + t_4 + \dots + t_{2n}} t_{2n+1}, \quad y_{2n} = k^{t_1 + t_3 + \dots + t_{2n-1}} t_{2n}.$$

The converse equations determining T when X and Y are known can then easily be written down, using (2).

It is evident from these two sets of equations that the correspondence between the ternary fractions T and (X, Y) is always $(1, 1)$. Thus the correspondence between the points T and (X, Y) is always $(1, 1)$, provided none of the corresponding ternary fractions terminate. It is easy, however, to prove the identity of the pair of formally different fractions (X_1, Y_1) and (X_2, Y_2), which correspond to a terminating T_1 and its equivalent fraction T_2 which ends in $\dot{2}$. On the other hand the fractions T_1 and T_2, or T_1, T_2, T_3, T_4, which correspond to a point (X, Y) with one or two terminating coordinates actually are distinct.

X (and similarly Y) is a continuous function of the variable T, having nowhere a differential coefficient.

(9)   p. 171.   Take n circles, each of which lies entirely outside all the others.   Let P be a point of the region R outside all the circles, and S the set of all the images of P, formed by inversions at all the circles, repeated in any manner whatever. It may be shewn [54] that this set of operations is the same as all the single inversions at all the circles which are images of the given circles: this set of circles we shall call *the set of circular images*; it may be constructed by inverting first each of the given circles with respect to the others, and then inverting each of the circles so obtained with respect to one another, and so on *ad infinitum*.   Hence we may shew that the images of the region R form a set of non-overlapping regions, each bounded by n circles.

It now follows at once that *the set S is an isolated and therefore a countable set*, since in each image of R there is one and only one image of P.   If $n=1$, S consists of a single point, otherwise S is a countably infinite set.   The first derived set S' of S consists of two points when $n=2$, these being the point circles of the coaxal system determined by the two circles.   If, however, $n>2$, S' *is a perfect set, dense nowhere in every region and on every curve*, and, when $n=3$, or $n>3$ and the n circles have an orthogonal circle, the set is of the type constructed in Ex. 1, p. 171, Fig. 20.

The points of S' may be shewn, in fact, to be the point-circles of the set of circular images, that is, each point L of S' is determined as the sole point internal to an infinite number of the circular images lying one inside the other, since the radii in this case decrease without limit.   In each such circular image there are at least two other circular images, and in each circular image at least one point-circle ; it follows, therefore, that S' *is dense in itself*.   Again, S' *is closed*, being a first derived set, and therefore is perfect. That it is *dense nowhere in every region* follows from the fact that the same is true of S.   That it is *dense nowhere on every curve* follows from the fact that there is no point of S' on any of the circumferences of the circular images. That the presence of an orthogonal circle reduces the set S' to the simpler form of Fig. 20 follows, since by inversion orthogonal circles remain orthogonal, and therefore the point-circles must all lie on the orthogonal circle of the given n circles.   These indications will suffice to enable the reader to complete the proofs.   This example was popularly sketched by Klein in his Lectures on the Application of the Calculus to Geometry (1902, lithographed by Teubner).   It occurs naturally in Geometry and in the Theory of Potential, and came into Klein's work on automorphic functions.   (Fricke, " Die Kreisbogenvierseite und das Princip der Symmetrie," *Math. Ann.* XLIV. pp. 565— 599.   Klein, Letter to Poincaré, *Comptes Rendus*, 1881.   Poincaré, *Acta Math.* III. p. 78 *seq.*   Klein-Fricke, *Modulfunktionen*, Vol. I. p. 103.   Klein-Fricke, *Automorphe Funktionen*, Vol. I. p. 428 *seq.*)

(10)   p. 187.   The idea of *span*, as defined in § 117, may be transferred to any set by substituting the word "set" for "closed region."   If the set is closed, it then follows that there is always at least one pair of points of the set whose distance apart is the span s, or the span s_L in any particular direction; the proofs are similar in character to that of Theorem 2, § 106, p. 175.

The following theorem, of which use is made on p. 281, at once suggests itself :—

The (outer) content of a set of span e is equal to, or less than $\frac{1}{4}\pi e^2$. In particular, *a circle of diameter e has the maximum area of any quadrable region of span e.*

To prove this theorem consider that, if the *distances* of any two points A and B from a point P are each not greater than e, the same is true of the distance of P from any point of the stretch AB. Thus *we do not alter the span of a set, if we add to the set all the points on all the stretches whose end-points belong to the set.* Further, *we do not alter the span by closing the set.* Neither of these processes decreases the (outer) content. Thus if the theorem can be proved for such a modified set, it follows generally.

Now these two processes of modification may be carried out in stages as follows. Take any point O, and any ray through O as axis of x. On each ordinate the points of any given set G determine a definite smallest segment containing all of them; the points of this segment, open at both ends, are precisely all the points between any two points of G on the ordinate. These points we adjoin to G and then close the set. Let this set be denoted by G'.

Let the greatest and least values of y for points of G' corresponding to any particular value of x be y_1 and y_2. Then y_1 *and* $-y_2$ *are both upper semi-continuous functions of* x (§ 160, p. 252). If on each ordinate we shift the closed segment, which is the section of G', so that that segment is bisected, the amount of shifting will be $\frac{1}{2}(y_1+y_2)$, which is a summable function (footnote †, p. 260), while the length of the segment is $\frac{1}{2}(y_1-y_2)$, which is an upper semi-continuous function. Hence we can shew that, after shifting, the set is still closed. For if we have any sequence of points of the new set $(x^{(i)}, y^{(i)})$, having (x, y) as limiting point,

$$-\tfrac{1}{2}\{y_1^{(i)}-y_2^{(i)}\} \leqslant y^{(i)} \leqslant \tfrac{1}{2}\{y_1^{(i)}-y_2^{(i)}\},$$

so that $$\tfrac{1}{2}\operatorname*{Lt}_{i=\infty}[-\{y_1^{(i)}-y_2^{(i)}\}] \leqslant y \leqslant \tfrac{1}{2}\operatorname{Lt}\{y_1^{(i)}-y_2^{(i)}\},$$

and therefore, (y_1-y_2) being an upper semi-continuous function,

$$-\tfrac{1}{2}\{y_1-y_2\} \leqslant y \leqslant \tfrac{1}{2}(y_1-y_2).$$

This last inequality shews that the limiting point (x, y) lies inside the segment which was the section of G', and therefore belongs to G' after shifting. Thus G' remaining a closed set we can apply the corollary on p. 259, and assert that *the content of G' has been unaffected by the shifting.*

It remains to shew that *the span of the shifted set is not greater than e.* For since

$$-\tfrac{1}{2}(y_1-y_2) \leqslant y \leqslant \tfrac{1}{2}(y_1-y_2),$$

and $$-\tfrac{1}{2}(y_1'-y_2') \leqslant y' \leqslant \tfrac{1}{2}(y_1'-y_2'),$$

the distance between the points (x, y) and (x', y') of the shifted set is not greater than

$$(x-x')^2 + \{\tfrac{1}{2}(y_1-y_2) + \tfrac{1}{2}(y_1'-y_2')\}^2,$$
$$\leqslant \tfrac{1}{4}\{4(x-x')^2 + (y_1-y_2)^2 + (y_1'-y_2')^2 + (y_1'-y_2)^2 + (y_1'-y_2')^2$$
$$- (y_1-y_1')^2 - (y_2-y_2')^2\},$$
$$\leqslant \tfrac{1}{4}\{4e^2 - (y_1-y_1')^2 - (y_2-y_2')^2\} \leqslant e^2.$$

Thus the span of the shifted set is less than, or equal to e; whence it follows that the whole shifted set lies inside the common part of two circles of radius e, whose centres lie at a distance $\frac{1}{2}e$ from O on the ray through O perpendicular to that chosen as axis of x; this latter is now an axis of symmetry of the set.

If we repeat this process, choosing now the ray perpendicular to the first as new axis of symmetry, and then choosing each of the two bisectors of the angles between the two axes of symmetry, and so on, the set, which becomes more and more symmetrical without increasing its span or altering its content, always lies inside the common part of the circles of radius e, whose centres lie on the rays through O perpendicular to the axes of symmetry and at a distance $\frac{1}{2}e$ from O.

Now this common part approaches as limit the circular region of radius $\frac{1}{2}e$ with O as centre. Thus, choosing h arbitrarily, we can determine k so that, after k shiftings, our set lies inside a region whose area is less than $\frac{1}{4}\pi e^2 + h$. This proves that the content of the set is not greater than $\frac{1}{4}\pi e^2$, which, since the area of a circle of span e is $\frac{1}{4}\pi e^2$, proves the theorem.

(11) p. 204. Jordan uses *parfait* for *closed*. Borel uses *parfait* sometimes in this sense, sometimes in the sense of perfect, differentiating the former as "relativement parfait" (Jordan), and the latter as "absolument parfait" (Cantor) (*Leçons sur la Théorie des Fonctions*, pp. 35, 36.) Lebesgue uses *parfait* in Cantor's sense, and uses *fermé* for *closed*.

(12) p. 207. Let ABC be one of the triangles of span $< e_i$. Produce the sides and take A_o, B_o, and C_o respectively in the angles vertically opposite to BAC, ABC, and BCA. Then the triangle $A_oB_oC_o$ will contain the triangle ABC. In order that the span of $A_oB_oC_o$ should be less than e_i, we only have to choose A_o inside the common part of the circular domains with centres B and C and radius e_i, and similarly restrict the choices of B_o and C_o to lie within certain domains. Finally, in order to insure that the triangle $A_oB_oC_o$ should have none of the inconveniences specified in the text, we only have to avoid choosing for A_o one of the $\frac{1}{2}3n(3n-1)$ intersections of sides of the n triangles, then choose B_o so as not to lie on any of the lines joining A_o to any of those points, and finally choose C_o so as not to lie on any of the lines joining either A_o or B_o to any of those points.

(13) p. 226, lines 17 and 21. If $n > 3$, R_{i-1} is not one of the regions which necessarily overlap with R_{i+1}; hence any common parts of R_{i-1} and R_{i+1} contain no points of the Jordan curve. Thus this case is the same as that touched on in lines 14—20. The reduction alluded to in line 17 is easily effected; it is not given in full in the text, since to do so would break the argument. If $U_{j,k}$ be the common part of R_j and R_k containing no point of the Jordan curve, we only have to interchange those parts of the boundaries of R_j and R_k which bound $U_{j,k}$, and then round off the common boundary points by small circles of radius d, where $2d$ is the minimum distance of $U_{j,k}$ from the Jordan curve.

(14) p. 227, line 3. The polygon Π is *simple* (p. 179, footnote †), since C_i and B_i can be joined by a simple polygonal line, lying in R_i and only intersecting the remaining regions R_j ($j \neq i$) at B_i and C_i. Thus we may speak of the inside and outside of Π (lines 14, 18, 20).

APPENDIX

(15) P. 7, line 2u. The choice of an interval sufficiently small for f(x) to be zero only at x = b is not necessary, since we may assume that f(x) = 0 has no rational root.

(16) P. 8, line 9u. Add the following footnote, giving Liouville's papers: Comptes Rendus, 18 (1844), pp. 883-885 and 910-911; and Journal des Mathématiques pures et appliquées, Ser. 1, 16 (1851), pp. 133-142.

(17) P. 19, lines 17d - 19d. Expand the sentence "A closed interval is a special case of a closed set, and an open interval of an open set" into the following: If we add to any set E all the non-included limiting points, it is easily shown that we get a closed set. This is said to be the set got by closing E. A closed interval is a special case of a closed set, and an open interval of an open set. Any point of a completely open interval is called an internal point of that interval (open or closed), and any point of a set which belongs to a completely open interval contained in the set is called an internal point of the set. Any internal point of the complementary set is called an external point of the interval or set, and the remaining points are called the boundary points of the set. In the case of an interval PQ, these are the end-points P and Q.

(18) P. 21, line 12u. After the definition, add the following examples:

1) All the points of $(0, 1)$.
2) All the rational points of a segment.
3) All the points of the segment of the form $2nM/2^m$.
4) Sierpiński's set. In the decimal notation, let

$$x = 0.x_1x_2x_3 \ldots ,$$

where

$$\text{Lt } (x_1 + x_2 + \ldots)/n = b,$$

b being any chosen number greater than zero and less than one. The set of points x is dense everywhere in $(0, 1)$.

Sierpiński, Bulletin de l'Académie des Sciences de Cracovie, Ser. A (1913), pp. 76-82.

(19) P. 22. Add the two following paragraphs to the end of Article 10:

A set, each point of which is a limit on one side only, may be said to be dense in itself on one side only, and if the side is the same at each point, to be dense in itself on the right, or on the left. If every point is a limiting point on both sides of the set, the set may be said to be dense in itself on both sides.

The rational limiting points of the Liouville numbers are an example of a set which is dense in itself on the left only. Taking any such number, reducing its last figure by 1, and affixing what may be called the maximum Liouville termination (that is, making all subsequent e's equal to 9 when the scale is decimal, or n - 1 when n is the base) we get a set which is dense in itself on the right only. The former set consists of the right-hand end-points, the latter of the left-hand end-points, of the black intervals of the Liouville points.

(20) P. 23, line 9u. Add the following: The set got by closing E (see Appendix (17) above) differs from E_1 in including the isolated points (if any) of E. If, and only if, E is dense in itself, the set got by closing E is the first derived set E_1.

(21) P. 30, line 16d. Precede the definition of a limiting point of a set of intervals with the following version:

A limiting point of a set of intervals is a limiting point of left-hand end-points of some of the intervals, and at the same

time a limiting point of the right-hand end-points of the same intervals; and a limiting interval of a set of intervals is an interval
whose left-hand end-point is a limiting point of the left-hand end-
points of some of the intervals, and at the same time the right-
hand end-point is a limiting point of the right-hand end-points of
the same intervals.

(22) P. 36. See Appendix (25) below.

(23) P. 41, line 25d. Before the sentence beginning with
'Therefore' insert the following: For if P be any point of the
given set and d_i the first of the given intervals having P as
internal point, P will, by construction, be separated by an interval from all the b's formed from the d's subsequent to d_i. On
the other hand, the b's formed from d_1, ... , d_i, being finite in
number, have no limiting points, so that P cannot be a limiting
point of the b's at all.

(24) P. 41, line 15u. After the proof of Theorem 6, add the
following corollary:

Cor.   The result still holds if the left-hand (right-hand)
end-points of the black intervals of the closed set are only right-
hand (left-hand) end-points of intervals of the given set.

For we can enlarge such an interval so as to include the end-
point in question and no further point of the set, and then apply
the theorem, subsequently curtailing any such interval which may
be among the finite number. This proves the corollary.

(25) P. 44, Theorem 12 to p. 46, end of Article 22. To be
transposed to p. 36, presumably at the end of Article 18.

(26) P. 55, line 3d.   Add the following footnote to Theorem
21:

The theorem commonly called the Cantor-Bendixson
Theorem is the Theorem of the Nucleus for closed sets. For
Bendixson's proof, see Acta Mathematica II (1883), p. 419. For
Cantor's, see Mathematische Annalen, 23 (1884), p. 468 (the
latter is also to be found in Cantor's Abhandlungen, p. 224).

(27) P. 55, line 5d. To the Corollary of Theorem 21, add,
in some form, the following: Very important theorem. Necessary and sufficient condition for a set to be countable is that it
contain no point of condensation.

(28) P. 62, line 11 u. Add the following proof of Theorem 29:

Let F denote any coherence, $F = Fa + Fc$.

Let E denote either the original set, or some coherence preceding F. Then any point of F is a point of E. Also $E = Ea + Ec$. Suppose P to be a point of F, but not a limiting point of Ea. Then we can describe an interval d round P, so as to contain no point of Ea. Hence every point of E in the interval d is a point of Ec, and as such, a limiting point of E. Thus the subset of E in d, which includes P, is dense in itself and therefore belongs to every coherence and, in particular, to Fc. Thus such a point P is not a point of Fa.

Hence if Q is a point of Fa, it must be a limiting point of Ea, that is, of any adherence preceding F. This proves the theorem.

The following are corollaries of Theorem 29.

Cor. If Q is a limiting point of an adherence, it is a limiting point of every adherence preceding it in the natural order.

Cor. If Q be a limiting point of the whole set, but not a limiting point of its ultimate coherence U, either it is a limiting point of all the adherences, or there is a definite adherence A such that Q is not a limiting point of A but is a limiting point of every adherence preceding A in the natural order.

(29) P. 64, line 9u and p. 65, line 17u.

"Blaricum, 4.4.13.

Dear Mrs. Young,

. . . .

It seems to me that in 'The Theory of Sets of Points' on p. 65 in cor. 2 the words 'and sufficient' must be destroyed, and that on p. 64 the proof of the lemma is incomplete.

Yours sincerely,

L. E. J. Brouwer."

(30) P. 66. A new proof of Theorem 33:

For E, being closed, contains its first derived set E_1; whence, by the preceding theorem, it follows that E is the inner limiting set of any sequence of sets of intervals such as are there considered.

(31) P. 69, line 15d. Add the following: In fact, if U be
any perfect set dense nowhere, whose right-handed limiting
points (for instance) are rational, U will be internal to the intervals at every stage and therefore will form part of the inner limiting set. We have seen that this is the case with the Liouville
points (that is, the example given in Appendix (19) above). Thus
the inner limiting set certainly includes all the Liouville points.

(32) P. 70, line 11 u. Add the following at the end of Article
33:
Any set which consists of all the points of a countably infinite set of closed sets is an ordinary outer limiting set, and any
set which consists of all points common to a countably infinite
number of sets of intervals is an ordinary inner limiting set.

(33) P. 71, line 3u. Add the following at the end of Article
34:
Typical set of the first category of potency c and content
zero.
In every black interval of Cantor's typical ternary set G_1
put a similar set, calling the whole set G_2. In every black interval of this new set put another set similar to G_1, and so on.
The final set G is dense everywhere and is therefore of the first
category, of potency c and content zero. Numerically:

in G_1, $x = e_1 e_2$ .... ($e_i = 0$ or 2).
G_2, $x = e_1 e_2$ .... ($e_1 = 0$ or 2, and only one of the e_i's
    may be a 1).
G_r, $x = e_1 e_2$ .... ($e_i = 0$ or 2, and n - 1 only of the
    e_i's may be a 1).
G, $x = e_1 e_2$ .... ($e_i = 0$ or 2, and a finite number
    of the e_i's may be 1's).

(34) P. 86, line 2u. Also in: Smith, Collected Mathematical
Papers (Oxford, 1894; repr. New York, 1965), Vol. 2, p. 90.

(35) P. 88, line 1u. The following theorem should be
added at the end of Article 47:

Given three positive quantities r, e and s, we can always
describe intervals of length less than or equal to 2r with the
points of a closed set G as centres, so that the content of the
intervals lies between I and I + e, I being the content of the
set, and so that the sum of the overlapping parts is less than s.

(36) P. 99, line 3d to p. 100, line 1u. Omit everything from Theorem 22 to Theorem 25 inclusive.

(37) P. 104, line 18d. Before Theorem 28, insert:
We then have the following important theorem, which is a generalisation of that given above, in Appendix (35).

Theorem. If G be a set of points of outer content I, and r, e and s are any three given positive numbers, we can describe intervals of length less than or equal to r with the points of G as centres, so that the content of the intervals lies between I and I + e, while the sum of the overlapping parts is less than s.

The proofs of this theorem and of the theorem of Appendix (35) are omitted. The proof of the latter entails the use of the generalised Heine-Borel Theorem (page 41) while the proof of the former uses Theorem 5 of page 38 instead. Also, where the latter uses a finite number of s_i's whose sum is equal to (or, less than or equal to) s, the former uses an infinite number of such s_i's which constitute a convergent series whose sum is equal to (or, less than or equal to) s.

(38) P. 109, line 3d. Combine Theorem 33 with Theorem 26 on page 101, and omit the last two paragraphs (lines 15u to 1u) on page 109.

(39) P. 111, line 10d. After Corollary 3, insert:

Ex. 4a. A set of the first category whose content is that of the continuum, and a set of the second category of content 0.

Let G_n be the set got by the method of Ex. 2, page 79, with 3^n as base m of the system of division. The content I_n of the set G_n has the limit 1 and therefore the outer limiting set has content 1. The complementary set is a set of the second category of content 0.

Another example is obtained by describing intervals s/2, $s/2^2$, ... round rational points and then making s describe a sequence with zero as limit.

(40) P. 147, line 14d. Add the following additional references at the end of the first footnote:
Zermelo, Mathematische Annalen, 65 (1908), pp. 261-281. Poincaré, 'Dernières Pensées,' pp. 102-139. Lebesgue, Journal de Mathématiques Pures et Appliquées, Ser. 6, 1 (1905), pp. 184-187.

(41) P. 151, line 14d. Add the footnote: The definition of multiplication for an infinite number of factors involves the axiom of Appendix (7) (pages 289-290). (In modern terminology, the Axiom of Choice.)

(42) P. 167, line 1u. Add the following reference to the footnote:

Jordan, 'Cours d'Analyse' (2nd ed.), Tome I, p. 90.

(43) P. 180, line 9u. To the definition of a domain add:

A favourite definition, lately adopted, of a domain, which is merely a form of the above, is that it is a set of points, each of which is an internal point of the set, and such that any two points can be joined by a broken line of such internal points.

(44) P. 184, line 6u. Eliminate the footnote and substitute the following proof of the corollary.

If P and Q are internal to a single triangle they can be joined by such a path, consisting of a single stretch. If not, let the number of triangles forming a transitive set joining P to Q be m, and omit from these any one, say d_1, which contains P. The common points of d_1 and the remaining m - 1 triangles form a closed set, not the null-set, since the m triangles form a transitive set. Let P_1 be the nearest point of this closed set to Q. Then repeat the reasoning with P_1 in place of P, remembering that P_1 is joined to Q by a transitive set containing at most m - 1 triangles, namely the given triangles omitting d_1, or a subset of these.

Thus after at most m - 1 such steps we arrive at a point P_k which lies in one of the triangles containing Q. Then PP_1, P_1P_2, P_2P_3, ... , P_kQ is a polygonal path with at most m sides all internal to the region. If this polygon cuts itself, we may omit the loop formed between the first and second passages through such a cross-point, and doing this as often as may be required as we pass from P to Q along our polygon, we reduce it to a simple one joining P to Q and lying inside the given region. This proves the corollary.

(45) P. 191, line 1u. Add the following reference to the footnote:

Bolzano, Abh. Böhm. Ges. Wiss. Prague. Ser. 3, V (1818); also in Ostwalds's Klassiker (ed. P. E. B. Jourdain), no. 153, pp. 3-43. (French translation in: Rev. Hist. Sci. Appl. 17 (1964), pp. 136-164.)

(46) P. 213, line 12d. This region is described by the triangles consisting of internal points (cf. Article 112, page 182).

(47) P. 213, line 2u. After the proof of Theorem 23, insert
the following:

Theorem. Any region which is the intersection of two simply
connected regions is simply connected.

For any boundary point of the region must be a boundary
point of one or the other of the simply connected regions, so that
in any circle containing that point there are points external to
that region and therefore to the given region. Thus every boundary point is a rim point. Again, any simple polygon consisting of
points internal to the given region consists of points internal to
both simply connected regions and therefore encloses no boundary point of either, and therefore no boundary point of the given
region. Thus the given region has no inner rims. Therefore
the given region is simply connected. Thus the common internal
points of two simply connected regions form simply connected
regions.

(48) P. 216, lines 11d and 16d. Add the following statement
to Theorem 25 and add the following proof to the proof of
Theorem 25. Delete all the corollaries and replace them with
the following corollary:

The converse is also true, provided the correspondence embraces all the points of the plane which come under consideration.

To prove that if G is closed so is g, let p be any limiting
point of g and P its corresponding point. Then, since the correspondence is continuous, if the correspondence embraces p,
in every interval containing p there are points of g. Therefore
in every circle with P as centre there are points of G, so that
P is a point or limiting point of G, and therefore certainly a
point of G, as G is closed. Therefore p is a point of g, so that
g is closed.

Cor. In a correspondence of the whole plane, the correspondents of any point form a closed set.

(49) P. 222, line 5d. Add the following footnote to the end
of Article 139:
Consult Brouwer, Mathematische Annalen, 69 (1910), pp.
169-175 and Carathéodory, Mathematische Annalen, 73 (1913),
pp. 314-320.

APPENDIX303

(50) P. 223, line 20d. Before Theorem 3, insert:

Take any two points P and Q of the curve, and first let P be the point corresponding to a and z, and let Q be the point corresponding to any other point q of (a, z). Then the points of the given Jordan curve corresponding to the points of (a, q) and (q, z) respectively form two Jordan curves, which are arcs of the given Jordan curve and together make it up completely. These two arcs have no common points except P and Q, since (a, q) and (b, q) have no common point but q, and a is regarded as identical with z. If R is any point other than P or Q of the given curve, its correspondent r is internal to one of the stretches (a, q) and (q, b) and is therefore not a limiting point of the other. Consequently R is a point of one arc and not a limiting point of the other. Thus P and Q divide the given Jordan curve into two arcs with no common limiting points except P and Q. If the point corresponding to a and z is not one of the two points P or Q, the reasoning is similar and the result the same. Thus the Jordan curve is a simple closed curve when a and z have the same correspondent.

(51) P. 242, line 4u. Add the following alternative definition of quadrable region:

A region is said to be quadrable if its points can be arranged in ordinates whose highest points (x, y) are such that y is summable with respect to x.

(52) P. 248, line 8u. Ex. 3 is not in the right place, and the preceding paragraph is not right. Ex. 3 belongs before the preceding remark on the falsity of the converse, which would then be demonstrated by re-constructing Ex. 3 with m^2 divisions at each stage.

(53) P. 257, line 5u. At the end of this line, insert the sentence:

These sets are closed (Article 160, page 252) because, by the first part of the theorem, X(x) is a semi-continuous function.

(54) P. 292, line 6d. This must be amended. The lines beginning with the words "It may be shewn that this set of operations" are misleading. This is not true of the images of P but only of the images of R as a whole.

BIBLIOGRAPHY*.

AMES, L. D. "On the theorem of Analysis Situs relating to the division of the plane or of space by a closed curve or surface," 1903. *Bull. Amer. Math. Soc.*, Ser. 2, X. pp. 301—305.

—— "An arithmetic treatment of some problems in Analysis Situs," 1905. *Amer. Journ. of Math.*, XXVII. Reproduced in Osgood's *Lehrbuch der Funktionentheorie*, 1906, pp. 130—141.

AMIGUES, E. "La théorie des ensembles et les nombres incommensurables," 1892. *Ann. de la Fac. des Sciences de Marseille*, II. pp. 33—43.

ARZELÀ, C. "Un teorema intorno alle serie di funzioni," 1885. *Rend. dell' Acc. dei Lincei*, Ser. 4, I. pp. 262—267.

—— "Funzioni di linee," 1889. *Rend. dell' Acc. dei Lincei*, Ser. 4, V. (semestre 1) pp. 342—348. "Sulle funzioni di linee," 1895. *Mem. R. Accad. di Bologna*, Ser. 5, V. pp. 225—244.

—— "Sulle serie di funzioni," 1900. *Mem. R. Accad. di Bologna*, Ser. 5, pp. 131—186, VIII. 701—744.

—— "Sulle serie di funzioni di variabili reali," 1902. *Rendiconti della R. Accad. di Bologna*.

—— "Sul secondo teorema della media per gli integrali doppi," 1903. *Mem. della R. Accad. di Bologna*, Ser. 5, X. pp. 90—100, and a number of other memoirs containing applications of the Theory of Sets.

ASCOLI, G. "Sulle serie di Fourier," 1875. *Ann. di Mat.*, Ser. 2, VI. pp. 21—71 and 298—351; see specially p. 56.

—— "Le curve limite di una varietà data di curve," 1884. *Mem. dell' Acc. dei Lincei*, Ser. 3, XVIII. pp. 521—586.

—— "Riassunto della mia memoria: 'Le curve limite di una varietà data di curve,' ed osservazioni critiche alla medesima," 1888. *Rend. dell' Ist. Lomb.*, Ser. 2, XXI. pp. 226—239, 257—265, 294— 300, 365—371.

BAIRE, R. "Sur les fonctions de variables réelles," 1899. *Ann. di Mat.*, Ser. 3, III. pp. 1—123.

—— "Sur la théorie des ensembles," 1899. *Comptes Rendus*, CXXIX. pp. 946—949.

BAIRE, R.; BOREL, E.; HADAMARD, J.; LEBESGUE, H. "Cinq lettres sur la théorie des ensembles," 1905. *Bull. de la Soc. Math. de France*, XXXIII. pp. 261—273.

BECKMAN, L. K. "Om dimensionsbegreppet och dess betdelse för matematiken," 1888. *Diss.* Upsala.

---

* The Roman numbers refer to the volume. The date is, as far as was possible, that affixed to the work, and is therefore sometimes anterior to that of publication. The volume number renders the date of publication unnecessary for purposes of identification and reference.

BENDIXSON, I. "Quelques théorèmes de la théorie des ensembles de points," 1883. *Acta Math.* II. pp. 415—429.

—— "Några studier öfver oändliga punktmängder," 1883. *Öfvers. af vet. akad. förhandlingar* (Stockholm), XL. No. 2, pp. 31—35.

—— "Sur la puissance des ensembles parfaits de points," 1884. *Bihang till vet. ak. handlingar* (Stockholm), IX. No. 6.

—— "Un théorème auxiliaire de la théorie des ensembles," 1884. *Bihang till vet. ak. handlingar* (Stockholm), IX. No. 7.

BERNSTEIN, F. "Ueber einen Schönflies'schen Satz der Theorie der stetigen Funktionen zweier reellen Veränderlichen," 1900. *Gött. Nachr.* pp. 98—102.

—— "Untersuchungen aus der Mengenlehre," 1901. *Diss.* (Göttingen). Reprinted in *Math. Ann.* LXI. pp. 117—155.

—— "Bemerkung zur Mengenlehre," 1904. *Gött. Nachr.* pp. 557—560.

—— "Ueber die Reihe der transfiniten Ordnungszahlen," 1904. *Math. Ann.* LX. pp. 187—193.

—— "Zum Kontinuumproblem," 1904. *Math. Ann.* LX. pp. 463—464.

—— "Zur Theorie der reellen Zahlen," 1905. *Jahresb. d. d. Mathvgg.* XIV. pp. 447—449.

BETTAZZI, R. "Teoria delle grandezze," 1890. Pisa.

—— "Su una corrispondenza fra un gruppo di punti ed un continuo ambedue lineari," 1888. *Ann. di Mat.*, Ser. 2, XVI. pp. 49—60.

—— "Sulla catena di un ente in un gruppo," 1896. *Atti dell' Acc. di Torino*, XXXI. pp. 446—456.

—— "Gruppi finiti ed infiniti di enti," 1896. *Atti dell' Acc. di Torino*, XXXI. pp. 506—512.

—— "Fondamenti per una teoria generale dei gruppi," 1896, 1897. *Periodico di mat.* XI. pp. 81—96, 112—142, 173—182 ; XII. pp. 40—42.

—— "Appendice ai Fondamenti per una teoria generale dei gruppi," 1897. *Periodico di mat.* XII. pp. 40—42.

—— "Sulla definizione d' infinito," 1897. *Periodico di mat.* XII. pp. 91, 92.

—— "Sulla definizione del gruppo finito," 1897. *Atti dell' Acc. di Torino*, XXXII. pp. 352—355.

—— "Sui punti di discontinuità delle funzioni di variabile reale," 1892. *Rend. del circ. mat. di Palermo*, VI. pp. 173—195.

BIASI, G. "Sulla definizione di infinito," 1897. *Periodico di mat:* XII. pp. 34—35.

BIERMANN, O. "Theorie der analytischen Functionen," 1887. Leipzig.

BINDONI, A. "Sui numeri infiniti ed infinitesimi attuali," 1902. *Rend. dell' Acc. dei Lincei*, Ser. 5, XI. 2° sem. pp. 205—209.

BLICHFELDT, H. F. "On the determination of the distance between two points in space of *n* dimensions," 1902. *Trans. Amer. Math. Soc.* III. pp. 467—481.

BLISS, G. A. "The exterior and interior of a plane curve," 1904. *Bull. Amer. Math. Soc.*, Ser. 2, X. pp. 398—404.

DU BOIS-REYMOND. "Versuch einer Classification der willkürlichen Functionen reeller Argumenten nach ihren Aenderungen in den kleinsten Intervallen," 1874. *Journ. f. Math.* LXXIX. pp. 21—37 (specially p. 36).

—— "Die Allgemeine Funktionentheorie," 1882. Tübingen. French translation by G. Milhaud and A. Girot, 1887, Paris.

BOLZANO, B. "Paradoxieen des Unendlichen," 1851; herausgegeben von F. Přihonský, Leipzig ; 2nd ed., Berlin, 1889.

BOREL, E. "Remarque relative à la communication de M. Hadamard," 1898. *Verh. des Math. Kongress in Zürich*, pp. 204—205.

—— "Leçons sur la théorie des fonctions," 1898. Paris, Gauthier-Villars.

—— "À propos de l'infini nouveau," 1899. *Revue phil.* XLVIII. pp. 383—390.

—— "L'antinomie du transfini," 1900. *Rev. phil.* XLIX. pp. 378—383.

—— "Contribution à l'analyse arithmétique du continu," 1903. *Journ. de math. pures et appl.*, Ser. 5, IX. pp. 329—375.

—— "Un théorème sur les ensembles mesurables," 1903. *Comptes Rendus*, CXXXVII. pp. 966—967*.

—— "Quelques remarques sur les ensembles de droites ou de plans," 1903. *Bull. de la Soc. Math. de France*, XXXI. pp. 272—275.

—— "Sur une propriété des ensembles fermés," 1905. *Comptes Rendus*, CXL. pp. 298—300.

—— "Quelques remarques sur les principes de la théorie des ensembles," 1905. *Math. Ann.* LX. pp. 194—195.

—— "Leçons sur les fonctions de variables réelles et les développements en séries de polynômes," 1905. Paris, Gauthier-Villars.

BORTOLOTTI, E. "Contributo alla teoria degli insiemi," 1902. *Rend. dell' Acc. dei Lincei*, Ser. 5, XI. 2° sem. pp. 45—52.

BRODÉN, T. "Beiträge zur Theorie der stetigen Functionen einer reellen Veränderlichen," 1896. *Journ. für Math.* CXVIII. pp. 1—60.

—— "Ueber das Weierstrass-Cantor'sche Condensationsverfahren," 1896. *Öfversigt af Kongl. Vet.-Akad. Förhandlingar*, 1896, No. 8, pp. 583—602.

—— "Ueber Grenzwerthe für Reihenquotienten," 1897. *Bihang till K. Svenska Vet.-Akad. Handlingar*, XXIII. Afd. I. No. 2, pp. 3—49.

—— "Reelle Functionen mit überalldicht liegenden Nullstellen," 1897. *Math. Ann.* LI. pp. 299—320.

—— "Funktionentheoretische Bemerkungen und Sätze," *Acta Univ. Lundensis*, XXXIII. p. 37.

—— "Functionentheoretische Bemerkungen und Sätze," 1897. *Acta Reg. Soc. Physiogr. Lund.* VIII. pp. 1—45.

—— "Wahrscheinlichkeitsbestimmungen bei der gewöhnlichen Kettenbruchentwickelung reeller Zahlen," 1900. *Öfversigt af K. Vet.-Akad. Förhandlingar*, Stockholm, 1900, No. 2, pp. 239—266.

—— "Derivirbare Functionen mit überall dichten Maxima und Minima," 1900. *Öfversigt af K. Vet.-Akad. Förhandlingar*, Stockholm, 1900, No. 4, pp. 423—441.

—— "Fortgesetzte Untersuchungen über derivirbare Funktionen mit überall dichten Maxima und Minima," 1900. *Öfversigt af K. Vet.-Akad. Förhandlingar*, Stockholm, 1900, No. 6, pp. 743—761.

—— "Bemerkungen über Mengenlehre und Wahrscheinlichkeitstheorie," and "Noch einmal die Gylden'sche Wahrscheinlichkeitsfrage," 1901. *Malmö. Skanska Lithografiska Aktiebolaget.*

BURALI-FORTI, C. "Sulle classi ordinate e i numeri transfiniti," 1894. *Rend. del circ. mat. di Palermo*, VIII. pp. 169—179.

* See W. H. Young, "Open sets and the theory of content," p. 16.

BURALI-FORTI, C. "Sulle classi derivate a destra e a sinistra," 1894. *Atti dell' Acc. di Torino*, XXIX. pp. 382—394.

—— "Sul limite delle classi variabili," 1895. *Atti dell' Acc. di Torino*, XXX. pp. 227—243.

—— "Sur quelques propriétés des ensembles d'ensembles et leurs applications à la limite d'un ensemble variable," 1896. *Math. Ann.* XLVII. pp. 20—32.

—— "Le classi finite," 1897. *Atti dell' Acc. di Torino*, XXXII. pp. 34—52.

—— "Sopra un teorema del Sig. G. Cantor," 1897. *Atti dell' Acc. di Torino*, XXXII. pp. 229—237.

—— "Una questione sui numeri transfiniti," 1897. *Rend. del circ. mat. di Palermo*, XI. pp. 154—164.

—— "Sulle classi ben ordinate," 1897. *Rend. del circ. mat. di Palermo*, XI. p. 260.

CANTOR, G. "Ueber die Ausdehnung eines Satzes aus der Theorie der trigonometrischen Reihen," 1872. *Math. Ann.* V. pp. 123—132. French trans., *Acta Math.* II. 1883, pp. 336—348.

—— "Ueber eine Eigenschaft des Inbegriffs aller reellen algebraischen Zahlen," 1874. *Journ. für Math.* LXXVII. pp. 258—262. French trans., *Acta Math.* II. 1883, pp. 305—310.

—— "Ein Beitrag zur Mannigfaltigkeitslehre," 1877. *Journ. f. Math.* LXXXIV. pp. 242—258. French trans., *Acta Math.* II. 1883, pp. 311—328.

—— "Ueber einen Satz aus der Theorie der stetigen Mannigfaltigkeiten," 1879. *Gött. Nachr.* pp. 127—135.

—— "Ueber unendliche, lineare Punktmannigfaltigkeiten," 1879—1883.

I. *Math. Ann.* XV. 1879, pp. 1—7.
II. *Math. Ann.* XVII. 1880, pp. 355—358.
III. *Math. Ann.* XX. 1882, pp. 113—121.
IV. *Math. Ann.* XXI. 1883, pp. 51—58.
V. *Math. Ann.* XXI. 1883, pp. 545—591.
VI. *Math. Ann.* XXIII. 1883, pp. 453—488.

French trans. of I—V, *Acta Math.* II. 1883, pp. 349—408. V. was previously published separately under the title "Grundlagen einer allgemeinen Mannigfaltigkeitslehre. Ein mathematisch-philosophischer Versuch in der Lehre des Unendlichen." Leipzig, 1883.

—— "Sur divers théorèmes de la théorie des ensembles de points situés dans l'espace continu à n dimensions. Première communication," 1883. *Acta Math.* II. pp. 409—414.

—— "De la puissance des ensembles parfaits de points," 1884. *Acta Math.* IV. pp. 381—392.

—— "Ueber verschiedene Theoreme aus der Theorie der Punktmengen in einem n-fach ausgedehnten stetigen Raume G_n," 1885. *Acta Math.* VII. pp. 105—124.

—— "Ueber die verschiedenen Ansichten in Bezug auf die actual unendlichen Zahlen," 1886. *Bihang till Vet.-Akad. Handlingar*, Stockholm, XI. No. 19. Partially reproduced in the *Zeitschr. f. Phil.* LXXXVIII. 1886, pp. 224—233, and in *Natur und Offenbarung* (Münster), XXXII. 1886, pp. 46—49. Part was published separately under the title "Zur Frage des actualen Unendlichen," 1886, and together with the next on this list, this memoir was published as a pamphlet with the title "Zur Lehre vom Transfiniten," 1890, Halle a/S.

CANTOR, G. "Mitteilungen zur Lehre vom Transfiniten," 1887. *Zeitschr. f. Phil.* XCI. and XCII.

—— "Ueber eine elementare Frage der Mannigfaltigkeitslehre," 1892. *Jahresb. d. d. Mathvgg.* I. pp. 75—78. Ital. trans., by G. Vivanti, *Riv. di mat.* II. 1892, pp. 165—167.

—— "Beitrage zur Begründung der transfiniten Mengenlehre."
I. *Math. Ann.* XLVI. 1895, pp. 481—512.
II. *Math. Ann.* XLIX. 1897, pp. 207—246.
French trans. by F. Marotte, Paris, 1899. Italian trans. of I. by F. Gerbaldi, *Riv. di Mat.* V. 1895, pp. 129—162.

—— "Sui numeri transfiniti," 1895. Extract from a letter to G. Vivanti. *Riv. di Mat.* V. pp. 104—108.

—— "Lettera a G. Peano," 1895. *Riv. di Mat.* V. pp. 108—109.

CESARO, E. "Sur la représentation analytique des régions et des courbes qui les remplissent," 1897. *Darb. Bull.*, Ser. 2, XXI. pp. 257—266.

COUTURAT, L. "De l'infini mathématique," 1896. Paris, Alcan.

—— "Sur la définition du continu," 1900. *Rev. de métaphysique et de morale*, VIII. pp. 157—168.

—— "Les principes des mathématiques," 1904. *Rev. de métaphysique et de morale*, XII. pp. 19—50, 211—240, 661—698, 810—844.

CRANZ, C. "Ueber den Unendlichkeitsbegriff in der Mathematik und Naturwissenschaft," 1895. *Phil. Studien*, XI. pp. 1—40.

DARBOUX, G. "Mémoire sur les fonctions discontinues," 1875. *Ann. de l'Éc. Norm.*, Ser. 2, IV. pp. 57—112.

DEDEKIND, R. "Was sind und was sollen die Zahlen?" 1888. Braunschweig. (2 ed., 1893.)

DEHN, M. "Ueber raumgleiche Polyeder," 1900. *Gött. Nachr.* pp. 345—354.

—— "Ueber den Rauminhalt," 1902. *Math. Ann.* LV. pp. 465—478.

—— "Zwei Anwendungen der Mengenlehre in der elementaren Geometrie," 1904. *Math. Ann.* LIX. pp. 84—88.

DICKSTEIN, S. "Pojęcia i metody matematyki," I. 1891. Warsaw.

DINI, U. "Fondamenti per la teorica delle funzioni di variabili reali," 1878. Pisa. German trans. by J. Lüroth and A. Schepp, 1892, Leipzig.

EMCH, H. "Some applications of the theory of assemblages," 1902. *The math. gazette*, II. pp. 173—175.

ENESTRÖM, G. "Om G. Cantor's uppsats: Ueber die verschiedenen Ansichten in Bezug auf die actual unendlichen Zahlen," 1885. *Öfversigt af K. Vet.-Akad. Förhandlingar*, Stockholm, XLII. No. 10, pp. 69—70.

EVELLIN. "Philosophie et mathématiques: l'infini nouveau," 1898. *Rev. phil.* XLV. pp. 113—119.

EVELLIN, et Z. "Philosophie et mathématiques: l'infini nouveau," 1898. *Rev. phil.* XLVI. pp. 473—486.

—— "L'infini nouveau," 1900. *Rev. phil.* XLIX. pp. 135—143.

FABER, G. "Ueber die Abzählbarkeit der rationalen Zahlen," 1905. *Math. Ann.* LX. pp. 196—203.

310 BIBLIOGRAPHY

FRÉCHET, M. "Sur l'écart de deux courbes et sur les courbes limites," 1905. *Trans. Amer. Math. Soc.* VI. pp. 435—449.

—— "Sur quelques points du Calcul Fonctionnel," 1906. *Rend. del circ. mat. di Palermo*, XXII. pp. 1—74.

GALDEANO, Z. G. de. "La moderna organización de la matemática. Conferencia segunda," 1899. *El progresso matemático*, Ser. 2, I. pp. 45—51, 77—87, 110—115, 154—156.

GARIBALDI, C. "Sull' estensione degli aggregati numerabili," 1894. *Rend. del circ. mat. di Palermo*, VIII. pp. 157—160.

—— "Un piccolo contributo alla teoria degli aggregati," 1895. *Rend. del circ. mat. di Palermo*, IX. pp. 198—201.

GENOCCHI, A. "Differentialrechnung und Grundzüge der Integralrechnung, herausgegeben von G. Peano." German trans. by G. Bohlmann and A. Schepp. 1898—1899. Leipzig, Teubner.

GIUDICE, F. "Subfiniti e transfiniti dal punto di vista di G. Cantor," 1892. *Rend. del circ. mat. di Palermo*, VI. pp. 161—164.

—— "Sui numeri dati mediante insiemi di numeri razionali," 1893. *Periodico di mat.* VIII. pp. 144—150, 172—180.

—— "Sulle successioni di numeri reali," 1903. *Rend. del circ. mat. di Palermo*, XVII. pp. 191—197.

GUICHARD, M. "Théorie des points singuliers essentiels," 1883. *Ann. de l'école norm. sup.*, Ser. 2, XII. pp. 301—394.

GUNDERSEN, C. "On the content or measure of assemblages of points," 1901. *Diss.*, New York.

GUTBERLET, C. "Das Problem des Unendlichen," 1885. *Zeitschr. f. Phil.* LXXXVIII. (Halle), pp. 179—223.

HADAMARD, J., and LOREY, W. "Question 765, Réponse," 1896. *L'Intermédiaire des mathématiciens*, III. pp. 53, 209.

HADAMARD, J. "Sur certaines applications possibles de la théorie des ensembles," 1898. *Verh. des Math.-Kongr. in Zürich*, pp. 200—202.

—— "Les surfaces à courbures opposées et leurs lignes géodésiques," 1898. *Journ. de math. pures et appl.* Ser. 5, IV. pp. 27—73.

See also BAIRE, BOREL and PINCHERLE.

HANKEL, H. "Untersuchungen über die unendlich oft oscillirenden und unstetigen Funktionen," 1870. Freiburg. Reprinted in *Math. Ann.* XX. 1882, pp. 63—112.

HANNEQUIN, A. "Essai critique sur l'hypotèse des atomes dans la science contemporaine," 1899. Paris, Alcan.

HARDY, G. H. "The cardinal number of a closed set of points," 1903. *Mess. of Math.*, New Series, XXXIII. No. 389, pp. 67—69.

—— "A theorem concerning the infinite cardinal numbers," 1903. *Quart. Journ. of Math.* XXXV. pp. 87—94.

—— "The continuum and the second number class," 1905. *Proc. Lond. Math. Soc.*, Ser. 2, IV. pp. 10—17.

HARNACK, A. "Vereinfachung der Beweise in der Theorie der Fourier'schen Reihen," 1881. *Math. Ann.* XIX. pp. 235—279.

—— "Notiz über die Abbildung einer stetigen linearen Mannigfaltigkeit auf eine Unstetige," 1883. *Math. Ann.* XXIII. pp. 285—288.

—— "Lehrsätze über die Integrale der Differentialquotienten, II." 1884. *Math. Ann.* XXIV. pp. 216—252.

HARNACK, A. "Ueber den Inhalt von Punktmengen," 1885. *Math. Ann.* XXV. pp. 241—250.

HAUSDORFF, F. "Über eine gewisse Art geordneter Mengen," 1901. *Leipzig. Ber.* pp. 460—475.

—— "Der Potenzbegriff in der Mengenlehre," 1904. *Jahresb. d. d. Mathvgg.* XIII. pp. 569—571.

HEINE, E. "Die Elemente der Functionenlehre," 1872. *Journ. für Math.* LXXIV. pp. 172—188.

HILBERT, D. "Ueber die stetige Abbildung einer Linie auf ein Flächenstück," 1890. *Verh. d. Naturf. u. Ärzte, Bremen,* pp. 11, 12. Reprinted in *Math. Ann.* XXXVIII. pp. 459—460.

—— "Grundlagen der Geometrie," 1899, second enlarged edition, 1903, Leipzig, Teubner. Reproduced in *Math. Ann.* LVI. 1902, pp. 381—422. English trans. by E. J. Townsend, 1902, Chicago, the Open Court Publishing Company. French trans. by L. Laugel, 1900, *Ann. de l'éc. norm.* Sér. 3, XVII. pp. 103—209.

HOBSON, E. W. "Inner limiting sets of points of a linear interval," 1904. *Proc. Lond. Math. Soc.,* Ser. 2, II. pp. 316—326*.

—— "The general theory of transfinite numbers," 1905. *Proc. Lond. Math. Soc.,* Ser. 2, III. pp. 170—188.

—— "On the arithmetic continuum," 1905. *Proc. Lond. Math. Soc.,* Ser. 2, IV. pp. 21—28.

HUNTINGDON, F. "The continuum as a type of order," 1905. *Ann. of Math.,* Ser. 2, VI. pp. 151—184, and VII. pp. 15—43.
See POINCARÉ.

HURWITZ, A. "Ueber die Entwickelung der allgemeinen Theorie der analytischen Functionen in neuerer Zeit," 1898. *Ver. des Math. Kongresses in Zürich,* pp. 91—112.

ILLIGENS, E. "Die unendliche Anzahl und die Mathematik," 1893. Münster, Theissing.

JANSSEN VAN RAAY, W. H. L. "Sur une classe de grandeurs transfinies," 1894, 1895. *Giorn. di mat.* XXXII. pp. 1—22, and XXXIII. pp. 329—360.

—— "De jongsten onderzoekingen betreffende het oneindig groote," 1897. *Handelingen van het 6de Nederl. Natuur- en Geneeskundig Congres,* pp. 211—218.

JORDAN, C. "Cours d'analyse de l'École Polytéchnique," 2nd ed. I. Paris, 1892—93.

JOURDAIN, P. "On unique, non-repeating integer functions," 1901. *Mess. of Math.*

—— "On the transfinite cardinal numbers of aggregates of functions," 1903. *Phil. Mag.,* Ser. 6, VI. pp. 323—326.

—— "The cardinal number of the aggregate of integrable functions," 1903. *Mess. of Math.* XXXIII. No. 389, pp. 78, 79.

—— "On functions, all of whose singularities are non-essential," 1903. *Mess. of Math.* XXXIII. No. 395, pp. 166—171.

---

* The reference to my work in the note on p. 323 is not quite complete. The inference that Theorem II. of Dr Hobson's paper is not contained in its full scope in my paper is incorrect. W. H. YOUNG.

JOURDAIN, P. "On the transfinite cardinal numbers of well-ordered ag-gregates," 1903. *Phil. Mag.*, Ser. 6, VII. 1904, pp. 61—75.

—— "On the transfinite cardinal numbers of number-classes in general," 1903. *Phil. Mag.*, Ser. 6, VII. 1904, pp. 294—303.

—— "On transfinite cardinal numbers of the exponential form," 1904. *Phil. Mag.*, Ser. 6, VIII. 1905, pp. 42—56.

—— "On the proof that every aggregate can be well-ordered," 1905. *Math. Ann.* LX. pp. 465—470.

—— "On the general theory of functions," 1905. *Journ. für Math.* CXXVIII. 1905, pp. 169—210.

JÜRGENS, E. "Der Begriff der *n*-fachen stetigen Mannigfaltigkeit," 1898. *Jahresb. d. d. Mathvgg.*, VII. pp. 50—55.

KERRY, B. "Ueber G. Cantor's Mannigfaltigkeitsuntersuchungen," 1885. *Vierteljahrschr. für wiss. Phil.* IX. pp. 191—232.

KEYSER, C. J. "Theorems concerning positive definitions of finite assemblage and infinite assemblage," 1901. *Bull. Amer. Math. Soc.*, Ser. 2, VII. pp. 218—226.

KILLING, W. "Bemerkungen über Veronese's transfinite Zahlen," 1895—96. *Index lectionum in Academia Monasteriensi.*

—— "Ueber transfinite Zahlen," 1896. *Math. Ann.* XLVIII. pp. 425—432.

—— "Einführung in die Grundlagen der Geometrie," II. 1898. Paderborn, Schöningh.

KLEIN, F. "Nicht-Euklidische Geometrie," 1889 ; (lithogr.) 2nd ed. 1893 (pp. 328 ff.). Leipzig, Teubner.

—— "Anwendung der Differential- und Integralrechnung auf Geometrie," 1901 (lithogr.). Leipzig, Teubner.

KÖNIG, J. "Zum Kontinuum-Problem," 1904. *Math. Ann.* LX. pp. 177—180 and 462. See also Bernstein, *id.* p. 463.

—— "Ueber die Grundlagen der Mengenlehre und das Kontinuumproblem," 1905. *Math. Ann.* LXI. pp. 156—160.

KÖPCKE, A. "Ueber Differentiirbarkeit und Anschaulichkeit der stetigen Functionen," 1887. *Math. Ann.* XXIX. pp. 123—140.

—— "Ueber eine durchaus differentiirbare, stetige Function mit Oscillationen in jedem Intervalle," 1889. *Math. Ann.* XXXIV. pp. 161—171.

—— "Nachtrag zu dem vorigen," 1890. *Math. Ann.* XXXV. pp. 104—109.

LEBESGUE, H. "Sur les fonctions de plusieurs variables," 1899. *Comptes rendus*, CXXVIII. pp. 811—813.

—— "Sur la définition de l'aire d'une surface," 1899. *Comptes rendus*, CXXIX. pp. 870—873.

—— "Sur quelques surfaces non réglées applicables sur le plan," 1899. *Comptes rendus*, CXXVIII. pp. 1502—1505.

—— "Sur la définition de certaines intégrales de surface," 1900. *Comptes rendus*, CXXXI. pp. 867—870.

—— "Sur le minimum de certaines intégrales," 1900. *Comptes rendus*, CXXXI. pp. 935—937.

—— "Sur une généralisation de l'intégrale définie," 1901. *Comptes rendus*, CXXXII. pp. 1025—1028.

—— "Intégrale, longueur, aire," 1902. Thèse présentée à la Faculté de Paris. *Annali di Mat.* Ser. 3, VII. pp. 231—359.

LEBESGUE, H. "Sur le problème des aires," 1903. *Bull. de la Soc. Math. de France*, XXXI. pp. 197—203.

—— "Leçons sur l'intégration et la recherche des fonctions primitives," 1904. Paris, Gauthier-Villars.

—— "Une propriété caractéristique des fonctions de classe 1," 1904. *Bull. de la Soc. Math. de France*, XXXII. pp. 1—14.

—— "Sur les fonctions représentables analytiquement," 1904. *Comptes rendus*, CXXXIX. pp. 29—31, 1905. *Journ. de Math. pures et appl.*, Ser. 6, I. pp. 139—216.

LENNES, N. J. "Volumes and Areas," 1905. *Trans. Am. Math. Soc.* VI. pp. 486—490.

LERCH, M. "Prispevek k nauce o mnozinach bodu v rovine," 1884. *Prag. Acad.* pp. 176—178.

—— "O soustavach bodu a jich vznamu v analisi," 1886. *Casopis pro pestovani math.* (Prag), XV. p. 211.

LEVI, BEPPO. "Sulla teoria delle funzioni e degli insiemi," 1900. *Rend. dell' Acc. dei Lincei*, Ser. 5, IX. 2° sem. pp. 72—79.

—— "Intorno alla teoria degli aggregati," 1902. *Rend. del R. Ist. Lomb.*, Ser. 2, XXXV. pp. 863—868.

LEVI-CIVITA, T. "Sugli infiniti ed infinitesimi attuali quali elementi analitici," 1893. *Atti dell' Ist. Veneto*, Ser. 7, IV. pp. 1765—1815.

—— "Sui numeri transfiniti," 1898. *Rend. dell' Acc. dei Lincei*, Ser. 5, VII. 1° som. pp. 91—96, 113—121.

LINDELÖF, E. "Sur quelques points de la théorie des ensembles," 1903. *Comptes rendus*, CXXXVII. pp. 697—700.

—— "Remarques sur un théorème fondamental de la théorie des ensembles," 1905. *Acta Math.* XXIX. pp. 183—190.

LORIA, G. "La definizione dello spazio a *n* dimensioni e l' ipotesi di continuità del nostro spazio secondo le ricerche di Giorgio Cantor," 1887. *Giorn. di mat.* XXV. pp. 97—108.

LÜROTH, J. "Ueber gegenseitig eindentige und stetige Abbildung von Mannigfaltigkeiten verschiedener Dimensionen auf einander," 1878, 1899. *Sitzungsberichte d. phys.-med. Soc. zu Erlangen*, X. pp. 190—195, XXXI. pp. 86—91.

MACCAFERRI, E. "Sugli insiemi continui e sugli insiemi connessi," 1894. *Rivista di Mat.* IV. pp. 97—103.

MAILLET, E. "Sur les équations différentielles et la théorie des ensembles," 1902. *Bull. de la Soc. math. de France*, XXX. pp. 195—201. *Comptes Rendus*, CXXXV. pp. 434—435.

MEYER, F. "Elemente der Arithmetik und Algebra," 1885. Halle.

MILESI, L. "Sulla impossibile coesistenza della univocità e della continuità nella corrispondenza che si può stabilire fra due spazi continui ad un numero differente di dimensioni," 1892. *Rivista di Mat.* II. pp. 103—106.

MILHAUD, G. "L'infini mathématique," 1897. *Revue philosoph.* XLIII. pp. 296—310.

MITTAG-LEFFLER, G. "Sur la théorie des fonctions uniformes d'une variable," 1882. *Comptes rendus*, XCIV. pp. 414—416, 511—514, 713—715, 781—783, 938—941, 1040—1042, 1105—1108, 1163—1166 ; XCV. pp. 335—336.

MITTAG-LEFFLER, G. "Sur la représentation analytique des fonctions monogènes uniformes d'une variable indépendante," 1884. *Acta Math.* IV. pp. 1—79.

MÖBIUS, A. F. "Barycentrische Calcul," 1827. Ch. VI. *Ges. Werke,* I. pp. 237 ff. (Projective Scale).

MOORE, E. H. "On certain crinkly curves," 1900. *Trans. Amer. Math. Soc.* I. pp. 72—90.

NETTO, E. "Beitrag zur Mannigfaltigkeitslehre," 1879. *Journ. f. Math.* LXXXVI. pp. 263—268.

OSGOOD, W. "Non-uniform convergence and the integration of series term by term," 1896. *Amer. Journ. of Math.* XIX. pp. 155—190.

—— "On the Existence of the Green's function for the most general simply connected plane region," 1900. *Trans. Amer. Math. Soc.* I. pp. 310—314.

—— "Zweite Note über analytische Functionen mehrerer Veränderlichen," 1900. *Math. Ann.* LIII. pp. 461—464.

—— "Ueber einen Satz des Herrn Schönflies aus der Theorie der Functionen zweier reeller Veränderlichen," 1900. *Gött. Nachr.* pp. 94—97.

—— "A Jordan curve of positive area," 1902. *Trans. Amer. Math. Soc.* IV. No. 1, pp. 107—112.

—— "Lehrbuch der Funktionentheorie," erster Band, erste Hälfte (1906). Leipzig und Berlin, Teubner.

—— "Selected topics in the general theory of functions," 1898. *Bull. Am. Math. Soc.* Ser. 2, V. pp. 59—87.

PAINLEVÉ, P. "Sur les lignes singulières des fonctions analytiques," 1887, Paris. *Ann. de Toulouse,* II. (1888).

DE PAOLIS, R. "Teoria dei gruppi geometrici e delle corrispondenze che si possono stabilire tra i loro elementi," 1890. *Memorie della Società Italiana delle Scienze* (detta dei XL.), Ser. 3, III.

PASCH, M. "Ueber einige Punkte der Functionentheorie," 1887. *Math. Ann.* XXX. pp. 132—154.

PEANO, G. "Arithmetices principia, nova methodo exposita," 1889, Turin.

—— "Sur une courbe qui remplit toute une aire plane," 1890. *Math. Ann.* XXXVI. pp. 157—160.

—— "Démonstration de l'intégrabilité des équations différentielles ordinaires," 1890. *Math. Ann.* XXXVII. pp. 182—228.

—— "Lezioni di analisi infinitesimale," 1893. Turin, Candeletti.

—— "Teoria dei gruppi di punti," 1894. Turin, Fodratti and Lecco. Extracted from *Formulaire de mathématiques* (Turin, 1895, Bocca-Clausen), pp. 58—64, and from *Rivista di mat.* 1894, IV. pp. 33—35.

PHRAGMÉN, E. "Beweis eines Satzes aus der Mannigfaltigkeitslehre," 1884. *Acta Math.* V. pp. 47—48.

—— "En ny sats inom teorien för punktmängder," 1884. *Öfvers. af vet. akad. förh.* (Stockholm), XLI. No. 1, pp. 121—124.

—— "Ueber die Begrenzungen von Continua," 1885. *Acta Math.* VII. pp. 43—48.

PINCHERLE, S. "Remarque relative à la communication de M. Hadamard," 1898. *Verh. des Math.-Kongr. in Zürich,* p. 203.

POINCARÉ, H. "Hilbert's Grundlagen der Geometrie," 1902. *Bull. des Sc.*

Math., Ser. 2, XXVI. pp. 249—279, and *Journ. des Sav.* pp. 252—271. English translation by E. V. Huntingdon, *Bull. Amer. Math. Soc.* 1904, Ser. 2, X. pp. 1—23.

PRINGSHEIM, A. "Grundlagen der allgemeinen Functionenlehre," 1899. *Encyklop. d. math. Wiss.* II. pp. 1—53.

RIEMANN, B. "Ueber die Darstellbarkeit einer Funktion durch eine trigonometrische Reihe," 1854. *Habilitationsschrift, Abh. d. k. Gesellschaft d. Wissenschaft zu Göttingen*, XIII. *Gesammelte Werke*, 2nd edition, pp. 227—264.

—— "Ueber die Hypothesen, welche der Geometrie zu Grunde liegen," 1854. *Habilitationsschrift, Abh. d. k. Ges. d. Wiss. zu Göttingen*, XIII. *Gesammelte Werke*, 2nd edition, pp. 272—287.

RIESZ, F. "Ueber einen Satz der Analysis Situs," 1904. *Math. Ann.* LIX. pp. 409—415*.

LE ROY, E., and VINCENT, G. "Sur l'idée de nombre," 1896. *Revue de métaphysique et de morale*, IV. pp. 738—755.

RUSSELL, B. "The principles of mathematics," I. 1903. Cambridge, Univ. Press.

SCHATUNOVSKY. "Ueber den Rauminhalt der Polyeder," 1903. *Math. Ann.* LVII. pp. 496—508.

SCHEEFFER, L. "Allgemeine Untersuchungen über Rectification der Curven," 1884. *Acta Math.* V. pp. 49—82.

—— "Zur Theorie der stetigen Functionen einer reellen Veränderlichen," 1884. *Acta Math.* V. pp. 183—194, 279—296.

SCHÖNFLIES, A. "Ueber die Abbildung von Würfeln verschiedener Dimensionen auf einander," 1806. *Gött. Nachr.* pp. 255—266.

—— "Transfinite Zahlen, das Axiom des Archimedes und die projectivische Geometrie," 1897. *Jahresb. d. d. Mathvgg.* V. pp. 75—81.

—— "Sur les nombres transfinis de M. Veronese," 1897. *Rend. dell' Acc. dei Lincei*, Ser. 5, VI. 2º sem. pp. 362—368.

—— "Ueber die Verteilung der Stetigkeits- und Unstetigkeitspunkte bei punktweise unstetigen Functionen einer reellen Variabeln," 1899. *Gött. Nachr.* pp. 187—194.

—— "Ueber einen Satz aus der Analysis Situs," 1899. *Gött. Nachr.* pp. 282—290.

—— "Mengenlehre," 1899. *Encyklop. d. math. Wiss.* I. pp. 184—207.

—— "Die Entwickelung der Lehre von den Punktmannigfaltigkeiten. Bericht erstattet der deutschen Mathematiker-Vereinigung" (Bericht über die Mengenlehre), 1900. *Jahresb. d. d. Mathvgg.* VIII. pp. 1—251.

—— "Ueber die überall oscillirenden differenzirbaren Functionen," 1901. *Math. Ann.* LIV. pp. 553—563.

—— "Ueber einen grundlegenden Satz der Analysis Situs," 1902. *Gött. Nachr.* pp. 185—192.

—— "Beiträge zur Theorie der Punktmengen," I. 1902, II. 1903, III. 1906. *Math. Ann.* LVIII. pp. 195—234, LIX. pp. 129—160, LXII. pp. 286—328.

* Containing a short proof of Schönflies's theorem referred to on p. 229 *supra*.

SCHÖNFLIES, A. " Ueber den Beweis eines Haupttheorems aus der Theorie der Punktmengen," 1903. *Gött. Nachr.* pp. 21—31.

—— " Ueber Stetigkeit und Unstetigkeit der Funktionen einer reellen Veränderlichen," 1904. *Wiener Ber.* CXIII. Abt. IIa*.

—— " Ueber wohlgeordnete Mengen," 1904. *Math. Ann.* LX. pp. 181—186.

—— " Bemerkung zu dem vorstehenden Aufsatz des Herrn Young," 1905. *Math. Ann.* LXI. pp. 287—288.

SCHRÖDER, E. " Ueber G. Cantor'sche Sätze," 1897. *Jahresb. d. d. Mathvgg.* v. pp. 81—82.

—— " Ueber zwei Definitionen der Endlichkeit und G. Cantor'sche Sätze," 1898. *Abh. d. Leop.-Car. Akad. d. Naturforscher,* LXXI. No. 6.

—— " Die selbstständige Definition der Mächtigkeiten 0, 1, 2, 3 und die explicite Gleichzahligkeitsbedingung," 1898. *Abh. d. Leop.-Car. Akad. der Naturforscher,* LXXI. No. 7.

SCHWARZ, H. " Ein Beitrag zur Theorie der Ordnungstypen," 1888. *Diss.* Halle.

SMITH, H. J. S. " On the integration of discontinuous functions," 1875. *Proc. Lond. Math. Soc.* VI. pp. 140—153.

STOLZ, O. " Ueber einen zu einer unendlichen Punktmenge gehörigen Grenzwerth," 1884. *Math. Ann.* XXIII. pp. 152—156.

—— " Zur Erklärung der Bogenlänge und des Inhalts einer krummen Fläche," 1902. *Trans. Amer. Math. Soc.* III. pp. 23—37.

TANNERY, J. " De l'infini mathématique," 1897. *Rev. gén. des sciences pures et appl.* VIII. pp. 129—140.

TANNERY, P. " Note sur la théorie des ensembles," 1884. *Bull. de la Soc. math. de France,* XII. pp. 90—96.

—— " Le concept scientifique du continu : Zénon d'Élée et Georg Cantor," 1885. *Rev. phil.* octobre, 1885.

—— " Sur le concept du transfini," 1894. *Rev. de métaphysique et de morale,* II. pp. 465—472.

THOMAE, J. " Sätze aus der Functionentheorie," 1878. *Gött. Nachr.* pp. 466—468.

—— " Beitrag zur Mannigfaltigkeitslehre," 1896. *Zeitschr. f. Math. und Phys.* XLI. pp. 231—232.

VEBLEN, O. " The Heine Borel Theorem," 1904. *Bull. Amer. Math. Soc.* Ser. 2, X. pp. 436—439.

—— " A system of axioms for Geometry," 1904. *Trans. Amer. Math. Soc.* V. pp. 343—384.

—— " Theory of Plane curves in non-metrical analysis situs," 1904. *Trans. Amer. Math. Soc.* VI. No. 1, pp. 83—98.

—— " Definition in terms of order alone in the linear continuum and in well-ordered sets," 1905. *Trans. Amer. Math. Soc.* VI. No. 2, pp. 165—171.

* At the end of this paper exception is taken, with justice, to the wording of the table in my paper in the *Wiener Bericht.* The intention was correct, and in the analogous paper for higher space—" Ordinary inner limiting sets in the plane and higher space "—the wording is amended. With the rest of Schönflies's remarks I find myself only in partial agreement. The reader is referred to my paper for the positive results contained in it. W. H. YOUNG.

VELTMANN, W. "Ueber die Anordnung unendlich vieler Singularitäten einer Function," 1882. *Zeitschr. f. Math. und Phys.* XXVII. pp. 176—179.

—— "Zur Theorie der Punktmengen," 1882. *Zeitschr. f. Math. und Phys.* XXVII. pp. 313—314.

—— "Die Fourier'sche Reihe," 1882. *Zeitschr. f. Math. und Phys.* XXVII. pp. 193—235.

VERONESE, G. "Fondamenti di geometria," 1891. Padua. German trans. by A. Schepp, Leipzig, 1894.

—— "Intorno ad alcune osservazioni sui segmenti infiniti e infinitesimi attuali," 1896. *Math. Ann.* XLVII. pp. 423—432.

—— "Sul postulato della continuità," 1897. *Rend. dell' Acc. dei Lincei,* Ser. 5, VI. 2° sem. pp. 161—168.

—— "Segmenti e numeri transfiniti," 1898. *Rend. dell' Acc. dei Lincei,* Ser. 5, VII. 1° sem. pp. 79—87.

VITALI, G. "Sui gruppi di punti," 1904. *Rend. del circ. mat. di Palermo,* XVIII. pp. 116—126.

—— "Una proprietà delle funzioni misurabili," 1905. *Rend. del. Ist. Lomb.,* Ser. 2, XXXVIII. pp. 599—603.

—— "Sulle funzioni ad integrale nullo," 1905. *Rend. del circ. mat. di Palermo,* XX. pp. 136—141*.

VIVANTI, G. "Fondamenti della teoria dei tipi ordinati," 1889. *Ann. di mat.* Ser. 2, XVII. pp. 1—35.

—— "Notice historique sur la théorie des ensembles," 1892. *Bibliotheca math.* Ser. 2, VI. pp. 9—25.

—— "Teoria degli aggregati," 1894. Turin, Fodratti and Lecco. Extracted from the *Formulaire de mathématiques* (Turin, 1895, Bocca-Clausen), I. pp. 65—74, and from the *Rivista di mat.* III. pp. 189—192, and IV. pp. 135—140†.

—— "Sugli aggregati perfetti," 1899. *Rend. del circ. mat. di Palermo,* XIII. pp. 86—88.

—— "Lezioni sulla teoria delle funzioni analitiche," 1899. Reggio Calabria, Massara (lithogr.).

—— "Teoria delle funzioni analitiche," 1901. Milano, *Manuali Hoepli,* 312—313. German translation by A. Gutzmer, 1906. Leipzig, Teubner.

VOLPI, R., and ZOCCOLI, E. G. "Di un' applicazione della teoria dei gruppi del Cantor al problema gnoseologico," 1896. Modena, Moneti.

VOLTERRA. "Alcune osservazioni sulle funzioni punteggiate discontinue," 1880. *Giorn. di mat.* XIX. pp. 76—86.

—— "Sulle funzioni analitiche polidrome," 1888. *Rend. dell' Acc. dei Lincei,* Ser. 4, IV. 2° sem. pp. 355—362.

* The author is clearly unacquainted with my papers, since he attributes to Lindelöf (*loc. cit. supra*) a theorem given first in my earlier paper on "Overlapping intervals." The same paper contains the generalised form of the Heine-Borel Theorem which is now generally used, and which Vitali, following Borel, refers to as due to Borel and Lebesgue. Lebesgue's ideas and my own have run on curiously parallel lines, and quite independently; as is well-known, the French are rarely well acquainted with English work. W. H. YOUNG.

† The list of literature in this pamphlet, as well as the supplementary "Lista bibliografica della teoria degli aggregati 1893—1899" (*Bibliotheca Math.,* Ser. 3, I. (1900), pp. 160—165, Leipzig), have been incorporated in the present one; in the original the arrangement is chronological.

WHITEHEAD, A. N. "On Cardinal Numbers," 1902. *Amer. Journ. of Math.*
XXIV. pp. 367—394.
—— "The logic of relations, logical substitution groups and cardinal numbers," 1903. *Amer. Journ. of Math.* XXV. pp. 157—178.
—— "Theorems on cardinal numbers," 1904. *Amer. Journ. of Math.* XXVI.
pp. 157, 158.

YOUNG, G. CHISHOLM. "On the form of a certain Jordan Curve," 1905.
Quart. Journ. of Math. XXXVII. No. 145, pp. 87—91.
YOUNG, W. H. "On the Density of Linear Sets of Points," 1902. *Proc.
Lond. Math. Soc.* XXXIV. pp. 285—290.
—— "Sets of Intervals on the Straight Line," 1902. *Proc. Lond. Math. Soc.*
XXXV. pp. 245—268.
—— "On Closed Sets of Points defined as the limit of a sequence of Closed
Sets of Points," 1902. *Proc. Lond. Math. Soc.* XXXV. pp. 269—282.
—— "A Note on Unclosed Sets of Points defined as the limit of a sequence
of Closed Sets of Points," 1902. *Proc. Lond. Math. Soc.* XXXV.
pp. 283—284.
—— "Overlapping Intervals," 1902. *Proc. Lond. Math. Soc.* XXXV. pp.
384—388.
—— "Sur l'intégration des séries," 1903. *Comptes rendus,* CXXXVI. pp.
1632—1634.
—— "Zur Lehre der nicht abgeschlossenen Punktmengen," 1903. *Leipz.
Ber.,* pp. 287—293.
—— "On the Analysis of Linear Sets of Points," 1903. *Quart. Journ.
of Math.* XXXV. No. 138, pp. 102—116.
—— "A Note on the Condition of Integrability of a Function of a Real
Variable," 1903. *Quart. Journ. of Math.* XXXV. No. 138, pp. 189—192.
—— "On Non-uniform Convergence and Term-by-term Integration of
Series," 1903. *Proc. Lond. Math. Soc.,* Ser. 2, I. pp. 89—102.
—— "On the Infinite Derivates of a Function of a Real Variable," 1903.
Archiv för Mat., Astr., och Fysik, I. pp. 201—204.
—— "On Closed Sets of Points, and Cantor's Numbers," 1903. *Proc. Lond.
Math. Soc.,* Ser. 2, I. pp. 230—248.
—— "Ueber die Einteilung der unstetigen Funktionen und die Verteilung
ihrer Stetigkeitspunkte," 1903. *Wiener Ber.* CXII. Abt. 11 a, pp. 1307
—1316.
—— "On Sequences of Sets of Intervals containing a given Set of Points,"
1903. *Proc. Lond. Math. Soc.,* Ser. 2, I. pp. 262—266.
—— "On the distribution of the points of non-uniform convergence of
a series of functions," 1903. *Proc. Lond. Math. Soc.,* Ser. 2, I. pp.
356—360.
—— "On an extension of the Heine-Borel theorem," 1904. *Mess. of Math.,*
New Ser. XXXIII. No. 393, pp. 129—132.
—— "Open sets and the theory of content," 1904. *Proc. Lond. Math. Soc.,*
Ser. 2, II. pp. 16—51.
—— "On upper and lower integration," 1904. *Proc. Lond. Math. Soc.,* Ser. 2,
II. pp. 52—66.
—— "The tile theorem," 1904. *Proc. Lond. Math. Soc.,* Ser. 2, II. pp.
67—69.

YOUNG, W. H. "The general theory of integration," 1904. *Phil. Trans.*, Ser. A, CCIV. pp. 221—252. Abstract in the *Proc. R. S.* LXXIII. pp. 445—449.

—— "On a perfect plane set," 1905. *Mess. of Math.*, New Ser. XXXIV. No. 406, p. 160. (See footnote, *Quart. Journ. of Math.* XXXVII. No. 145, p. 91.)

—— "The potencies of closed and perfect sets," 1905. *Quart. Journ. of Math.* XXXVI. No. 143, pp. 280—284.

—— "On regions and sets of regions," 1905. *Quart. Journ. of Math.* XXXVII. No. 145, pp. 1—35.

—— "Ordinary inner limiting sets in the plane or higher space," 1905. *Proc. Lond. Math. Soc.*, Ser. 2, III. pp. 371—380.

—— "Linear content of a plane set of points," 1905. *Proc. Lond. Math. Soc.*, Ser. 2, III. pp. 461—477.

—— "Zur Theorie der nirgends dichten Punktmengen in der Ebene," 1905. *Math. Ann.* LXI. pp. 281—286.

—— "A note on sets of overlapping intervals," 1906. *Rend. del circ. mat. di Palermo*, XXI. pp. 125—127.

ZERMELO, E. "Ueber die Addition transfiniter Kardinalzahlen," 1901. *Gött. Nachr.* pp. 34—38.

—— "Beweis dass jede Menge wohlgeordnet werden kann," 1904. *Math. Ann.* LIX. pp. 514—516.

ZORETTI, L. "Sur les ensembles parfaits et les fonctions uniformes," 1904. *Comptes Rendus*, CXXXVIII. pp. 674—676.

INDEX OF PROPER NAMES.

(The references are to pages.)

GENERAL INDEX.

(The references are to pages.)

Abut, 39 ff., 268
Actual, 287
Addendum, 158
Addition, 37, 150; ordinal, 154, 157; theorem, 81, 84, 101, 107, 116, 241, 242
Additive class, 108, 110, 118; inner, 100, 108, 109; outer, 108, 116, 117
Adherence, 57, 67, 234, 284
Adjacent, 166
After, 122
Aleph, 136, 145, 153, 157, 290
Algebraic number, 6, 8, 44, 146
Amalgamate, 42
Analyse, 55
Analysis, 284 ff.
Arbitrary, choice, 147, 289; constant, 170
Arc, 171, 212, 219, 222, 227, 229
Area, 173, 178, 242, 259, 281
Arithmetic, 37; continuum, 161
Arithmetical, 225, 291
Associative law, 37, 158
Augendum, 158
Automorphic function, 292
Axiom, 14, 290

Band, 245 ff.
Beads, 244
Bean-shaped region, 275 ff.
Before, 122
Belonging to, 284
Between, 122
Binary, fraction, 20, 21, 51, 64, 128 ff., 131, 134, 163, 236, 290; notation, 17; numbers, 25
Bisect, bisection, 19, 23, 293
Black, arcs, 171; intervals, 19 ff., 48, 52, 83, 112, 133, 151, 287; region, 173, 203, 243
Boundary, 169, 170, 178, 184, 185, 189, 196, 198, 209, 227, 239, 242
Branch, 220

Calculation, of content, 250 ff.; of length, 267 ff.

Calculus, 146, 292; of potencies, 150
Cantor-Bernstein-Schroeder Theorem, 147 ff.
Cantor-Dedekind axiom, 14, 16, 134
Cantor's, numbers, 25, 37, 145 ff., 284; theorem, 38, 47, 76, 177, 199; theorem of deduction, 26, 89, 93, 194, 228, 285; typical ternary set, 20, 48, 60, 171 ff.
Cardinal number, 1, 93, 145 ff., 153, 157
Cartesian coordinates, 161
Choice, 147, 289 ff.
Chord, 268, 269
Chow, 184, 213, 221 (Ex. 3)
Circle, 164 ff., 170, 177, 181, 185, 190, 192, 195, 196, 228, 262, 271 ff., 292
Circumference, 171, 181, 182, 190, 278 ff.
Class, of Cantor's numbers, 155, 156, 284. (*See also* Set)
Clockwise, 277
Close order, 127 ff., 134
Closed, 15, 17, 19, 26, 27, 44, 52, 76, 97, 103, 116, 135, 171, 178, 187, 188, 192, 195, 196, 201, 211, 218, 252, 284; connected set, 207; ordinally, 127, 134, 287; curve, 213, 220, 221
Coefficient, differential, 292
Coherence, 57, 116, 234, 284; deduced, 59; ultimate, 60, 61, 67, 116
Commutative law, 37, 158
Compare, 4
Complementary, 215, 229, 261
Component, 16, 23, 35, 134, 149, 150, 157, 158, 242, 288, 290; proper, 36, 149, 150, 284, 288; closed, 89 ff., 249 ff., 287
Concentrated, 18
Condensation, point of, 32. (*See also* Limiting point)
Connected, 195, 204 ff., 207, 219; doubly, 227; *r*-ply, 208; simply, 208, 212, 227, 244
Consecutive, 23, 264, 268, 277

CHELSEA

SCIENTIFIC

BOOKS

LECTURES ON ERGODIC THEORY
By P. R. HALMOS

CONTENTS: Introduction. Recurrence. Mean Convergence. Pointwise Convergence. Ergodicity. Mixing. Measure Algebras. Discrete Spectrum. Automorphisms of Compact Groups. Generalized Proper Values. Weak Topology. Weak Approximation. Uniform Topology. Uniform Approximation. Category. Invariant Measures. Generalized Ergodic Theorems. Unsolved Problems.

"Written in the pleasant, relaxed, and clear style usually associated with the author. The material is organized very well and painlessly presented."
—*Bulletin of the A.M.S.*

—1956-60. viii + 101 pp. 5⅜x8.   8284-0142-X.

ELEMENTS OF QUATERNIONS
By W. R. HAMILTON

Sir William Rowan Hamilton's last major work, and the second of his two treatises on quaternions.

—3rd ed. 1899/1901-68. 1,185 pp. 6x9. 8284-0219-1.
Two vol. set.

RAMANUJAN:
Twelve Lectures on His Life and Works
By G. H. HARDY

The book is somewhat more than an account of the mathematical work and personality of Ramanujan; it is one of the very few full-length books of "shop talk" by an important mathematician.

—1940-59. viii + 236 pp. 6x9.   8284-0136-5.

GRUNDZUEGE DER MENGENLEHRE
By F. HAUSDORFF

The original, 1914 edition of this famous work contains many topics that had to be omitted from later editions, notably, the theories of content, measure, and discussion of the Lebesgue integral. Also, general topological spaces, Euclidean spaces, special methods applicable in the Euclidean plane, the algebra of sets, partially ordered sets, etc.

—1914-49. 484 pp. 5⅜x8.   8284-0061-X.

SET THEORY
By F. HAUSDORFF

Hausdorff's classic text-book is an inspiration and a delight. The translation is from the Third (latest) German edition.

"We wish to state without qualification that this is an indispensable book for all those interested in the theory of sets and the allied branches of real variable theory."—*Bulletin of A. M. S.*

—2nd ed. 1962. 352 pp. 6x9.   8284-0119-5.

Grundzüge Einer Allgemeinen Theorie der
LINEAREN INTEGRALGLEICHUNGEN
By D. HILBERT
—1912-53. 306 pp. 5½x8½. 8284-0091-1.

GEOMETRY AND THE IMAGINATION
By D. HILBERT and S. COHN-VOSSEN
Translated from the German by P. NEMENYI.

"A fascinating tour of the 20th century mathematical zoo. . . . Anyone who would like to see proof of the fact that a sphere with a hole can always be bent (no matter how small the hole), learn the theorems about Klein's bottle—a bottle with no edges, no inside, and no outside—and meet other strange creatures of modern geometry will be delighted with Hilbert and Cohn-Vossen's book."
—*Scientific American.*

"Should provided stimulus and inspiration to every student and teacher of geometry."—*Nature.*

"A mathematical classic. . . . The purpose is to make the reader *see* and *feel* the proofs. . . . readers can penetrate into higher mathematics with . . . pleasure instead of the usual laborious study."
—*American Scientist.*

"Students, particularly, would benefit very much by reading this book . . . they will experience the sensation of being taken into the friendly confidence of a great mathematician and being shown the real significance of things."—*Science Progress.*

"A person with a minimum of formal training can follow the reasoning. . . . an important [book]."
—*The Mathematics Teacher.*

—1952-68. 358 pp. 6x9. 8284-0087-3.

GESAMMELTE ABHANDLUNGEN
(Collected Papers)
By D. HILBERT

Volume I (Number Theory) contains Hilbert's papers on Number Theory, including his long paper on Algebraic Numbers. Volume II (Algebra, Invariant Theory, Geometry) covers not only the topics indicated in the sub-title but also papers on Diophantine Equations. Volume III carries the sub-title: Analysis, Foundation of Mathematics, Physics, and Miscellaneous Papers.

—1932/35-66. 1,457 pp. 6x9. 8284-0195-0.   Each vol.
Three vol. set

PRINCIPLES OF MATHEMATICAL LOGIC
By D. HILBERT and W. ACKERMANN

"As a text the book has become a classic . . . the best introduction for the student who seriously wants to master the technique. Some of the features which give it this status are as follows:

"The first feature is its extraordinary lucidity. A second is the intuitive approach, with the introduction of formalization only after a full discussion of motivation. Again, the argument is rigorous and exact . . . A fourth feature is the emphasis on general extra-formal principles . . . Finally, the work is relatively free from bias . . . All together, the book still bears the stamp of the genius of one of the great mathematicians of modern times."—*Bulletin of the A.M.S.*

—1950-68. xii + 172 pp. 6x9. 8284-0069-5.

SQUARING THE CIRCLE, and other Monographs
By HOBSON, HUDSON, SINGH, and KEMPE
FOUR VOLUMES IN ONE.

SQUARING THE CIRCLE, by *Hobson*. A fascinating account of one of the three famous problems of antiquity, its significance, its history, the mathematical work it inspired in modern times, and its eventual solution in the closing years of the last century.

RULER AND COMPASSES, by *Hudson*. "An analytical and geometrical investigation of how far Euclidean constructions can take us. It is as thoroughgoing as it is constructive."—*Sci. Monthly*.

THE THEORY AND CONSTRUCTION OF NON-DIFFERENTIABLE FUNCTIONS, by *Singh*. I. Functions Defined by Series. II. Functions Defined Geometrically. III. Functions Defined Arithmetically. IV. Properties of Non-Differentiable Functions.

HOW TO DRAW A STRAIGHT LINE, by *Kempe*. An intriguing monograph on linkages. Describes, among other things, a linkage that will trisect any angle.

"Intriguing, meaty."—*Scientific American*.

—388 pp. 4½x7½. 8284-0095-4. Four vols. in one.

SPHERICAL AND ELLIPSOIDAL HARMONICS
By E. W. HOBSON

"A comprehensive treatise . . . and the standard reference in its field."—*Bulletin of the A. M. S.*
—1931-65. xi + 500 pp. 5⅜x8. 8284-0104-7.

ELASTOKINETIK: Die Methoden zur Angenäherten Lösung von Eigenwertproblemen in der Elastokinetik
By K. HOHENEMSER

—(Erg. der Math.) 1932-49. 89 pp. 5½x8½. 8284-0055-5.

RULER AND COMPASSES, by H. P. HUDSON.
See HOBSON

PHYSIKALISCH-MATHEMATISCHE MONOGRAPHIEN
By W. v. IGNATOWSKY, et al.
THREE VOLUMES IN ONE.

CONTENTS: 1. *Untersuchungen einiger Integrale mit Besselschen Funktionen und ihre Anwendung auf Beugungserscheinungen*, by Ignatowsky. 2. *Kreisscheibenkondensator*, by Ignatowsky. 3. *Table of a Special Function*, by Bursian and Fock.
—1932-66. 16 + 232 pp. 6¼x9¼. 8284-0201-9. Three vols. in one.

VORLESUNGEN UEBER DYNAMIK
By C. G. J. JACOBI

This is Volume 8 of the *Gesammelte Werke*.
—2nd ed. 1884-68. viii+300 pp. 6½x9¼. 8284-0227-2.

THE MATHEMATICAL THEORY OF THE TOP,
by F. KLEIN. See SIERPINSKI

FAMOUS PROBLEMS, and other monographs
By KLEIN, SHEPPARD, MacMAHON, and MORDELL

FOUR VOLUMES IN ONE.

FAMOUS PROBLEMS OF ELEMENTARY GEOMETRY, by *Klein.* A fascinating little book. A simple, easily understandable, account of the famous problems of Geometry—The Duplication of the Cube, Trisection of the Angle, Squaring of the Circle—and the proofs that these cannot be solved by ruler and compass—presentable, say, before an undergraduate math club (no calculus required). Also, the modern problems about transcendental numbers, the existence of such numbers, and proofs of the transcendence of *e.*

FROM DETERMINANT TO TENSOR, by *Sheppard.* A novel and charming introduction. Written with the utmost simplicity. PT I. Origin of Determinants. II. Properties of Determinants. III. Solution of Simultaneous Equations. IV. Properties. V. Tensor Notation. PT II. VI. Sets. VII. Cogredience, etc. VIII. Examples from Statistics. IX. Tensors in Theory of Relativity.

INTRODUCTION TO COMBINATORY ANALYSIS, by *MacMahon.* A concise introduction to this field. Written as introduction to the author's two-volume work.

THREE LECTURES ON FERMAT'S LAST THEOREM, by *Mordell.* This famous problem is so easy that a high-school student might not unreasonably hope to solve it; it is so difficult that tens of thousands of amateur and professional mathematicians, Euler and Gauss among them, have failed to find a complete solution. Mordell's very small book begins with an impartial investigation of whether Fermat himself had a solution (as he said he did) and explains what has been accomplished. This is one of the masterpieces of mathematical exposition.

—2nd ed. 1962. 350 pp. 5⅜x8. Four vols. in one.

        8284-0108-X.   Cloth
        8284-0166-7.   Paper

VORLESUNGEN UEBER NICHT-EUKLIDISCHE GEOMETRIE
By F. KLEIN

—1928-59. xii + 326 pp. 5x8.   8284-0129-2.

ENTWICKLUNG DER MATHEMATIK IM 19. JAHRHUNDERT
By F. KLEIN

TWO VOLUMES IN ONE.

Vol. I treats of the various branches of advanced mathematics of the prolific 19th century; Klein himself was in the forefront of the mathematical activity of latter part of the 19th and early part of the 20th centuries.

Vol. II deals with the mathematics of relativity theory.

—1926/27-67. 616 pp. 5¼x8. 8284-0074-1. Two vols. in one.

FOUNDATIONS OF ANALYSIS

By E. LANDAU

"Certainly no clearer treatment of the foundations of the number system can be offered. . . . One can only be thankful to the author for this fundamental piece of exposition, which is alive with his vitality and genius."—*J. F. Ritt, Amer. Math. Monthly.*

—2nd ed. 1960. xiv + 136 pp. 6x9. 8284-0079-2.

ELEMENTARE ZAHLENTHEORIE

By E. LANDAU

"Interest is enlisted at once and sustained by the accuracy, skill, and enthusiasm with which Landau marshals . . . facts and simplifies . . . details."
                    —*G. D. Birkhoff, Bulletin of the A. M. S.*

—1927-50. vii + 180 + iv pp. 5½x8½. 8284-0026-1.

ELEMENTARY NUMBER THEORY

By E. LANDAU

The present work is a translation of Prof. Landau's famous *Elementare Zahlentheorie*, with added exercises by Prof. Paul T. Bateman.

—2nd ed. 1966. 256 pp. 6x9.   8284-0125-X.

Einführung in die Elementare und Analytische Theorie der ALGEBRAISCHE ZAHLEN

By E. LANDAU

—2nd ed. 1927-49. vii + 147 pp. 5⅜x8. 8284-0062-8

NEUERE FUNKTIONENTHEORIE, by E. LANDAU.
See WEYL

Mémoires sur la Théorie des SYSTEMES DES EQUATIONS DIFFERENTIELLES LINEAIRES, Vols. I, II, III

By J. A. LAPPO-DANILEVSKII

THREE VOLUMES IN ONE.

A reprint, in one volume, of Volumes 6, 7, and 8 of the monographs of the Steklov Institute of Mathematics in Moscow.

"The theory of [systems of linear differential equations] is treated with elegance and generality by the author, and his contributions constitute an important addition to the field of differential equations."—*Applied Mechanics Reviews.*

—1934/5/6-53. 689 pp. 5⅜x8.   8284-0094-6.
                    Three vols. in one.

VORLESUNGEN UEBER DIE ALGEBRA DER LOGIK
By E. SCHRÖDER

One of the classics of logic.

The present edition includes, as an addendum to the third volume, the complete text of the short two-volume work *Abriss der Algebra der Logik*.

—2nd ed. 1966. (1st ed.: 1890-1905; 1909/10.) 2,192 pp. 6x9. 5 vols. in 3. 8284-0171-3. Three vol. set. **$35.00**

GESAMMELTE MATHEMATISCHE ABHANDLUNGEN
By H. A. SCHWARZ

—1890-1971. 726 pp. 6x9. Two vols. in one.

AN INTRODUCTION TO THE OPERATIONS WITH SERIES
By I. J. SCHWATT

Many useful methods for operations on series, methods for expansions of functions, methods for the summation of many types of series, and a wealth of explicit results are contained in this book. The only prerequisite is knowledge of the Calculus.

—1924-62. x + 287 pp. 5⅜x8. 8284-0158-6.

PROJECTIVE METHODS IN PLANE ANALYTICAL GEOMETRY
By C. A. SCOTT

CHAPTER HEADINGS: I. Point and Line Co-ordinates. II. Infinity. Transformation of Coordinates. III. Figures Determined by Four Elements. IV. The Principle of Duality. V. Descriptive Properties of Curves. VI. Metric Properties of Curves; Line at Infinity. VII. Metric Properties of Curves; Circular Points. VIII. Unicursal (Rational) Curves. Tracing of Curves. IX. Cross-Ratio, Homography, and Involution. X. Projection and Linear Transformation. XI. Theory of Correspondence. XII. The Absolute. XIII. Invariants and Covariants.

—3rd ed. 1923-61. xii + 290 pp. 5⅜x8. 8284-0146-2

LEHRBUCH DER TOPOLOGIE
By H. SEIFERT and W. THRELFALL

This famous book is a modern work on *combinatorial topology* addressed to the student as well as to the specialist. It is almost indispensable to the mathematician who wishes to gain a knowledge of this important field.

"The exposition proceeds by easy stages **with examples and illustrations at every turn.**"

—*Bulletin of the A. M. S.*

—1934-68. vii + 353 pp. 5⅜x8. 8284-0031-8.

SET TOPOLOGY
By R. VAIDYANATHASWAMY

In this text on Topology, the first edition of which was published in India, the concept of partial order has been made the unifying theme.

Over 500 exercises for the reader enrich the text.

CHAPTER HEADINGS: I. Algebra of Subsets of a Set. II. Rings and Fields of Sets. III. Algebra of Partial Order. IV. The Closure Function. V. Neighborhood Topology. VI. Open and Closed Sets. VII. Topological Maps. VIII. The Derived Set in T_1 Space. IX. The Topological Product. X. Convergence in Metrical Space. XI. Convergence Topology.

—2nd ed. 1960. vi + 305 pp. 6x9.   [139]

LECTURES ON THE GENERAL THEORY OF INTEGRAL FUNCTIONS
By G. VALIRON

—1923. xii + 208 pp. 5¼x8.   [56]

GRUPPEN VON LINEAREN TRANSFORMATIONEN
By B. L. VAN DER WAERDEN

— (Ergeb. der Math.) 1935. 94 pp. 5½x8½.   [45]

THE LOGIC OF CHANCE
By J. VENN

One of the classics of the theory of probability. Venn's book remains unsurpassed for clarity, readability, and sheer charm of exposition. No mathematics is required.

CONTENTS: PART ONE: Physical Foundations of the Science of Probability. CHAP. I. The Series of Probability. II. Formation of the Series, III. Origin, or Causation, of the Series. IV. How to Discover and Prove the Series. V. The Concept of Randomness. PART TWO: Logical Superstructure on the Above Physical Foundations. VI. Gradations of Belief. VII. The Rules of Inference in Probability. VIII. The Rule of Succession. IX. Induction. X. Causation and Design. XI. Material and Formal Logic . . . XIV. Fallacies. PART THREE: Applications. XV. Insurance and Gambling. XVI. Application to Testimony. XVII. Credibility of Extraordinary Stories. XVIII. The Nature and Use of an Average as a Means of Approximation to the Truth.

—Repr. of 3rd ed. xxix+508 pp. 5⅜x8.   [173]   Cloth
                                          [169]   Paper